Trench Rescue

Principles and Practice to NFPA 1006 and 1670

FOURTH EDITION

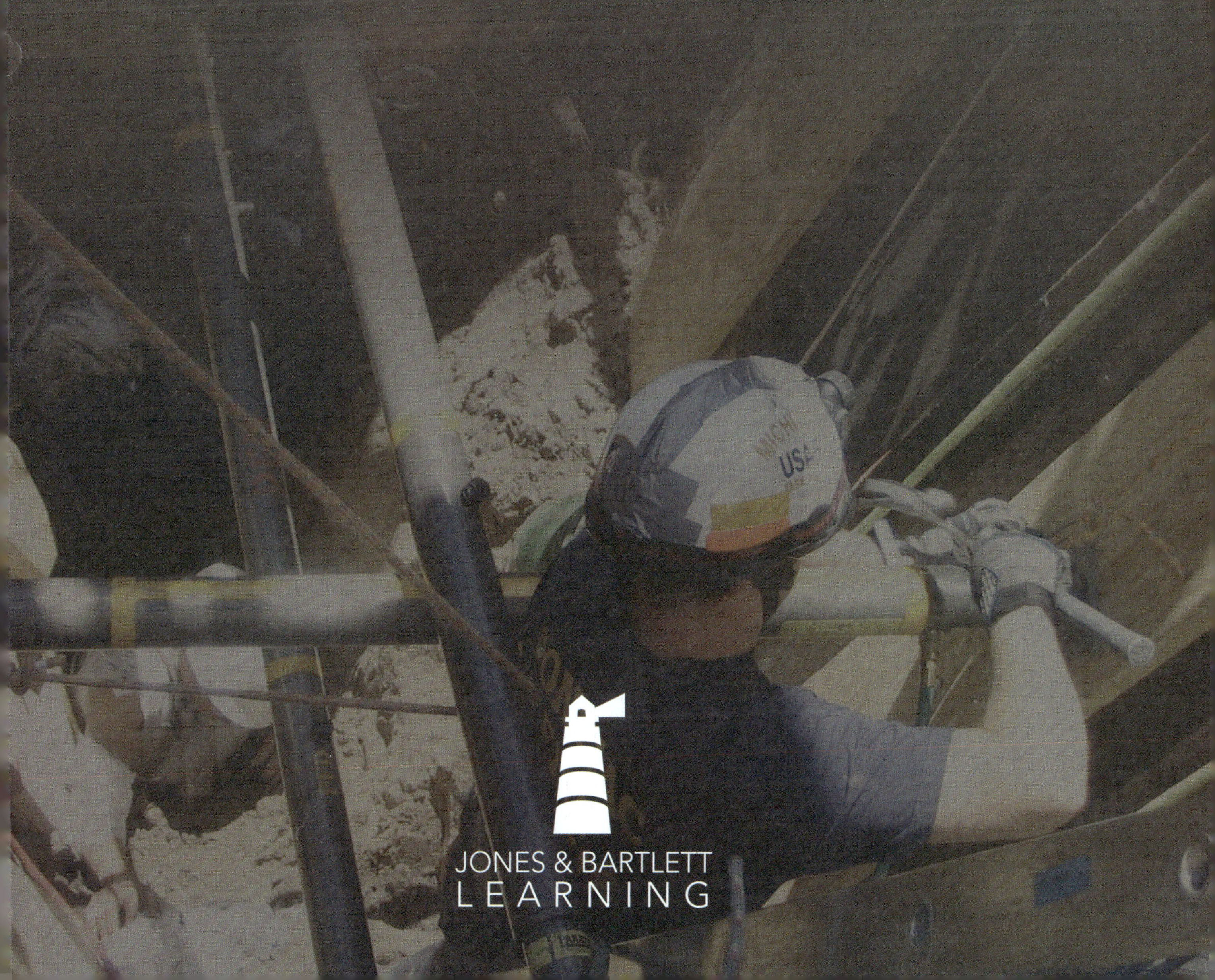

JONES & BARTLETT
LEARNING

Trench Rescue
Principles and Practice to NFPA 1006 and 1670
FOURTH EDITION

Ron "Z" Zawlocki, BS
Battalion Chief (retired), City of Pontiac
Rescue Team Manager, MI-TF1-US&R
Detroit, Michigan

Craig Dashner, PE
Construction Manager, OHM Advisors, Livonia, Michigan
Lead Structures Specialist, MI-TF1-US&R
Detroit, Michigan

Jones & Bartlett Learning
World Headquarters
25 Mall Road, Suite 600
Burlington, MA 01803
978-443-5000
info@jblearning.com
www.jblearning.com
www.psglearning.com

Jones & Bartlett Learning books and products are available through most bookstores and online booksellers. To contact the Jones & Bartlett Learning Public Safety Group directly, call 800-832-0034, fax 978-443-8000, or visit our website, www.psglearning.com.

Substantial discounts on bulk quantities of Jones & Bartlett Learning publications are available to corporations, professional associations, and other qualified organizations. For details and specific discount information, contact the special sales department at Jones & Bartlett Learning via the above contact information or send an email to specialsales@jblearning.com.

Copyright © 2022 by Jones & Bartlett Learning, LLC, an Ascend Learning Company

All rights reserved. No part of the material protected by this copyright may be reproduced or utilized in any form, electronic or mechanical, including photocopying, recording, or by any information storage and retrieval system, without written permission from the copyright owner.

The content, statements, views, and opinions herein are the sole expression of the respective authors and not that of Jones & Bartlett Learning, LLC. Reference herein to any specific commercial product, process, or service by trade name, trademark, manufacturer, or otherwise does not constitute or imply its endorsement or recommendation by Jones & Bartlett Learning, LLC and such reference shall not be used for advertising or product endorsement purposes. All trademarks displayed are the trademarks of the parties noted herein. *Trench Rescue: Principles and Practice to NFPA 1006 and 1670, Fourth Edition* is an independent publication and has not been authorized, sponsored, or otherwise approved by the owners of the trademarks or service marks referenced in this product.

There may be images in this book that feature models; these models do not necessarily endorse, represent, or participate in the activities represented in the images. Any screenshots in this product are for educational and instructive purposes only. Any individuals and scenarios featured in the case studies throughout this product may be real or fictitious but are used for instructional purposes only.

The publisher has made every effort to ensure that contributors to *Trench Rescue: Principles and Practice to NFPA 1006 and 1670, Fourth Edition* materials are knowledgeable authorities in their fields. Readers are nevertheless advised that the statements and opinions are provided as guidelines and should not be construed as official policy. The recommendations in this publication or the accompanying resources do not indicate an exclusive course of action. Variations taking into account the individual circumstances and local protocols may be appropriate. The publisher disclaims any liability or responsibility for the consequences of any action taken in reliance on these statements or opinions.

21620-2

Production Credits
VP, Content Strategy and Implementation: Christine Emerton
Director of Product Management: Bill Larkin
Content Strategist: Jennifer Deforge-Kling
Project Manager: Kristen Rogers
Senior Project Specialist: Alex Schab
Digital Project Specialist: Rachel DiMaggio
Director of Marketing: Brian Rooney
Content Services Manager: Colleen Lamy

VP, Manufacturing and Inventory Control: Therese Connell
Composition: S4Carlisle Publishing Services
Cover Design: Scott Moden
Text Design: Scott Moden
Media Development Editor: Faith Brosnan
Rights Specialist: Liz Kincaid
Cover Image (Title Page, Part Opener, Chapter Opener): © Ron Zawlocki.
Printing and Binding: LSC Communications

Library of Congress Cataloging-in-Publication Data
Names: Dashner, Craig, author. | Zawlocki, Ron, author.
Title: Trench rescue : principles and practice to NFPA 1006 and 1670 / Craig Dashner, Ron Zawlocki.
Description: Fourth edition. | Burlington, MA : Jones & Bartlett Learning, [2022] | Revision of: Trench rescue / Cecil V. "Buddy" Martinette, Jr., MA, CFO, EFO, Cheif, Wilmington Fire Department, Wilmington, North Carolina, Ron "Z" Zawlocki, BS, Battalion Cheif (Retired), City of Pontiac, Task Force Leader, MI-TF1-US&R, Detroit, Michigan. 2017. Third edition. | Includes bibliographical references and index.
Identifiers: LCCN 2021022719 | ISBN 9781284216202 (paperback)
Subjects: LCSH: Excavation--Safety measures--Textbooks. | Rescue work--Textbooks. | Trenches--Textbooks. | Search and rescue operations--Textbooks.
Classification: LCC TA730 .M273 2022 | DDC 628.9/2--dc23
LC record available at https://lccn.loc.gov/2021022719

6048

Printed in the United States of America
25 24 23 22 21   10 9 8 7 6 5 4 3 2 1

Brief Contents

SECTION **1**
Awareness Level   1

CHAPTER **1** Introduction to Trench Rescue   2

CHAPTER **2** Soil and Collapse Mechanics   17

CHAPTER **3** Initial Actions   42

CHAPTER **4** Personal Protective Equipment and Equipment Basics   64

SECTION **2**
Operations Level   93

CHAPTER **5** Hazard Mitigation   94

CHAPTER **6** Managing the Trench Incident   115

CHAPTER **7** Operations Level Trench Rescue Shoring   127

CHAPTER **8** Victim Care and Extrication   185

SECTION **3**
Technician Level   205

CHAPTER **9** Lifting and Load Stabilization   206

CHAPTER **10** Technician Level Trench Rescue Shoring   235

APPENDIX **A:** Advanced Trench Rescue Shoring   274

APPENDIX **B:** NFPA 1006 and 1670 Correlation Guide   298

APPENDIX **C:** T-L Method – A Metric Guide   304

GLOSSARY   313

INDEX   318

Contents

Skill Drills	xii
Acknowledgments	xiii
About the Authors	xv
Foreword	xvii

SECTION 1
Awareness Level — 1

CHAPTER 1
Introduction to Trench Rescue — 2

Introduction	3
Incidence of Trench Collapse	4
OSHA CFR 1926 Subpart P, Excavations	5
OSHA and Trench Rescue	5
National Fire Protection Association (NFPA)	6
NFPA 1670: Trench Rescue Levels	7
Awareness Level	7
Operations Level	7
Technician Level	8
Soil Basics and Collapse Mechanics	9
Trench Rescue Response Systems	9
Trench and Excavation Hazards	10
Sample OSHA Safety Measures	11
Collapse (Cave-In)	11
Falling Objects	11
Fall Hazards	11
Utility Hazards	11
Hazardous Atmospheres	12
Physical Hazards	12
Ingress or Egress	15

CHAPTER 2
Soil and Collapse Mechanics — 17

Introduction	18
OSHA-29 CFR 1926.651 and 1926.652	19
OSHA Soil Classification	19
Stable Rock	21
Type A	21
Type B	21
Type C	22
Other Soil Classifications	22
Soil Mechanics	22
Soil Types	22
Forces Associated with Soil	22
Vertical Force	22
Horizontal Force	25
Soil Strength	25
Soil Friction	25
Cohesion	26
Moisture Content	27
Unconfined Compressive Strength	27
Trench Collapse	27
Types of Trench Collapse	27
Spoil Pile Slide	27
Lip Shear Failures	28
Slough Failure	29
Rotational Failures	29
Toe Failure	30
Wall Shear Failures	30
Wedge Failures	31
Collapse Patterns	32

Conditions and Factors That Lead to Collapse	33	Establish Operational Zones	53
Water	33	Prepare the Site for Incoming Resources	55
Water Table	33	Hazard Management	55
Severe Environmental Conditions	33	Defensive Measures	55
Varying Soil Profiles	34	Defensive Mitigation Actions for Underground Utilities	56
Disturbed Soils	34	Unbroken Utility Lines	56
Superimposed (Surcharge) Loads	34	Nonentry Rescue and Victim Self-Rescue	57
Vibration	35	Nonentry Rescue	61
Signs of Impending Collapse	35		
Visible Cracks	36		
Visible Bulging on Walls or Floors	36		
Water	36		
Undercut or Blown Out Trench Walls	36		

CHAPTER 3
Initial Actions — 42

CHAPTER 4
Personal Protective Equipment and Equipment Basics — 64

Introduction	43	Introduction	65
Scene Management	43	The Importance of Proper Equipment	65
Incident Command	43	Development of a Safety Culture	66
Command Post	44	Personal Protective Equipment for Trench Rescue	66
Staging Area	44	Personal-Issue PPE	66
Standard Operating Guidelines (SOGs) and Incident Action Plans	45	Torso, Arm, and Leg Protection (Coat and Trouser or Coverall)	66
Developing the Incident Action Plan	45	Head Protection (Helmet)	66
Size-Up	46	Hand Protection (Gloves)	67
Situation Assessment	46	Foot Protection (Boots)	68
Dispatch Information	46	Eye Protection (Safety Glasses)	68
Initial Assessments	47	Firefighting Turnout Gear	69
Interview Techniques	47	Team-Issued PPE	69
Classifying the Emergency	47	Respiratory Protection	69
Hazard Assessment	47	Hearing Protection	69
Victim Assessment	48	Cutting Tool Protection	69
Trench Assessment	49	Trench Rescue Shoring Equipment Overview	70
Resource Assessment	50	Lip Protection	70
Determine the Scope of the Rescue	51	Lip Protection Best Practices	72
Evaluation	51	Trench Rescue Panels	72
Summon Resources	52	How Panels Work	73
Tier 1 Response	52	Panel Materials	74
Tier 2 Resources	52	Best Practices for Panels	74
Site Control	53	Panel Ropes	75
Shut Down Traffic	53	Wales	76
		Best Practices for Wales	76
		Wale Hangers	78
		Wale Ropes	78
		Backfill	78

Backfill Options	78
Struts	80
Best Practices for Struts	81
Pneumatic Struts	81
Timber Struts	81
Screw Jack Struts	82
Hydraulic Struts	82
Tools and Appliances	83
Strut Collar Locking Tools	83
Tape Measure and Angle Finder	83
Pike Poles	84
Pickets and Sledge Hammers	84
Shovels	84
Hammers and Nails	84
Chainsaw	85
Ventilation Equipment	85
Ladders	85
Scene Lighting	85
Dewatering Devices	85
Utility Control	86
Victim Removal Equipment	86
Lifting and Stabilizing Heavy Objects	86
Rapid Intervention Team Equipment	86
Cutting Station	86

SECTION 2
Operations Level   93

CHAPTER 5
Hazard Mitigation   94

Introduction	95
Hazard Identification	95
Types of Hazards	95
Trench Collapse	95
Utilities	96
Traffic	96
Physical Hazards	97
Water	97
Severe Environmental Conditions	97
Biological Hazards	97
Hazardous Materials	97
Atmospheric	97
Hazard Control Plan	98
Hazard Mitigation PPE	98
Briefing	98
Hazard Mitigation	99
Hazard Control Equipment	99
Atmospheric Testing Equipment	99
Dewatering Devices	100
Sewer and Water Control Equipment	100
Electric Control Equipment	101
Ventilation Equipment	101
Natural Gas Control Equipment	102
Potential Offensive Mitigation Techniques	102
Atmospheric Monitoring	102
Dewatering Devices	102
Underground Electrical Wires	106
Gas	106
Ventilation	106
Rescue Ventilation	107
Hazardous Materials	107
Severe Environmental Conditions	107
Enhanced Hazard Mitigation	113
Ongoing Hazard Mitigation	113

CHAPTER 6
Managing the Trench Incident   115

Introduction	116
Size-Up	116
Size-Up and Expanding the Incident Command Structure	117
Command	117
Logistics	118
Operations	119
Incident Management Tools	120
Incident Termination	121

CHAPTER 7
Operations Level Trench Rescue Shoring   127

Introduction	128
The History of Trench Rescue Shoring	129

Comparing Rescue Shoring to Construction Shoring	129
Purpose	129
Soil Conditions	129
Time	130
Planning	130
Principles of Trench Rescue Shoring	131
Trench Rescue Shoring Essentials	131
Collect/Distribute Loads	131
Transfer Loads (Struts)	132
Resist Loads	132
Lateral Soil Forces and Firefighters	132
T-L Method	133
Soil Classification	133
Soil Forces	133
How To Use The T-L Method for Trench Rescue	134
Surcharge Loads (ScL)	135
Determining the Spoil Pile Surcharge	135
Adding the Spoil Pile Surcharge	135
Determining the Equipment Surcharge	136
Adding the Equipment Surcharge	136
Trench Depth to L Conversion	136
T-L Method in Action	136
Tabulated Data for Shoring Equipment	137
Prescriptive Shoring Designs	144
Criteria for Safe Zone in a Trench	144
Trench Rescue Shoring Plan	145
Shoring Plan Briefing	145
Procedure for the Trench Rescue Shoring Plan	145
Step 1: Trench Size-Up	145
Step 2: Primary Shoring	147
Step 3: Secondary Shoring	147
Step 4: Complete Shoring	147
Step 5: Shoring Performance Assessment	147
Emergency Procedures	147
Trench Rescue Equipment	147
Struts	148
Strut Force	148
Pneumatic Struts	149
Pneumatic Strut Pressure	149
Pneumatic Strut Placement	150
Pneumatic Strut Spacing	150
Hydraulic Struts	150
Screw Jack Struts	150
Timber Struts	151
Timber Strut Placement	151
Timber Strut Activation Forces	151
Trench Rescue Panels	152
Wales	153
Shoring Voids	154
Void Assessments	154
Shielding Systems	155
Practices of Trench Rescue Shoring	156
Lip Protection	157
Trench Rescue Shoring	160
Entry Shoring Overview	161
Non-entry Shoring Overview	162
Panel Installation	163
Installing Pneumatic Shores–Non-Entry	167
Installing Pneumatic Shores–Entry	167
Installing Timber Shores	171
Installing Inside Wales	172
Shoring Voids	174
Air Bag Backfill	174
Wood Backfill	175
Buttress	175
Soil Backfill	176
Open Lip Void Shoring Procedure with Entry Techniques	176
Shoring System Disassembly and Removal	176
Removal Methods	177
Machine Removal	177
Manual Removal	177
Machine Tear Out	178
Shoring System Disassembly/Removal Plan	178
Assessment	178
Briefing	178
Machine Removal	178
Personnel	178

CHAPTER 8
Victim Care and Extrication — 185

Introduction	186
Mechanism of Injury	186
Cave-In Incidents	186
Incidents Without a Cave-In	187
Non-entry Rescue and Victim Self-Rescue	187
Entry Operations	189
Pre-entry Briefing	189
Personal Protective Equipment for Entry Team Operations	189
Entry Team Duties	189
Extrication	189
Soil Entrapment	190
Vacuum Systems	191
Vacuum Trucks	191
Vacuum Truck Response	193
Operations	193
Victim Care Considerations	194
Providing Victim Care	197
Victim Assessment and Initial Care	197
Victim Stabilization	198
Victim Care Involving a Collapse	198
Special Considerations	199
Removing a Victim from a Trench	201
Victim Packaging Equipment	202
Victim Harness	202
Litter Basket	202

SECTION 3
Technician Level — 205

CHAPTER 9
Lifting and Load Stabilization — 206

Introduction	207
Lifting Mechanics	207
Levers	209
Pulleys	209
Rope-Based Systems	210
Basic Techniques	210
Load Assessment	210
Weight	210
Type of Material	210
Center of Gravity	210
Forces Acting on the Load	212
Hazards Associated with Heavy Objects	212
Load Stabilization	213
Mechanics of Load Stabilization	213
Stabilization Plan	214
Compression-Based Stabilization	214
Cribbing	214
Box Crib System	214
Cribbing Rules	215
Cribbing Details	215
Cribbing Angled Objects	216
Wedges and Shims	216
Struts	216
Wheel Chocks	217
Tension-Based Stabilization	217
Inspecting Stabilization Equipment	218
Inspecting Pneumatic Struts	218
Inspecting Tiebacks	218
Synthetic Rope	218
Webbing Inspection	218
Lever Hoist Inspection	218
Inspecting Rigging Equipment	219
Wire Rope Inspection	219
Synthetic Web Slings and Polyester Round Sling Inspection	219
Alloy Steel Chain Slings Inspection	219
Rigging Hardware: Hook, Shackle, Turnbuckle, and Carabiner Inspection	219
Lifting Plan	219
Lifting Techniques	220
Surface-Based Lifting Techniques	220
Types of Surface-Based Lifting Tools	220
Lifting Tool Capacities	220
Surface-Based Lifting Tool Inspection	221
Overhead Lifting Techniques	222
Critical Angles	222
Types of Overhead Lifts Used for Trench Rescues	224
Performing Bridge Lifts	225
Lifting Bridge Capacities	225

Air Bag Bridge Lift	225
Performing Bipod Lifts	225
Bipod Components	226
Bipod Capacities	226
Hand Signals for Bridge Lifts and Bipod Lifts	226
Overhead Lifting Tools	226
Coordinating the Use of Heavy Equipment	226
Equipment Operator	227
Communications	227
Heavy Equipment Terminology	227
Lifting Capacities	228
Hand Signals Needed for Heavy Equipment Lifting	228

CHAPTER 10
Technician Level Trench Rescue Shoring — 235

Introduction	236
Intersecting Trenches	237
Deep Trenches	237
Sheeting and Shoring for the Technician Level	238
Supplemental Sheeting and Shoring	238
Spot Shoring	238
Rescue Plan	238
Risk versus Benefit Analysis	238
Shoring Plan	238
Shoring Performance Assessment	239
Victim Release	239
Path for Removal	239
Pre-entry Briefing	239
Shoring Intersecting Trenches	239
Plan Application for an Intersecting Trench	242
Precautions for an Intersecting Trench	242
Preparation to Stabilize an Intersecting Trench	242
Intersecting Trench–Specific Shoring Equipment	243
Standard Shoring Equipment for Intersecting Trenches	243
Personnel for Intersecting Trenches	243
L-Trench Shoring Plan	243
Shoring Size-Up for an Intersecting Trench	244
Walkthrough of Shoring a L-Trench	244
Primary Shoring	244
Secondary Shoring	245
Expand the Safe Zone	246
Putting It All Together: Shoring the L-Trench	246
T-Trench Shoring Plan	246
Walkthrough of Shoring a T-Trench	250
Primary Shoring	250
Secondary Shoring	251
Complete the Shoring	252
Putting It All Together: Shoring a T-Trench	252
Deep Trench	260
Application for Deep Trenches	260
Precautions for Deep Trenches	260
Preparation Tasks for Shoring Deep Trenches	260
Shoring Equipment for Deep Trenches	260
Personnel for Deep Trenches	261
Shoring Plan for a Deep Trench	261
Size-Up	261
Walkthrough of Shoring a Deep Trench	261
Primary Shoring	261
Primary Shoring (Top Panel Set)	261
Secondary Shoring (Bottom Panels)	262
Secondary Shoring (Top Panels)	262
Complete Shoring	263
Putting It All Together: Shoring a Deep Trench	263
Utilizing Supplemental Shoring	263
Application for Supplemental Shoring	267
Precautions for Supplemental Shoring	268
Preparation for Supplemental Shoring	268
Walkthrough of Installing Supplemental Shoring	268
Putting It All Together: Supplemental Shoring	268
Spot Shoring	269
Single-Point Shoring	270
Severe Environmental Conditions	270
Victim Removal Review	271

APPENDIX A: Advanced Trench Rescue Shoring — 274
APPENDIX B: NFPA 1006 and 1670 Correlation Guide — 298
APPENDIX C: T-L Method – A Metric Guide — 304

Glossary	313
Index	318

Skill Drills

CHAPTER **4**

SKILL DRILL 4-1   76
Roping the Panels

SKILL DRILL 4-2   87
Setting Up a Cutting Station Table

CHAPTER **5**

SKILL DRILL 5-1   103
Utilizing an Atmospheric Monitor

SKILL DRILL 5-2   104
Dewater a Trench

SKILL DRILL 5-3   108
Supply Ventilation

SKILL DRILL 5-4   109
Exhaust Ventilation

SKILL DRILL 5-5   110
Rescue Ventilation

CHAPTER **7**

SKILL DRILL 7-1   157
Installing Ground Pads

SKILL DRILL 7-2   159
Installing Lip Bridges

SKILL DRILL 7-3   163
One-Side Panel Installation

SKILL DRILL 7-4   166
Two-Side Panel Installation

SKILL DRILL 7-5   167
Installing Pneumatic Struts Without Entering the Trench

SKILL DRILL 7-6   169
Installing Pneumatic Shores with Entry Operations

SKILL DRILL 7-7   172
Installing Timber Shores

SKILL DRILL 7-8   177
Shoring Voids with Struts that Require Entry

CHAPTER **10**

SKILL DRILL 10-1   247
Shoring the L-Trench

SKILL DRILL 10-2   253
Shoring the T-Trench

SKILL DRILL 10-3   264
Shoring a Deep Trench

SKILL DRILL 10-4   269
Supplemental Shoring

Acknowledgments

Reviewers

Richard Atwood
Trench Rescue Instructor
Los Angeles County Fire Department
Los Angeles, California

Mark Bagley
NPQ Instructor II
Georgia Fire Academy
Battalion Chief
Cartersville Fire Department
Cartersville, Georgia

Aaron Bolinger
Assistant Chief
Warsaw-Wayne Fire Territory
Akron, Indiana

Sean Broyles
President, Tech-ResQ Training Specialists
Captain/Paramedic (retired)
Kingsport Fire Department
Kingsport, Tennessee

Carrie Erbe
Lieutenant/Paramedic
Independence Township Fire Department
Independence Township, Michigan
MI-TF1
US&R Medical Specialist

Brian S. Gettemeier
Captain
Cottleville Fire Protection District
St. Charles, Missouri

Michael A. Gorsuch
Captain
Columbia River Fire and Rescue
West Linn, Oregon

Chuck Guy
Toronto Fire Services
Whitby, Ontario, Canada

Jeff Hakola
Merced City Fire Department
Atwater, California

Tommy Hatfield, AS
Clayton County Fire and Emergency Services
World Wide Industrial Training Solutions
McDonough, Georgia

Jason Hoover
Martinsburg Fire Department
Martinsburg, West Virginia

Walter D. Idol, MS/EMT-P
Program Manager: Health Safety and Preparedness
University of Tennessee Institute for Public Service
Dandridge, Tennessee

Robert Lindstedt
Assistant Professor
Southern Maine Community College
Portland, Maine

Jose Mendia
Lieutenant
Technical Rescue Coordinator
TACO/InterContinental Fire Training Academy
Pembroke Pines, Florida

Ernesto Ojeda, CSCS, MEd
California State University, Dominguez Hills
San Pedro, California

John O'Neal
BS Fire Administration (in progress)
Columbia Southern University
Indiana-certified Trench Rescue Technician and Instructor 3
Noblesville Fire Department
Noblesville, Indiana

Jonathan A. Rigolo
Virginia Beach Fire Department
Virginia Beach, Virginia

Todd W. Saunders
Lieutenant
Raleigh Fire Department
Raleigh, Virginia

Gary Seidel
Fire Chief (retired)
Hillsboro, Oregon

Oliver-Denzil S. Taylor, PhD, PE
U.S. Army Engineer Research and Development Center
Vicksburg, Mississippi

Mike Tesarski
Toronto, Ontario, Canada

Mike Ulibarri
Unified Fire Authority Battalion Chief (retired)
Paratech Incorporated
West Jordan, Utah

Darin J. Virag
Captain
Charleston Fire Department
Charleston, West Virginia

William W. Watts
Assistant Chief
Arkansas State USAR Task Force Leader
Fayetteville Fire Department
Fayetteville, Arkansas

Keith Wilson
Lancaster County Fire Rescue
Lancaster County, South Carolina

About the Authors

Ron "Z" Zawlocki

Ron Zawlocki's career in the fire and rescue service spans five decades. He began as a firefighter with the City of Detroit in 1974 and retired as a battalion chief with the City of Pontiac in 2007. Chief Zawlocki continues to serve as a rescue team manager with Michigan's Urban Search and Rescue Task Force (MI-TF1). During his career, he has responded to several thousand fire, emergency medical, and technical rescue incidents. The technical rescue discipline that has been his most frequent response is trench rescue. As a result, he has developed a passion for trench rescue shoring that has led him to new levels of testing, research, and development in that discipline. Including training, exercises, and actual rescue incidents, Chief Zawlocki has shored more than 900 live trenches and has made a point to learn from every one of them.

Chief Zawlocki has a bachelor of science degree in secondary education from Eastern Michigan University. He has written the student manuals and instructor guides for trench rescue courses that were developed by Michigan State University and funded by the Michigan Occupational Safety and Health Administration. He is currently the lead trench rescue instructor for the Michigan Urban and Rescue Training (MUSAR) Foundation. Chief Zawlocki has taught rescue programs across the United States and in Canada, Mexico, and the Persian Gulf. He has had rescue-related articles published in *Fire Engineering, Firehouse, Fire Chief,* and *Fire and Rescue* magazines and has been both a classroom and a hands-on instructor at the Fire Department Instructor's Conference (FDIC).

Chief Zawlocki would like to make the following acknowledgments:

First and foremost, thanks to my friend and former coauthor, Cecil "Buddy" Martinette. Buddy is both an avid student and an exceptional teacher of trench rescue and responder safety. Without his dedication and support this book never would have happened.

In this edition we feature a new co-author, Craig Dashner. Craig is a professional engineer who is licensed in three states and is the lead structures specialist for Michigan's Urban Search and Rescue Task Force, MI-TF1. Craig has worked with me on improving rescue tactics and safety for over two decades. His tireless pro-bono research, testing, and advice are greatly appreciated by those of us who are lucky enough to work with him. Through this book we are now able to share his expertise with rescuers around the world.

I would like to acknowledge the people and organizations that have invested their time and energy into an ongoing effort to find a better way: MUSAR Training Foundation; Michigan State University/ERS-Professor Scott Tobey and Chief (retired) Don Fisher (Ann Arbor FD); International Union of Operating Engineers (Local#324) John Garcia, Lee Graham, Bill "Bear" Nelson, Pat Zinser and the entire staff at the training center in Howell, Michigan; Paratech Rescue International—Nigel Leatherby, Tom Gavin, Mike Ulibarri, and Bill Teach; Chief Robert Lamson, Battalion Chief Mike Nye, and the members of the City of Pontiac Fire Department; Chief John Cieslik and the members of the City of Rochester Fire Department.

I also would like to acknowledge my "band of brothers" (MUSAR instructors) who continue to prompt, support, question, and test everything we do: Aaron Osborn (Summit FD), Mike DeCreane (St. Clair Shores FD), Carl Hein (Ann Arbor FD), Chad Godfrey (Southfield FD), Mark Laux (Midland FD), Greg Payeur (Pittsfield Township FD), Jason Hendrie (Redford FD), Kevin Caldwell (Westland FD), Marcus Boudreaux (Detroit FD), Shawn Skelly (Canton FD), and Dave VanHolstyn (Grand Rapids FD).

The rescue instructors who guided and inspired me include: Mike McGroarty (LaHabra FD), Tim Gallagher (Phoenix FD), Ray Downey (FDNY), John O'Connell (FDNY), Nick Giordano (FDNY), David Hammond (FEMA), Chase Sargent (VBFD), Mike Brown (VBFD), Jon Rigolo (VBFD), Ron Winchester (Detroit FD), and Alan Zsido, Dave Potter, and Tim Campbell (Pontiac FD).

I offer a special tribute to my fellow instructors and friends who have passed but will not be forgotten: Chief Ray Downey (FDNY), Lt. Joey DiBernardo (FDNY), Lt. Andrew Fredricks (FDNY), FF Christopher Blackwell (FDNY), Fire Chief Mike McGroarty (LaHabra FD), David Knisley (Grand Rapids FD), and the LODD death of FF Tracy Williamson (Pontiac FD) that prompted a change in rescue operations in Michigan. I hope this book helps transfer their dedication and passion to the next generation of firefighters.

Finally and most importantly, I would like to acknowledge my family: my wife Jeri and my daughters Erin and Lauren. Their love, patience, and support have enabled me to continue to pursue my dreams.

Craig Dashner, PE

Craig Dashner's career in engineering is in its third decade. He began as a bridge engineer in 1993 and grew through his positions to his current role as a construction manager and partner at OHM

Advisors in Livonia, Michigan. Around 1998, Ron "Z" Zawlocki reached out to the Structural Engineers Association of Michigan, of which Dashner was the Emergency Response Committee chair at the time. This contact resulted in an unexpected friendship between Dashner and Z, one that pulled Dashner into the world of Urban Search and Rescue and then into trench rescue. This contact resulted in Dashner's position on Michigan Task Force 1 (MI-TF1) as the lead structures specialist, since the inception of the organization.

Dashner has worked with Z for many years developing new and improved methods for shoring collapsed trenches. He brings engineering expertise into the existing trench shoring teachings and ensures that the shoring can both stand up to the potential forces and can be installed safely from outside the trench. Countless hours have been spent in the MUSAR Training Foundation site, building, testing, and refining new shoring concepts to determine their strength, stability, and suitability for use. New materials and methods have been brought into the shoring designs to provide better strength and reliability. All of that effort has culminated in Dashner partnering with Z to write this book, which combines the expertise of an engineer with the experience of a firefighter. He hopes that this book, with the additions that he and Z have made, can make things better and safer for today's fire service personnel and the future generations that follow. They deserve whatever support that can be provided to them.

Dashner has a bachelor of science degree in civil and environmental engineering from Michigan State University. He is a subject matter expert for the MUSAR Training Foundation and frequent guest speaker at MUSAR Training Foundation Trench classes. In partnership with Z, Dashner has coauthored articles, presented at several rescue conferences across the country, including FDIC, and participates in an online group that provides answers to trench-related questions.

Mr. Dashner would like to make the following acknowledgments:

First and foremost, I must thank my coauthor and friend, Ron "Z" Zawlocki. Without that fateful contact in 1998, I may not have ever entered this world that Z calls rescue engineering. He is single-handedly responsible for a major portion of my career outside of my job. His introduction into the urban search and rescue world, followed by trench shoring, spawned an interest that has persisted for over 20 years. I would not be where I am now if it were not for that contact.

I would like to thank my partners, managers, and coworkers at OHM Advisors. They have supported my work with MI-TF1, covered for me in my absence when I am deployed, and provided the flexibility for me to use my engineering skills to make improvements for the fire service.

I would like to acknowledge the group of MUSAR instructors who have accepted me; respect me; and listen to, support, and question me; and build and test everything we do: Aaron Osborn (Summit FD), Mike DeCreane (St. Clair Shores FD), Carl Hein (Ann Arbor FD), Chad Godfrey (Southfield FD), Mark Laux (Midland FD), Greg Payeur (Pittsfield Township FD), Jason Hendrie (Redford FD), Kevin Caldwell (Westland FD), Marcus Boudreaux (Detroit FD), Shawn Skelly (Canton FD), and Dave VanHolstyn (Grand Rapids FD).

I would like to acknowledge my parents, Larry and Joyce, who have always supported me and my education. They supported me through college, and they are a big part of the person I am today.

Finally and most important, I would like to thank my family: my wife Heather, daughter Jordan, and son Nicholas. I will never understand how it must feel for them when I walk out the door with my bags, headed into a disaster only knowing that I "should" be home in 14 days. Never once have they held me back in my ability to serve my newfound partners in the fire service while assisting strangers in their time of greatest need. I thank them for their love, support, understanding, and patience which has allowed me to spend the time I have devoted to MI-TF1 and trench rescue shoring.

Foreword

Trench rescue is a multifaceted discipline. The response begins with the local first responders. The National Fire Protection Association suggests that first responders are proficient at the skills, knowledge, and abilities found in the trench rescue awareness sections of the NFPA 1670, *Standard on Operations and Training for Technical Search and Rescue Incidents* and NFPA 1006, *Standard for Technical Rescue Personnel Professional Qualifications*. Personnel trained at the awareness level must size up the incident, control the site, manage the scene, identify and summon the needed resources, and perform non-entry rescues when the conditions are favorable for those operations. When non-entry rescues are not feasible, additional resources are needed. Those resources must be proficient at mitigating the hazards and providing safe entry into the trench in order to perform rescue or recovery operations. Those operations include shoring trench walls to prevent collapse, extricating victims from entrapment, treating and packaging patients, and safely removing them from the trench. Responders trained at the trench rescue operations level have the knowledge, skills, and abilities needed to perform those duties at relatively shallow (8 feet deep or less) and simple (nonintersecting) trench configurations. As the depth, width, and complexity of the trenches increase, trench rescue technician level and beyond (advanced level) knowledge, skills, and abilities are required to resolve the incidents safely and efficiently.

In her study of trench rescue shoring practices (2009) Dr. S. Marie LaBaw found that there has been a lack of engineering principles and analysis regarding the shoring designs and equipment used by first responders and trench rescue teams. Additionally, Dr. LaBaw concluded that "underground construction shoring and trench rescue shoring are significantly different." Because of those significant differences, many of the theories, designs, and practices being used for construction shoring are not applicable to trench rescue shoring.

Previous trench rescue texts have been published using shoring designs, practices, and tabulated data that lacked engineering analysis. More recently, the trend has been to publish trench rescue books that use engineered shoring designs, methods, and tabulated data that were developed for use by construction workers. Unfortunately, the conditions, operational periods, scope, and purpose of shoring trenches for construction and repairs are significantly different than the conditions, operational periods, scope, and purpose at trench rescue incidents, and as a result, shoring designs, methods, and tabulated data developed for the construction industry are often inappropriate for trench rescue operations. This edition of *Trench Rescue* contains shoring designs, methods, and tabulated data that have been engineered specifically for use by first responders trained as trench rescuers. This innovative approach was made possible by a group of professional engineers who are also members (structures specialists) of federal and state urban search and rescue task force teams.

First, a unique method was developed that allows first responders to determine worst-case lateral soil forces rapidly and accurately. Second, tabulated data was engineered specifically for those worst-case soil conditions. The method (T-L method) was developed by Dr. Denzel-Oliver Taylor, PE (PhD) and Dr. S. Marie LaBaw, PE (PhD). The tabulated data was developed by Craig Dashner, PE and Dr. Janos Gergely, PE (PhD). Their work was coordinated and reviewed by George "Donnie" Barrier, PE who is a geotechnical forensic engineer. Our hope is that by providing engineered principles and practices that are developed specifically for trench rescue conditions, we will enhance rescuer safety and efficiency.

Ron "Z" Zawlocki, BS
Battalion Chief (retired), City of Pontiac
Rescue Team Manager, MI-TF1-US&R
Detroit, Michigan

Craig Dashner, PE
Construction Manager, OHM Advisors, Livonia, Michigan
Lead Structures Specialist, MI-TF1-US&R
Detroit, Michigan

Header image: Courtesy of Ron Zawlocki.

SECTION 1

Awareness Level

CHAPTER **1** **Introduction to Trench Rescue**

CHAPTER **2** **Soil and Collapse Mechanics**

CHAPTER **3** **Initial Actions**

CHAPTER **4** **Personal Protective Equipment and Equipment Basics**

CHAPTER 1

Awareness Level

Introduction to Trench Rescue

KNOWLEDGE OBJECTIVES

After studying this chapter, you should be able to:

- Define the terms *trench* and an *excavation*.
- Describe typical types of accidents at trench and excavation work sites. (**NFPA 1006: 12.1.6**, pp. 4–5)
- Describe the use of Occupational Safety and Health Agency (OSHA) regulations for trench rescue operations.
- Identify trench rescue references. (**NFPA 1006: 12.1.4**, pp. 5–9)
- Define a competent person per OSHA. (**NFPA 1006: 12.1.1**, p. 5)
- Identify the levels of response for trench rescue (awareness, operations, and technician), and describe the scope of each level. (**NFPA 1006: 12.1.6**, pp. 7–8)
- Recognize basic soil and collapse mechanics.
- Recognize and describe trench rescue terms and definitions.
- Identify the components of a trench rescue response system. (**NFPA 1006: 12.1.6**, pp. 9–10)
- Identify hazards frequently found at trench sites. (**NFPA 1006: 12.1.3**, pp. 10–12, 15)

SKILL OBJECTIVES

There are no skill objectives for this chapter.

You Are the Rescuer

Your fire department is dispatched to a construction site for a man trapped in a trench. The department is experienced with emergency medical incidents and fires but has never responded to a trench emergency. Upon arrival at the construction site, you see several pieces of heavy equipment, including excavators, trench boxes, front end loaders, dump trucks, and bulldozers. Several trenches have been dug and are ready for the installation of underground utilities that will serve the large housing complex that is under construction. Several mounds of dirt are piled up around the site. This is an unfamiliar environment to you and your crew, to say the least. As your engine pulls into the site, you are directed to a trench at the southeast corner of the complex.

You begin assessing the conditions and hazards as your crew walks to the end of the trench. Your first look at the trench reveals three construction workers and one police officer in an unprotected trench attempting to dig out the buried victim. An excavator is positioned at the other end of the trench and is being directed to dig out the collapsed soil. The operator is the brother of the man trapped in the trench. The digging has uncovered the victim's chest, and he is now buried to about his waist. The man is conscious, and the construction workers are telling you that they will have him out in about 5 minutes. One of your firefighters who has attended a trench rescue class tells you that it will likely take at least 30 minutes to free the victim from four feet of heavy, wet, compacted clay soil.

The construction crew supervisor informs you that this is Class B soil, and it has been stable since early this morning. The construction crew suspended work for approximately 2 hours this afternoon during a heavy thunderstorm. When they started work after the storm, the trench collapsed suddenly. The trench is 13 feet (4 meters [m]) deep and 5 feet (1.5 m) wide, with a cave-in from one side wall. Conditions are wet and muddy, with a couple of inches of water on the trench floor. As the first-arriving officer on the scene, your responsibility will be to decide on a safe and effective rescue plan.

1. What is the risk versus gain analysis for either continuing the current rescue attempt or changing the rescue tactics?
2. What needs to be done to protect the victim and rescuers from a secondary collapse?
3. What resources are needed to conduct a safe and efficient rescue operation in an unprotected 13-foot-deep (4-meter-deep) trench?

NAVIGATE Access Navigate for more practice activities.

Introduction

As a potential responder to trench rescue incidents, you need to have an understanding of trench accidents, the related laws and standards, the resources necessary to conduct safe and efficient rescue operations, and the related terminology. To begin, we will define both a trench and an excavation. The Occupational Safety and Health Administration (OSHA), the agency that ensures workplace safety, defines an **excavation** as any human-made cut, cavity, trench, or depression in the earth's surface and formed by earth removal **FIGURE 1-1**. A **trench** is defined as a narrow underground excavation that is deeper than it is wide and is no wider than 15 feet (4.5 m) **FIGURE 1-2**. Although a trench is an excavation, an excavation is not always a trench.

Trenches and excavations are built for a wide variety of reasons, including large construction projects, site balancing earthwork, road building projects, inground pools, septic fields, graves, and sand and gravel quarries **FIGURE 1-3**. However, most trench fatalities occur in trenches that are dug to install new

FIGURE 1-1 An excavation is any trench, cavity, cut or depression in the earth's surface.
© Richard Thornton/Shutterstock.

FIGURE 1-2 A trench is deeper than it is wide but not wider than 15 feet (4.5 m).

FIGURE 1-4 Trenches that are dug to repair or tap into existing underground utilities typically have four walls that must be protected.
© ShaunWilkinson/Shutterstock.

FIGURE 1-3 Trenches are often dug to install new underground utilities.

FIGURE 1-5 This worker was lucky to leave this trench unharmed.

underground utilities, to repair existing underground utilities, and to create and repair foundations for a variety of structures **FIGURE 1-4**.

Incidence of Trench Collapse

According to OSHA, hundreds of injuries and dozens of fatalities are caused by trench collapses each year. Due to an aging infrastructure and an increase in building construction, more work is being performed on underground utilities. With the increase in excavation and trench work, OSHA has deemed this work to be the most hazardous of all construction work. The Laborers' Health & Safety Fund of North America also found trench fatalities to be a serious problem in construction. According to data provided by the Bureau of Labor Statistics, cave-ins cause about three out of every four (75 percent) fatalities at construction sites.

> **TIP**
>
> As a result of repairs to underground utilities and new building construction, open trenches and excavations are a common sight in every type of community (urban, suburban, and rural). Thousands of people enter trenches every day of the year. Whether there will be a trench-related emergency depends on the soil conditions of the trenches, how closely the safety regulations were followed and, in some cases, luck **FIGURE 1-5**.

Due to the number of fatalities at construction sites, OSHA investigated the causes and found that lack of proper protection was the main reason that

FIGURE 1-6 The majority of trench rescue incidents involve collapse.

FIGURE 1-7 Hazards in this trench include disrupted utilities and collapse (unprotected trench walls).
© Aisyaqilumaranas/Shutterstock.

trenches collapse and either injure or kill workers. Most trench fatalities occur at residential work sites and the majority of these fatalities occur in trenches less than 9 feet (2.7 m) deep. Almost 50 percent of the time, there was no protection in place for workers and only 35 percent of the workers were offered trench safety training.

Firefighters responding to trench- and excavation-related accidents are exposed to the same hazards as the victims they are attempting to rescue. Firefighters can simplify the assessment of trench collapses by categorizing the situation as either collapse or noncollapse:

- Collapse situations: Collapse (**cave-in**) is the detachment of a mass of soil in the side (wall) of a trench or excavation and the sudden displacement that represents a hazard to any person inside or on the trench lip **FIGURE 1-6**.
- Noncollapse situations: These less common trench rescue incidents include victims struck by mobile equipment or falling loads, electrocutions, medical issues for construction workers in the trench, injuries by falls, and those overcome by hazardous atmospheres in trenches and excavations that have not collapsed.

For their own safety and for the safety of all persons on scene, firefighters must be able to recognize and mitigate all hazards associated with trench and excavation sites **FIGURE 1-7**.

OSHA CFR 1926 Subpart P, Excavations

The OSHA standard for excavations (29 CFR 1926 Subpart P) was created for the construction industry. As previously discussed, excavation and trenching are among the most hazardous construction operations. OSHA's excavation standards contain requirements for excavation and trenching operations that provide safe working practices for employees digging and working in and around trenches and excavations. OSHA standards are not merely recommendations. OSHA CFR 1926 Subpart P is a federal law. States that do not use federal OSHA regulations may develop their own laws as long as they meet or exceed federal OSHA requirements. Each excavation site is required to have a competent person who is responsible for safe practices and compliance with OSHA regulations. A **competent person** is defined as an individual, designated by the employer, who is capable of identifying existing and predictable hazards in the surroundings or working conditions that are unsanitary, hazardous, or dangerous to workers and who is authorized to take prompt corrective measures to eliminate them.

OSHA and Trench Rescue

It is important for firefighters to recognize that OSHA CFR 1926 Subpart P was not written for and does not address rescue operations. OSHA CFR 1926 Subpart P was written to ensure workers' safety at open excavations in new construction. The OSHA excavation standard is applied to rescue operations quite differently from one state to another. In fact, whether or not any parts of the regulation will be enforced during a rescue usually depends on the interpretation of the law by the local OSHA enforcement officer, thus creating an ad hoc and inconsistent approach for rescue operations. The authors recommend that members of organizations that are responsible for responding to trench emergencies read both their own state and federal OSHA standards. They should then set up a meeting with representatives of their OSHA compliance agency to determine what, if any, parts of the excavation (construction) standard will be enforced during a rescue operation.

An example of the differences in practices used by construction workers versus those used by rescuers is apparent with shoring. Shoring is a protective method used to prevent collapse or the unravelling of the trench wall. The sheeting and shoring guidance within the OSHA standard is designed for the installation that occurs shortly after a trench is dug and before the soil becomes active or starts to move. This is unlike the soil conditions found at most trench rescue incidents when the soil has collapsed and is likely to continue to collapse as the shoring is being installed. These differences in soil conditions require significantly different shoring techniques. Furthermore, rescuers shore trenches with people trapped in them. Construction workers do not shore trenches with anyone trapped in them. Victim considerations, such as protection from additional collapse and safety during shoring installation, require rescuers to use many specialized shoring techniques that are not included in the OSHA standard.

The shoring guidance found in OSHA CFR 1926 Subpart P is based on soil classifications and lateral soil forces created by OSHA. Type A-25, Type B-45, and Type C-60/C-80 soil classifications are not accurate representations of the soil forces found at trenches with soil failures and collapsing walls. OSHA specifically states that their classifications cannot be simply applied to theoretical (predicted) failure surfaces to develop a worst-case scenario for shoring designs and use. In 1994, OSHA cautioned against using Type A-25, Type B-45, and Type C-80 soil classifications for shoring system designs, instead stating that shoring should be designed by *a competent person*. This left firefighters without a simple, rapid, and accurate standard from which lateral soil forces and subsequent shoring designs could be calculated for rescue operations where time is critical and competent persons may not be readily available.

Additionally, the OSHA shoring guidance is designed to be able to resist soil forces for long durations (days and sometimes weeks) associated with many construction projects and not the interim or immediate forces generated in temporary or collapse conditions. Therefore, OSHA guidelines have different design criteria by necessity. As such, these permanent or semipermanent design guidelines require the use of very strong and heavy shoring equipment with shoring installation to be concurrent with trenching operations. Installing shoring via the OSHA guidelines often requires the use of heavy equipment, such as excavators, backhoes, etc., and operators to install the shoring that are typically unavailable in a rescue scenario.

In summary, the OSHA shoring guidance for braced (shored) trenches is not meant for rescue shoring operations. In the past, firefighters have attempted to create shoring systems that are more appropriate for rescue situations. Without the input of an experienced engineer, those shoring designs would not have the engineering principles needed to ensure safety. OSHA has provided an alternative to using the shoring guidance that the standard gives to construction workers. This alternative can be found in the Selection of Protective Systems (1926 Subpart P App. F) Option 4 Professional engineer design: Spec.1926.652 (c) (4), which requires the excavation to be designed by a registered professional engineer. It is important to use a professional engineer with subject matter expertise in both geotechnical engineering and trench rescue shoring principles and practices.

TIP

In addition to reading state and federal OSHA standards, rescuers need to be able to read the technical references that may be at a trench site, such site plans. Site plans detail the engineering specifications for the trench, including the length, depth, and width of the trench and the support structures. Awareness level rescuers should be able to gather key pieces of information from references, such as the depth of the trench and its purpose.

National Fire Protection Association (NFPA)

The National Fire Protection Association (NFPA) is the organization that creates standards for the fire service. Unlike OSHA regulations, which are laws, NFPA regulations are recommendations. However, unlike the OSHA trench and excavation regulations that are designed for the underground construction industry, NFPA recommendations are designed for the fire service and first responders, including firefighters. The two most trench rescue–specific standards are found in NFPA 1670: *Standard on*

Operations and Training for Technical Search and Rescue Incidents, and NFPA 1006: *Standard for Technical Rescue Personnel Professional Qualifications*. The purpose of NFPA 1670 is to assist the **authority having jurisdiction (AHJ)** in assessing a technical search and rescue hazard within the response area, to identify the level of operational capability, and to establish operational criteria. The purpose of NFPA 1006 is to specify the job performance requirements (JPRs) for service as technical rescue personnel. In most cases, the AHJ for events requiring search and rescue is the local fire department or public safety department. NFPA 1670 lists the minimum procedures that an AHJ needs to support for awareness, operations, or technician level service. NFPA 1006 lists the recommended JPRs that members of the organization (fire or public safety departments) need to be capable of to provide the level of service that their AHJ has decided to provide to the community. It is important to note that the NFPA outlines skills and knowledge necessary to safely shore a trench but does not provide prescriptive means and methods to do so. This chapter will discuss the operational criteria requirements (NFPA 1670), and throughout the book, we will describe the JPRs (NFPA 1006) for personnel performing at the trench rescue awareness, operations, and technician levels.

NFPA 1670: Trench Rescue Levels

NFPA 1670 outlines the three levels of response to trench rescue incidents. The AHJ can use this outline to determine the most appropriate level of response for their community. At a minimum, all first responders who will be sent to a trench emergency must be trained to the awareness level. Those first responders include all fire department line personnel, police officers, and members of the emergency medical services. It is common for fire departments to provide trench rescue awareness training to all personnel while designating select personnel with higher levels of training as trench rescue or technical rescue team members. While large fire departments may have their own trench rescue team capability, most fire departments rely on mutual aid agreements for those resources. The NFPA 1670 awareness, operations, and technician level response capabilities for organizations that are responsible for search and rescue operations are outlined next.

Awareness Level

According to NFPA 1670, awareness level responders are required to recognize incidents, control access to the site, and implement a response. Responsibilities include the following:

- Understand how often trenches are utilized in their area, and know how to obtain resources to respond to a trench collapse.
- Recognize the potential site hazards and signs of collapse, and control the emergency site.
- Set up trench lip protection around the site and perform non-entry rescues of mobile victims **FIGURE 1-8**. A **trench lip** is the area around the top edge of the trench face.

> **TIP**
>
> Per NFPA 1006, personnel trained to the awareness level need to be able to do the following:
> - Understand the basic definitions of terms used at trench rescue incidents.
> - Identify the types of trench or excavation incidents and the hazards associated with them.
> - Recognize the capabilities and limitations of awareness, operations, and technician level responders.
> - Implement the initial actions at a trench emergency (detailed in Chapter 3, *Initial Actions*).
> - Understand the components of their local standard operating procedures for awareness, operations, and technician level incidents (detailed in Chapter 6, *Managing the Trench Incident*).
> - Identify the resources needed at a trench or excavation rescue incident and the methods to summon the required resources.

Operations Level

Organizations that respond at an operations level should be able to identify and mitigate trench and

FIGURE 1-8 Trenches are dug in every community to install and repair underground utilities.
© Maksim Safaniuk/Shutterstock.

excavation hazards, in addition to providing collapse protection at a nonintersecting trench. Nonintersecting trenches are 8 feet (2.4 m) or less and the collapse protection only utilizes sheeting and shoring (panels and struts). Sheeting is designed to hold back debris and a strut is designed to carry force from one side of the trench to the other. In addition, organizations responding at the operations level should be able to do the following **FIGURE 1-9**:

- Perform scene size-up and identify unstable conditions.
- Identify the location of victims.
- Enter a trench safely for rescue and recovery operations.
- Install and remove sheeting and shoring and identify the safe use of a trench shield (**trench box**) and sloping and benching systems. (Sloping and benching will be covered in detail in Chapter 2, *Soil and Collapse Mechanics*, and trench shields and boxes will be covered in detail in Chapter 7, *Operations Level Trench Rescue Shoring*.)
- Obtain the assistance of utility services and provide ventilation as needed.
- Identify bell-bottom shaped excavations and install trench lip protection around the **rescue area**.
- Provide safe paths for entry and egress, and extricate a trapped victim.
- Hold a pre-entry briefing and document entry operations.
- List the duties and responsibilities of panel teams, entry teams, and shoring teams.
- Determine the mechanism of victim entrapment and provide methods for victim removal.

Technician Level

Organizations operating at the technician level at trench and excavation emergencies should be able to do the following **FIGURE 1-10**:

- Identify all trench hazards associated with individual (nonintersecting) and **intersecting trenches** with initial depths of more than 8 feet (2.4 m).
- Perform rescue and recovery operations, conduct digging and victim removal operations, and utilize isolation systems and associated rigging and heavy equipment to protect victims from collapse.
- Provide protection from collapse through the use of manufactured protective systems and evaluate the capacities of protective systems and supplemental shoring systems.
- Perform **atmospheric monitoring** for oxygen content, flammability, and toxicity within the affected rescue area.

FIGURE 1-9 The limited depth and configuration of trenches meeting the operations level criteria result in relatively few collapse-related emergencies.
© ChiccoDodiFC/Shutterstock.

FIGURE 1-10 Deep trenches and intersecting trenches have a high potential for collapse.

> **TIP**
>
> **Other NFPA Standards Related to Trench Rescue**
> - NFPA 472, *Standard for Competence of Responders to Hazardous Materials/Weapons of Mass Destruction Incidents*, 2013 edition
> - NFPA 1500, *Standard on Fire Department Occupational Safety and Health Program*, 2013 edition
> - NFPA 1561, *Standard on Emergency Services Incident Management System and Command Safety*, 2014 edition
> - NFPA 1951, *Standard on Protective Ensembles for Technical Rescue Incidents*, 2020 edition

Soil Basics and Collapse Mechanics

A basic knowledge of soil mechanics will help in understanding why soil fails and why trench and excavation walls collapse. Soil is composed of rock-based material (grains), minerals, water (moisture), organic material, and air. Because soil is made up of such diverse materials, it is often divided into three types based on the size of its particles. The basic soil types are sand, silt, and clay, and most soils have a combination of the three types. Sand has the largest particles (seen with the naked eye) and clay and silt have the smallest (seen only with a microscope). The size and shape of soil particles, which partly determines the soil friction, has an effect on a trench/excavation wall's collapse probability. Soil will be covered in detail in Chapter 2, *Soil and Collapse Mechanics*.

Gravity is a physical force of nature that draws everything to the center of the earth. When a trench is dug, it leaves trench walls that have unopposed forces, resulting from the pull of gravity. The pressure or force on the wall is called lateral earth pressure. All soils and all trench walls have lateral earth pressure that is trying to push the wall into the hole in the ground. The amount of lateral pressure that certain soils exert is a result of the weight of soil pushing down and the amount of resistance, or internal strength, that the soil has. The soil's strength is dependent on soil friction, cohesion, and moisture content. If the internal strength of the soil is greater than the lateral force, movement will not occur, and the trench wall will remain standing (for a time). If the lateral force is greater than the soil's internal strength, then the soil will become active and the wall will collapse unless the wall is properly shored. (See Chapter 2, *Soil and Collapse Mechanics*, for more information on soil mechanics and collapse.)

External forces and environmental conditions can weaken the soil's ability to resist lateral forces and can cause a trench wall to collapse. Sometimes just the weight of a human walking on a trench lip is enough to turn a standing wall into a collapse pile on the trench floor. The strength of the soil can change after a trench sits open for a period of time—the longer the trench sits open, the weaker it becomes. When trench walls collapse, they leave voids in the walls. The voids typically result in distinguishable patterns called collapse patterns. The most common collapse patterns are shown in **FIGURE 1-11**.

Every trench wall should be considered a potential collapse hazard. A trench that has collapsed has demonstrated that the soil's internal strength is not sufficient to resist the current lateral earth pressure. A trench with fissures, sloughing, raveling, or bulging has already started to fail. Contrary to folklore, a trench that has collapsed is highly likely to collapse again (secondary collapse). Although every trench collapse will not be preceded by visual warning signs, there are visual clues for secondary collapse. These will be covered in detail in Chapter 2, *Soil and Collapse Mechanics*.

A trench collapse commonly results in several cubic yards of dirt falling from the trench wall to the trench floor. One cubic yard of dirt consists of 27 cubic feet (3 ft × 3 ft × 3 ft). On average, a cubic foot of dirt weighs about 120 pounds (54 kilograms [kg]). That means that the average cubic yard of dirt weighs 3,240 pounds (1470 kg), which is about the weight of a mid-size car. The weight of soil and the manner in which it surrounds the body results in an entrapping mechanism. Just 1 cubic yard of dirt on top of a victim's chest is more than enough weight to mechanically limit the expansion of the lungs, resulting in death (compressive asphyxia). The weight of soil can also cause potentially fatal crush and compartment injuries. The force created by large falling chunks of soil can hit victims in the trench and cause impact-related injuries to the head, neck, spine, pelvis, extremities, and organs. Soil particles that enter and block both the upper and lower airways can cause death from suffocation.

Trench Rescue Response Systems

It is vitally important that the people responsible for planning and implementing a response to a trench emergency understand the resources that are needed

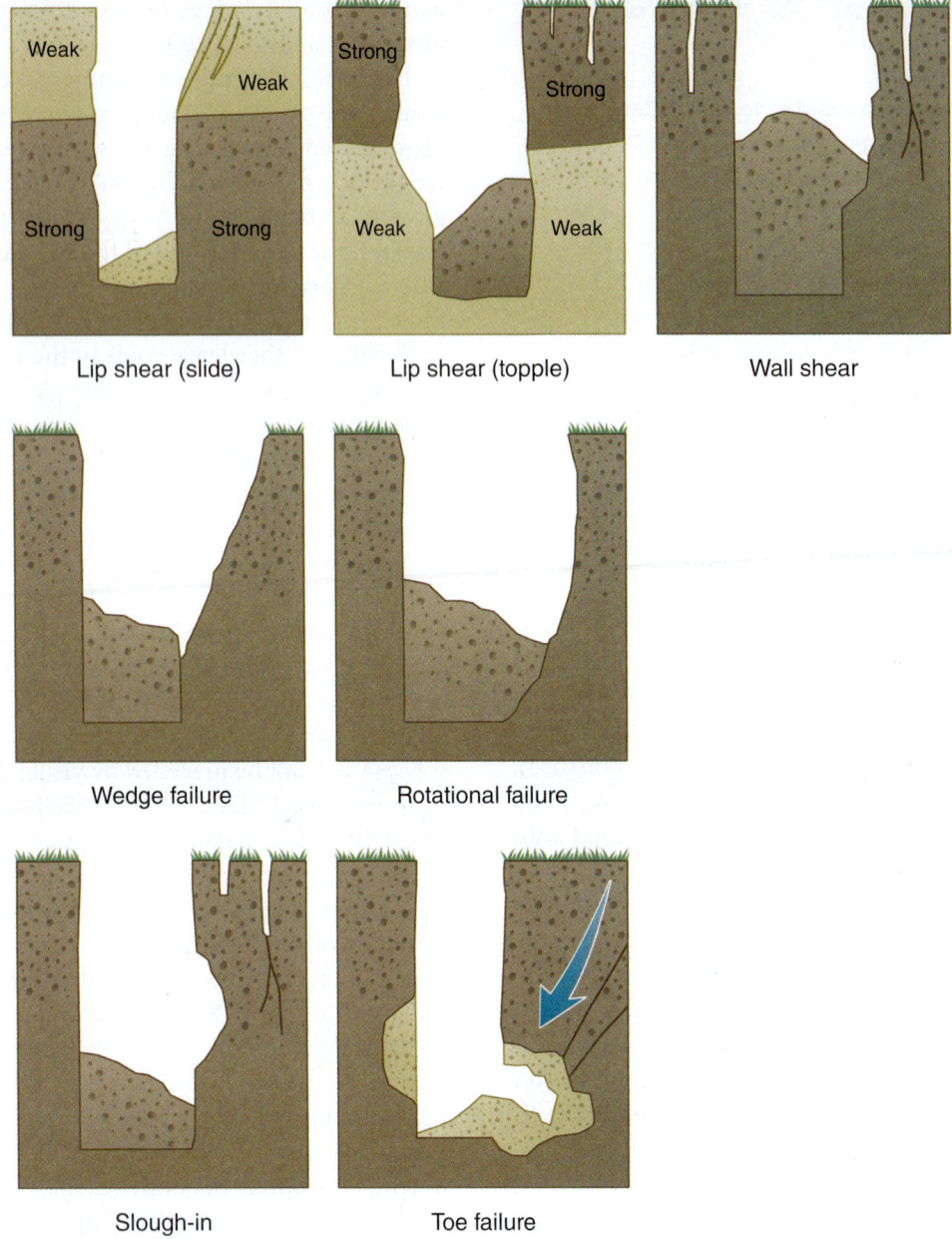

FIGURE 1-11 Common trench collapse patterns.

to conduct trench rescues safely and efficiently. While a trench rescue team is an important part of a trench rescue response system, it must be understood that it is only a part of a multifaceted system. A trench rescue response system is a community endeavor.

A well-prepared AHJ must develop a list of resources that are needed to mitigate the majority of trench incidents, such as utility companies and other fire departments. A well-prepared AHJ will establish agreements with each agency on the list long before the need for them to respond occurs. In addition to the resources that are frequently needed at trench rescue incidents, the AHJ should develop a list and establish agreements with resources for trench and excavation incidents that occur less frequently. The services and equipment that each resource brings to a trench rescue incident are detailed in Chapter 4, *Personal Protective Equipment and Equipment Basics*.

Trench and Excavation Hazards

Trenching and excavation work presents serious hazards to everyone engaged in that environment. The same hazards that confront workers at an excavation site also confront first responders and technical rescuers. First responders and trench rescue teams must be able to recognize, identify, and mitigate the hazards associated with trench and excavation operations.

Although the OSHA excavation standard was not designed for rescue operations, it does identify the hazards associated with the trench environment and it provides hazard mitigation guidance to construction workers. Although the methods and the degree of mitigation used by workers while performing underground construction work can be significantly different than the methods and degree of mitigation used by a first responder while rescuing a live victim, hazards must always be mitigated to an appropriate level that is based on a comprehensive risk versus gain analysis **FIGURE 1-12**. Risk versus gain analysis is a decision made by a first responder based on assessing the situation for hazards and weighing the risks likely to be taken against the benefits to be gained for taking those risks. This concept will be covered in detail in Chapter 5, *Hazard Mitigation*.

An understanding of the OSHA hazard control requirements provides rescuers with a standard point of reference to determine whether adequate safety measures have been implemented by the construction workers. Additionally, that understanding of hazard control is important to rescue personnel because it provides a safety guideline, which can help responders determine appropriate tactical direction.

The most common hazards associated with trench and excavation sites include collapse (cave-in), struck by objects, underground utilities (natural gas, water, sewer, electrical, etc.), injured by falls, hazardous atmosphere (carbon monoxide, hydrogen sulfide, explosive, oxygen deficient), hazardous materials (commonly gasoline, diesel fuel, solvents fluids), and physical hazards (commonly construction and rescue equipment on the trench lip).

Sample OSHA Safety Measures
Collapse (Cave-In)
Any time there are conditions where there is a potential for a trench collapse, protective systems are improperly installed, or hazardous conditions are suspected, every person must be removed from the area until the problems can be safely mitigated. In addition, no worker can enter a trench 5 feet (1.5 m) deep or greater until proper protection is installed, or the trench walls are sloped to proper backslopes or steps and it is declared safe by a competent person.

Falling Objects
Materials, equipment, and excavated soil should be securely placed at least 2 feet (0.6 m) back from the lip of the trench. Materials, such as pipes, should utilize restraining devices, or even better, be placed at 90 degrees on the ground beside the trench to prevent pipes from rolling into the trench. Any time that materials are being lowered into a trench, workers must place themselves away from the work so that the load cannot fall on them or pin them against a wall or other materials in the trench.

Fall Hazards
Trip hazards must be monitored and removed from the trench work zone **FIGURE 1-13**. Fall protection is not required unless the excavations are not readily seen because of plant growth or other visual barriers.

Utility Hazards
Utility identification and marking services should be utilized before digging begins to mark the location of

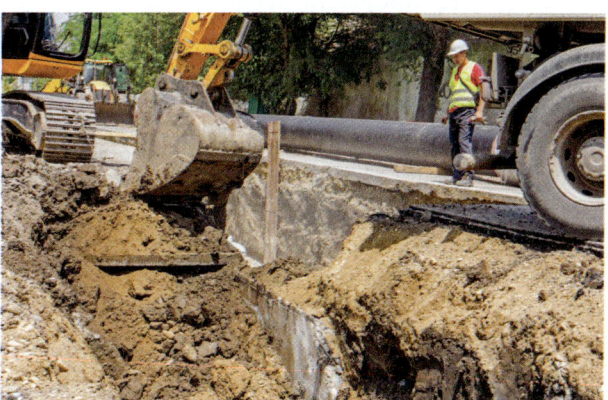

FIGURE 1-13 Slippery conditions (mud, water, ice, and snow) and trip hazards are prevalent at trench sites.
© roman023/Getty Images.

FIGURE 1-12 Evaluate the risks and gains before deciding the course of action.

FIGURE 1-14 Exposed or broken underground utilities can injure and kill trapped workers and responders.
Courtesy of Ron Zawlocki.

utilities in the area of the planned trench. The utilities may include communications, electric, natural gas, oil pipelines, water mains, and sanitary sewers. The exact location of utilities may be determined by trained professionals through hand digging, non-metallic probing, hydro excavating, or other safe methods. Any utilities that are within the excavation area should be removed or secured by a trained professional to protect all workers from the hazards **FIGURE 1-14**.

Hazardous Atmospheres

Air quality in trenches must be tested and monitored to ensure safe levels of oxygen, typically in or near landfills, contaminated ground, or where substances are stored that could sink into the trench. The need for air monitoring is a responsibility of the employer's designated competent person. If the competent person determines that hazardous atmospheres could or do exist, the trench should be evacuated immediately.

Atmospheres with less than 19.5 percent oxygen are hazardous and require safety precautions such as proper ventilation and air supplies. Atmospheres with flammable gases in concentrations measuring 20 percent of the lower flammable limit require precautions such as ventilation with fresh air **FIGURE 1-15**. If ventilation or air supplies are being used in the trench, it is important that the air continue to be monitored for hazardous atmospheric conditions.

FIGURE 1-15 Oxygen and flammable and toxic atmospheric levels must be evaluated prior to workers or rescuers entering a trench.
© Jones and Bartlett Publishers. Courtesy of MIEMSS.

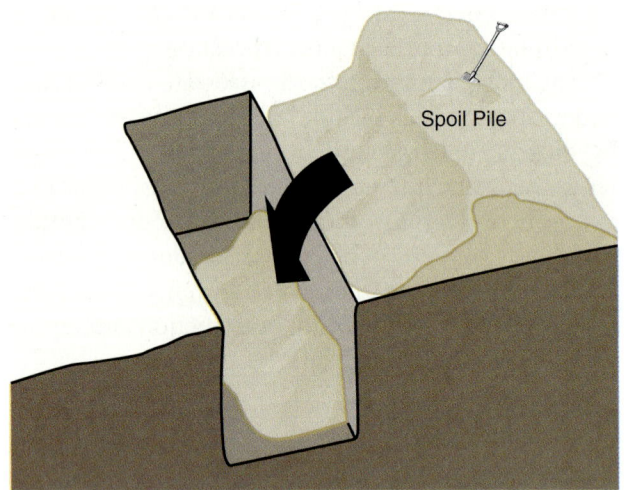

FIGURE 1-16 Spoil piles are hazards that must be keep at least 2 feet (0.6 m) from the trench walls.

Physical Hazards

The **spoil pile** and other materials must have a 2-foot (0.6 m) setback from the lip of the trench to avoid spilling into the trench **FIGURE 1-16**. In addition, heavy equipment must be kept away from trench walls.

Voices of Experience

A contractor was in the process of making a sewer hookup on a new residential parcel in a developing subdivision. The ground was undisturbed clay, and the trench was being dug by an excavator while a worker was in the trench. The worker was about 12 feet (3.7 m) down in the bottom of the trench performing the hookup when the sidewall sloughed in (collapsed) and trapped him. The operator of the excavator quickly moved his bucket against the sidewall and supported the section of wall that had not collapsed yet. Once the excavator bucket was pinned against the wall, the operator jumped in and began digging to free his co-worker.

A call went out to the local fire department for a trapped worker in a trench. The fire department had done some trench shoring training using timber struts but had limited experience and shoring equipment. As crews arrived, contact was made with the trapped worker, who was on his hands and knees, and the collapse had buried him to his chest, trapping his lower body and one of his arms. The initial actions of the responding team included placing ground pads on the lip of the trench. Once lip protection was completed, they added a ladder into the end of the trench for access and egress.

Shoring began by placing sheets of plywood supported by timber struts into the trench. Wooden wedges were used to pressurize the struts. The shoring design was loosely based on OSHA Type A Shoring Charts. Unfortunately, this was not Type A soil. Any trench that has collapsed should be considered Class C (unstable) soil at best.

The first pieces of shoring placed into the trench were two sheets ¾-inch CDX plywood (not trench rescue panels) placed horizontally just above the cave-in pile on the trench floor. The plywood covered the trench walls from the cave-in pile upward for 4 feet (1.2 m). Two timber struts were then installed to support the plywood sheets. This left about 4 feet (1.2 m) of unprotected trench wall above the plywood. The timber struts (4 × 4-inch and 2 × 6-inch) were spaced approximately 6 feet (15 cm) on center (horizontally).

In an attempt to add capacity to the shoring in place, 2 × 12-inch uprights were slid in vertically between the trench wall and the plywood. No additional shoring was added above the horizontal plywood sheets to support the top 4 feet (1.2 m) of exposed trench walls. This shoring plan was the result of minimally trained and equipped first responders working under the duress of a rescue situation and hazardous conditions. Neither an engineered prescriptive shoring design nor rescue shoring protocols were used. That resulted in a shoring system that was ultimately inadequate for the soil conditions that we faced. The reality is that we were just lucky that day.

The trapped worker was located below the bucket of the excavator. Shoring continued around the excavator bucket and timber struts were used to pin the excavator bucket to the trench wall. The bucket placement and the shoring resulted in only a small space for rescuers to operate. Once the initial shoring was installed, paramedics entered and began patient care that included crush syndrome protocols.

In addition to the challenging shoring issues that the rescuers were dealing with, they were also faced with the cold temperatures that are typical for December in Michigan. Partially frozen and compact clay trapped the worker and made the extrication difficult. Confined space ventilation with heater attachments and ductwork (tubing) was brought to the scene by the local telephone and power companies to prevent the victim and the rescuers in the trench from experiencing hypothermia. Because rescuers were working underneath an excavator bucket, the standard-length shovels that the team used were problematic. Eventually, small hand tools were

Voices of Experience

acquired to remove the soil. Reciprocating saws were used to cut large pieces of frozen soil into smaller pieces that could be removed manually.

Rescuers were not experienced with supplemental shoring and digging procedures. That led to numerous secondary collapse situations throughout the extrication process. A secondary collapse buried the victim above his shoulders and buried the rescuers up to their knees. Additionally, the spoil pile on the lip was not removed, thus resulting in spoil sliding in from above. Although progress was being made in the extrication of the victim, rescuers were exposed to hazardous conditions. After about 6 hours, the worker was finally uncovered. He was placed into a Stokes basket and removed from the trench. The worker sustained a dislocated hip and bruising from the incident.

Although ultimately this was a successful rescue, over the years the rescue team has taken every opportunity to better prepare for trench and excavation emergencies.

Lessons Learned

- We have procured better trench rescue shoring equipment, and additional training is being conducted. Team members have attended certified training from industry leaders in the trench rescue community. They have learned safer practices and honed their skills and equipment to be better prepared.
- We now have trench rescue shoring protocols (prescriptive engineered shoring designs), and more realistic trench training scenarios are being conducted to prepare our team for a variety of situations that they may encounter. Since this incident, we have learned about creating safe zones using engineered shoring equipment and shoring system designs.
- Partnerships have been created and nurtured from neighboring jurisdictions; this provides additional trained personnel and rescue equipment. Partnerships and mutual aid agreements have lessened the financial impact of equipment purchasing on any one organization and has distributed equipment costs across the response system. By training together, neighboring teams have increased the efficiency and safety of our trench rescue response.
- Paratech pneumatic strut equipment has been purchased to replace less efficient timber shores. Although timber shoring has not been eliminated as an option, pneumatic non-entry shoring has become the go-to trench rescue shoring method.
- Trench rescue panels with composite construction made with premium arctic birch plywood (Finform) were purchased and have greatly increased the safety factors of our shoring systems.

Trench rescue preparedness is a huge part of the progression of a response team. Working with professional engineers and conducting destructive testing of shoring designs are needed to validate the safety of trench rescue shoring systems. Repetition of shoring system installations will make it fluid when the emergency arrives. Understand your equipment, know why you do what you do, and continue to attend training programs outside of your team that will keep your mind sharp and current. We can all look back at how things were done, and we should all look forward to how to better prepare for the next call to action.

Mark Laux
Lieutenant, Midland Fire Department
Technical Rescue Response Team Leader, Michigan Region 3
Rescue Specialist, MI-TF1

Ingress or Egress

Easily accessible means of ingress (entrance) and egress (exit) via stairways, ladders, or ramps must be provided for people entering trenches. Trenches 4 feet (1.2 m) or greater in depth must have a stairway, ladder, or ramp a maximum of every 25 feet (7.6 m) **FIGURE 1-17**.

SAFETY TIP

Biological hazards, such as blood, saliva, urine, feces, and emesis are commonly associated with trench-related traumatic injuries. This hazard is not addressed in OSHA excavation standards but must be mitigated by rescuers by using universal precautions. (See Chapter 4, *Personal Protective Equipment and Equipment Basics*, and Chapter 8, *Victim Care and Extrication*, for more information.)

FIGURE 1-17 A rapid means of egress must be provided and maintained.
Courtesy of Greg Payeur.

After-Action

IN SUMMARY

- Excavation and trenching are among the most hazardous construction operations. According to OSHA, hundreds of injuries and dozens of fatalities are caused by trench collapses each year.
- Most trench fatalities occur in trenches that are dug to install new underground utilities, to repair existing underground utilities, and to create and repair foundations for a variety of structures.
- OSHA has established laws for safe working practices for workers engaged in excavation and trench work.
- The OSHA standard for excavations (29 CFR 1926 Subpart P) was created for the construction industry and contains requirements for excavation and trenching operations that provide safe working practices for employees digging and working in and around trenches and excavations.
- NFPA 1670 sets forth the minimum procedures that an AHJ needs to provide for awareness, operations, or technician level service.
- Awareness level rescuers are required to recognize incidents, control access to the site, and implement a response.
- Operations level rescuers should be able to identify and mitigate trench and excavation hazards, in addition to providing collapse protection at a nonintersecting trench.
- Technician level rescuers should be able to identify and mitigate trench and excavation hazards, in addition to providing collapse protection at both nonintersecting and intersecting trenches.
- The purpose of NFPA 1006 is to specify the JPRs for technical rescue personnel.
- OSHA specifies shoring for construction projects that are not suitable to emergency response, and NFPA specifies skills that shoring team members should possess to respond to an incident. Neither specify a prescriptive shoring standard to be safely used at an incident.
- The most common hazards associated with trench and excavation sites include collapse (cave-in), struck by objects, injured by falls, utility hazards, hazardous atmospheres, and physical hazards.

KEY TERMS

Atmospheric monitoring A method of evaluating the ambient atmosphere of a space, including but not limited to its oxygen content, flammability, and toxicity.

Authority having jurisdiction (AHJ) The local authority that draws on the power of law to make rules and regulations for the organization.

Cave-in The separation of a mass of soil or rock material from the side of an excavation or trench, or the loss of soil from under a trench shield or support system, and its sudden movement into the excavation, either by falling or sliding, in sufficient quantity so that it could entrap, bury, or otherwise injure and immobilize a person.

Competent person An individual, designated by the employer, who is capable of identifying existing and predictable hazards in the surroundings or working conditions that are unsanitary, hazardous, or dangerous to workers, and who is authorized to take prompt corrective measures to eliminate them.

Excavation Any human-made cut, cavity, trench, or depression in an earth surface formed by the earth's removal. In practical terms, when a hole is more than 15 feet (4.5 m) wide at its base, it is called an excavation. Overall, an excavation is wider than it is deep.

Rescue area The area located immediately surrounding the rescue site.

Spoil pile A pile of excavated soil next to the excavation or trench.

Strut A compression element used in the support of structures, excavation openings, or other loads.

Trench A narrow excavation (in relationship to its length) made below the surface of the ground. In general, the depth is greater than the width, but its width measured at the bottom does not exceed 15 feet (4.5 m).

Trench box A premade shielding system that can be placed and lowered as digging operations continue.

Trench floor The bottom of the trench.

Trench lip The area 2 feet horizontal and 2 feet vertical (0.6 m and 0.6 m) from the top edge of the trench face.

Trench walls The vertical or inclined earth surfaces formed as a result of excavation work.

REFERENCES

Laborers' Health & Safety Fund of North America. n.d. "Trenches and Excavations." https://www.lhsfna.org/index.cfm/occupational-safety-and-health/trenches-and-excavations/.

Occupational Safety and Health Agency (OSHA). n.d. "OSHA Fact Sheet: Trenching and Excavation Safety." https://www.osha.gov/Publications/trench_excavation_fs.html. Accessed January 25, 2021.

Occupational Safety and Health Agency (OSHA). n.d. "Trenching and Excavation." https://www.osha.gov/Publications/trench_excavation_fs.html#:~:text=A%20trench%20is%20defined%20as,15%20feet%20(4.5%20meters).&text=Cave%2Dins%20pose%20the%20greatest,to%20result%20in%20worker%20fatalities.

On Scene

1. What are the differences and similarities in the approach to trench emergencies and excavation emergencies?

2. What is the role of an awareness level rescuer at a trench rescue?

3. What is the role of an operations level rescuer at a trench rescue?

4. What is the role of a technician level rescuer at a trench rescue?

CHAPTER 2

Awareness Level

Soil and Collapse Mechanics

KNOWLEDGE OBJECTIVES

After studying this chapter, you should be able to:

- Identify characteristics associated with soil types. (**NFPA 1006: 12.1.2, 12.1.3**, pp. 19–22)
- Describe rescue concerns related to each soil type. (**NFPA 1006: 12.1.2, 12.1.3**, pp. 19–22)
- Identify forces associated with soil. (**NFPA 1006: 12.1.2, 12.1.3**, pp. 22–25)
- Describe major factors that influence soil strength. (**NFPA 1006: 12.1.2, 12.1.3**, pp. 25–27)
- Describe types of trench collapse. (**NFPA 1006: 12.1.2, 12.1.3**, pp. 27–33)
- Identify conditions and factors that may lead to trench collapse. (**NFPA 1006: 12.1.1, 12.1.2, 12.1.3, 12.1.5**, pp. 33–35)
- Recognize signs of impending collapse. (**NFPA 1006: 12.1.2, 12.1.3, 12.1.5**, pp. 35–36)

SKILL OBJECTIVES

There are no skills for this chapter.

You Are the Rescuer

You are called to the scene of a trapped worker. As usual, details are scarce. You arrive on the site to see a recent trench excavation with three workers looking into the hole with an additional worker sitting on a 60-year-old backhoe. The workers tell you that no one is trapped. They managed to pull the laborer out of the trench right after the wall collapsed, burying the worker to his knees. They begin discussing who will go into the hole to finish replacing the sewer line and hoping that the driveway does not crack because the homeowner will make them replace it at their cost.

You realize that you were not too far from needing to perform a rescue at this site and realize that you may be called back in an hour. You start to think about what you would have needed to do if the worker was still trapped.

You survey the scene and note the following characteristics: The trench is about 30 feet (9 meters [m]) long, 4 feet (1.2 m) wide, 10 feet (3 m) deep and looks to be moist clay soil. The trench extends from the front of the house and is right next to and parallel to a concrete driveway. The other side of the trench is open yard and the spoils are placed right up to the lip of the trench. The trench wall at the driveway failed in a slough failure. The slough is deepest at the mid depth of the trench and appears to be approximately 3 feet (0.9 m) into the original trench wall and about 15 feet (4.5 m) long.

Based on your assessment, you should already know the following in information if you are called back to this site to perform a rescue:

- The classification of the soil.
- The additional force that the spoil pile will apply to your shoring.
- The type of collapse that is most likely to occur and the resulting collapse pattern (voids).
- The response level (operations or technician) needed to perform a rescue in these soil conditions.

In the event you are called back to the site, consider the following questions:

1. How would you mitigate the risk of a concrete driveway slab that overhangs the failed area of the trench? What methods and equipment would you need?
2. If the excavation is right at the edge of the concrete driveway, do you need to place ground pads on the concrete driveway?
3. How will you determine the surcharge loading from the spoils pile and the driveway for use in determining what trench rescue shoring components to use?

 *Access Navigate for more practice activities.*

Introduction

It is important for trench rescuers to understand soil and collapse mechanics because collapse (cave-in) also poses the greatest risk to rescuers at trench rescue incidents. The Occupational Safety and Health Administration (OSHA) requires that employers create workplaces that have no recognized hazards. Recognized hazards in trenches include the risk of collapse or additional collapse, materials outside of the trench falling into the trench, hazardous atmospheres within the trench, utilities within the trench, and equipment failures that can injure or kill workers in the trench. The most common hazard is the risk of collapse or additional collapse, followed by the presence of utilities in the trench and materials falling into the trench.

OSHA has devoted standards 29 Code of Federal Regulations (CFR) 1926.650, 29 CFR 1926.651, and 29 CFR 1926.652 to this topic. However, OSHA and OSHA-approved state plans are not written for rescue operations. As a result, many of the soil assumptions, terminology, and shoring practices found in the OSHA standards have limited application to rescuers. However, many hazards and safety regulations found in the standard are applicable to rescuers working at a trench or excavation. Therefore, aspiring trench rescuers should read the OSHA-29 CFR 1926.650, OSHA-29 CFR 1926.651, and 29 CFR 1926.652 standards.

In this chapter, we will compare the key soil knowledge requirements found in the OSHA standards to the soil knowledge requirements that a rescuer must have to make safe and effective decisions at a rescue or recovery operation. We will examine soil mechanics, types of collapse, and resulting collapse patterns, and we will also identify signs of impending collapse.

OSHA-29 CFR 1926.651 and 1926.652

OSHA excavation standards, 29 CFR Part 1926, Subpart P contains the requirements for excavation and trenching operations for construction work. These are the requirements that employers must follow while performing trenching and excavation. To ensure scene safety, contractors and employers must follow the trenching and excavation requirements in 29 CFR 1926.651 and 1926.652 or comparable OSHA-approved state plan requirements.

As previously stated, OSHA defines a trench as a narrow underground excavation that is deeper than it is wide and is no wider than 15 feet (4.5 m). An excavation is defined as any human-made cut, cavity, trench, or depression in the earth's surface formed by earth removal. Therefore, a trench is an excavation, but an excavation is not necessarily a trench. Both excavations and trenches have similar safety concerns with vertical or near vertical walls that can fail with little to no notice. An excavation is more likely to offer a path for escape of a collapse, because the environment is not as confined.

Some important requirements of the OSHA standard are as follows:

- Trenches must be properly shored, sloped, or benched.
- Means of egress need to be provided within 25 feet (7.6 m) of any worker.
- Workers must be given adequate protection from the hazards presented by loose soil or rock that may roll or fall from an excavation face.
- Adequate protection must be provided from excavated or other materials or equipment that could fall or roll into the excavation.

These OSHA standards also provide a basis of soil pressure to be used in the design of construction shoring. This recommended soil pressure is dependent upon soil classification by a competent person. With the proper soil classification, the timber shoring systems to be utilized are provided in Appendix C of Subpart P.

Rescuers should recognize that the shoring provided in the OSHA standard is not designed for trench rescue and that the soil pressures provided by OSHA are possibly excessive for the short-term nature of trench rescue shoring. It is important to recognize that there is no nationally recognized method for trench rescue shoring. Past practices in trench rescue shoring are a patchwork of ideas and methods that have been handed down through the generations with no comprehensive engineering input at the system level. The authors hope that the engineering aspects included in this textbook will change those past practices and replace them with recommended best practices based on evidence.

OSHA Soil Classification

OSHA requires that soil be classified before digging begins. The classification must be performed by a competent person. A competent person must have the training and experience that enables them to conduct prescribed visual and manual soil tests. The tests conducted by a competent person in the field are not precise, however there is a margin of error built into the system that maintains a reasonable factor of safety. With few exceptions, firefighters do not have the experience and competency needed to determine soil classifications. This is primarily because firefighters will only practice competent person skills during training exercises, which may occur quarterly, at best. This does not provide enough regular use of the skills to maintain competency. That lack of experience and competency can result in a dangerous error. In Chapter 7, *Operations Level Trench Rescue Shoring*, we discuss the T-L method. Designed by civil engineers (geotechnical specialists), the T-L method allows firefighters to rapidly assess soil and to apply the results of that assessment to trench rescue shoring data to help select the proper shoring components. Understanding soil conditions will enable rescuers to interview and speak intelligently to a competent person at the scene of a trench emergency.

SAFETY TIP

Remember your safety and your team's safety is on the line, and trusting a competent person who overlooked the issues that resulted in collapse may not be the best individual from whom to take advice.

The system of classifying soils used by underground construction workers is a hierarchical approach to determine the performance of a soil based on a decreasing order of stability. In a nutshell, some general assumptions are made about which products are in the soil and how they can be expected to behave when excavated.

OSHA requires all classifications to be determined based on one visual test and one manual test performed by a competent person. Many of the manual tests recognized by OSHA are not practical for use at rescue incidents because of the skills, experience, equipment, and time needed to obtain accurate results.

TABLE 2-1 provides characteristics and concerns about soil, as classified by OSHA. Each type of soil represents a varying degree of danger based on the characteristics that make it a part of that class. When multiple layers are present in the soil, the classification will be determined by the layer that is normally least stable **FIGURE 2-1**.

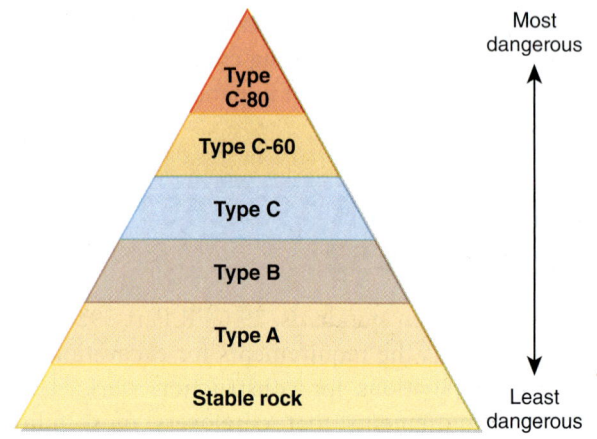

FIGURE 2-1 Stable rock is the least dangerous type of soil, and Type C-80 is the most dangerous.

TABLE 2-1 Soil Types

Type	Characteristics	Concerns
Type A	■ Soil is clay or a mix with mostly clay. Soil plasticity is present. Fissures and other signs of movement are not present. Spoil piles maintain steeper (greater than 56 degrees) angles of repose. ■ Unconfined compressive strength of 1.5 tons per square foot (tsf) or greater.	■ Unusual conditions for a rescue response because the soil is stable and less likely to collapse. ■ If soil shows signs of movement (fissures, sloughing, raveling, or collapse), or if the trench has been subjected to vibrations, install a shoring system that is strong enough to support C-60 soil.
Type B	■ Mixed soil with less clay. Includes granular soil with more than 15% clay. Fissures and other signs of **active soil** (sloughing or raveling) may be present. Includes previously excavated soils that are not Type C soils. Spoil pile angle of repose is 34–55 degrees. ■ Includes cohesive soil with unconfined compressive strength of 0.5–1.5 tsf.	■ The most common type of soil for a rescue response. Soil often looks fairly compact and stable when it is first dug, but becomes active with time. If soil shows signs of movement (fissures, sloughing, raveling, or collapse), or if the trench has been subjected to vibrations, install a shoring system that is strong enough to support C-60 soil.
Type C	■ Granular, sand, and sandy loam (mix). Can include submerged soil. Spoil pile angle of repose less than 34 degrees. ■ Includes cohesive soils with unconfined compressive strength less than 0.5 tsf.	■ Unstable and heavy soil that is further categorized as either C-60 or C-80. The numbers 60 and 80 represent the lateral pressure (per square foot of exposed wall) times depth.
C-60	■ Includes weak/unstable soil types that will stand long enough to install shoring and have a water level at or below the bottom of the excavation.	■ In Type C soils and in all soils whenever a cave-in has occurred, rescuers should install shoring systems that are strong enough to support C-60 soil unless the soil has C-80 characteristics.
C-80	■ Soil that will not stand up long enough to install shoring. Water level above the bottom (floor) of the trench. Moving soil that looks like wet concrete, mud, or quicksand.	■ Conventional trench rescue shoring (panels/struts and wales) is not effective in C-80 soils. Trench boxes, sloping, and sheet piling are commonly used techniques.

FIGURE 2-2 Stable rock.
© Gilles Paire/Shutterstock.

FIGURE 2-3 Type A soil.
Courtesy of Ron Zawlocki.

Stable Rock

The least dangerous soil type, from a collapse perspective, is stable rock (**FIGURE 2-2**). This kind of soil is a natural solid material that can remain standing after excavation. The danger associated with stable rock excavations generally comes from anything except a collapse. This does not mean, however, that some excavated products cannot fall on a worker. Accidents in this environment usually involve worker falls or equipment failures that cause entrapments.

Type A

Type A soil is cohesive material with an unconfined compressive strength of 1.5 tons per square foot (tsf) or greater. Examples of this type of soil include **clay**, silty clay, clay loam, and sandy clay loam (**FIGURE 2-3**). Cemented soils are also considered Type A. Soil is not considered Type A if any of the following conditions are present:

- The soil is fissured.
- The soil is subject to vibration.
- The soil has been previously disturbed.
- The soil is part of a sloped soil layer that is steeper than 4 horizontal units to 1 vertical unit.
- The material is subject to other factors that would require it to be classified as a less stable material.

FIGURE 2-4 Type B soil.
Courtesy of Ron Zawlocki.

Type B

Type B soil comprises cohesive material with an unconfined compressive strength greater than 0.5 tsf but less than 1.5 tsf, or granular cohesionless materials, including angular gravel, **silt**, silt loam, sandy loam, and sandy clay loam (**FIGURE 2-4**). Type B may also be a previously disturbed soil, unless it would otherwise be classified as Type C. Alternatively, it may be a soil that, while meeting the unconfined compressive force requirements for Type A, is fissured, or is subject to vibration from an external force, such as vehicles traveling on a roadway. In addition, it could be a material that is part of a sloped system steeper than 4 horizontal units to 1 vertical unit.

FIGURE 2-5 Type C soil.
Courtesy of Ron Zawlocki.

Type C

Type C soil is cohesive materials with an unconfined compressive strength of 0.5 tsf or less. They include granular soils, sand, and sandy loam (**FIGURE 2-5**). Type C soils also encompass submerged soils, soils from which water is freely flowing, and submerged rock that is not stable. Additionally, this soil type includes sloped or layered systems where the layers dip into the excavation at a slope of 4 horizontal units to 1 vertical unit, or steeper.

Type C soils are generally broken down into two subcategories. Type C-60 comprises soil that is either moist and cohesive or moist and granular, but that does not fit into the Type A or Type B classification and is not flowing or submerged. Type C-60 soil can be cut nearly vertical, and a trench in this soil will stand unsupported long enough to allow shoring to be installed. Type C-80 soil consists of moving or running soil that will not stand up long enough for shoring to be installed. It is often found below the water table or in heavily saturated areas.

In underground construction, after a soil has been classified and conditions that determined the original classification change, a reclassification must be done by a competent person. This may necessitate a change in the type of protective system selected to accomplish the rescue or at least a change in the risk versus gain analysis of the rescue attempt.

Other Soil Classifications

OSHA recognizes the use of soil classifications other than those provided in the OSHA standard. The use of alternative classifications is permitted only if the tabulated data are approved by a registered professional engineer for use in the design and construction of the protective system. The key here is the term *registered professional engineer*. Based on this OSHA interpretation, you may encounter additional subcategories of recognized soil types. In fact, you will encounter a new soil classification, called T-L soil, which has been designed by professional engineers. This soil type is discussed later in Chapter 7, *Operations Level Trench Rescue Shoring*. T-L soil is soil that has been engineered to meet worst-case soil conditions.

Soil Mechanics

Gaining a basic knowledge of soil mechanics will help provide an understanding of why soil fails and why trench walls collapse. One of the keys to success in trench rescue is recognition of how the various physical elements interact and then contribute to a collapse situation. Many of these factors, either by themselves or in conjunction with other physical factors, need to be evaluated before a trench rescue intervention is attempted. In this section, we will discuss soil types, forces associated with soil, and soil strength.

Soil Types

Soil is made up of rock-based material (grains), minerals, water (moisture), organic material, and air. Because soil is made up of such diverse materials, it is often divided into three basic types based on the size of the particles it contains. The basic soil types are sand, silt, and clay, and most soils have a combination of these soil types. Sand has the largest particles and clay has the smallest. Sand particles can be seen by the naked eye. Clay and silt particles may only be seen with a microscope **FIGURE 2-6**.

Soil is a multiphase system composed of soil particles, air, and fluid **FIGURE 2-7**. The ultimate strength (resistance) of soil depends on the interaction of these phases or components. Typically, the more fluid (water) in the soil, the weaker (lower the resistance) it is and the heavier it becomes. The heavier the soil is, the larger the potential lateral force (typically) at failure. Clay material is typically lighter in weight than sandy soil but can yield larger failure surfaces along a trench face.

Forces Associated with Soil

Vertical Force

One of the most important physical forces of nature that determines whether something stands or falls is the force that draws everything to the center of the

FIGURE 2-6 **A.** Clay. **B.** Silt. **C.** Sand.
A: © ognennaja/Shutterstock; **B:** © Fabio Lamanna/Shutterstock; **C:** © Viktoriya Pavliuk/Shutterstock.

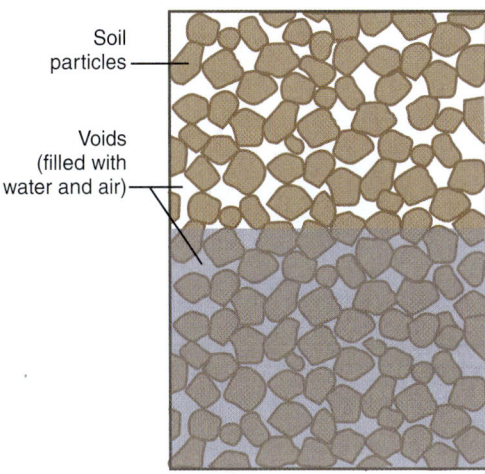

FIGURE 2-7 Soil particle.

FIGURE 2-8 Mass, as it relates to other objects of mass, determines the effects of gravity.

earth—**gravity**. On the most basic scale, if we dig a hole and then leave, the hole will eventually fill itself back in. This tendency is simply nature's way of reaching the lowest energy state. In reality, a hole in the ground (trench or excavation) is preventing the earth from reaching the lowest gravitational energy state, which is an overall spherical planetary shape **FIGURE 2-8**. Gravity is discussed in more depth in Chapter 9, *Lifting and Load Stabilization*.

Compounding the effects of gravity is the concept of **hydrostatic pressure**. Hydrostatic pressure is increased pressure caused by the addition of water to the soil profile. Soil is both porous and absorbent. Dry soil can weigh between 80 and 120 pounds per cubic foot (pcf; 1281–1922 kilograms per cubic meter [kg/m^3]). When the weight of water is added to an absorbent soil, the resulting weight can be astounding. In fact, in some cases, water-saturated soil can weigh as much as 150 pcf (2403 kg/m^3) **FIGURE 2-9**.

Of all the physical factors associated with a trench collapse, probably none is as poorly understood as the total weight of a volume of soil, or **soil density**. Many people simply cannot relate to dirt as a volume that has a great mass and weight that can hurt them. To prove this point, think about this question: Have you ever seen someone stand under a piano being lifted to

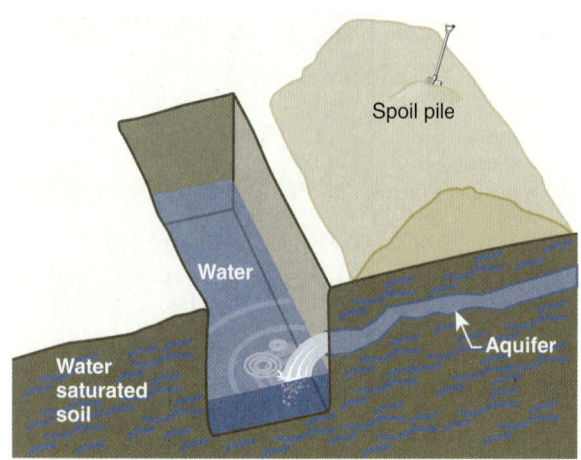

FIGURE 2-9 Water can change the dynamics of soil stability very quickly during rescue operations.

the upper floor in a building? No! If the piano were to fall, it would crush the individual. In contrast, many people have no qualms about standing in a trench beside a volume of potentially collapsible soil, which can be just as dangerous as a falling piano. More people die from trench collapse than falling pianos, and fortunately, both are preventable **FIGURE 2-10**.

The weight of 1 cubic foot of most soil is approximately 120 pounds (54 kg). That amounts to a 1 × 1 × 1-foot box of soil. (That figure could be less or more depending on the type of soil and its moisture content.) Certainly, an even greater volume than 1 cubic foot of soil would fall on you in any significant trench collapse.

To determine the weight of soil scientifically, we can make a few observations. In general terms, dry soil is one-half soil and one-half air. The specific gravity of rock is about 2.65, which means that it is 2.65 times heavier than water. If water weighs 62.4 pcf (999.6 kg/m^3), then rock would weigh 62.4 × 2.65, which equals 165.4 pcf (2649.5 kg/m^3).

Soil structure is composed of solids and voids. The solids are rock fragments, and the voids are the space between them. The voids can fill with water, and how much water is a function of the soil's porosity, or the number of voids. Soil is typically more porous than rock. If the soil you are dealing with is very dry and highly porous with one-half rock solids and one-half air, and assuming the specific gravity of 2.65, it weighs 82.7 pcf (1324.7 kg/m^3). If the soil is one-half rock and one-half water (saturated), each cubic foot would weigh 113.9 pounds (51.7 kg). (We arrived at 113.9 pounds (51.7 kg) by adding together one-half cubic foot of water at 31.2 pounds (14.2 kg) and one-half cubic foot of rock at 82.7 pounds (37.5 kg). Generally, soil weighs between 85 and 150 pcf (1361.6 and

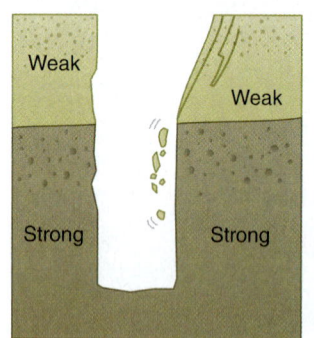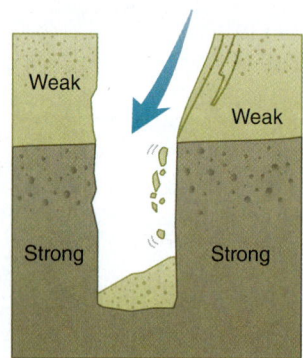

FIGURE 2-10 Most people would never stand under a grand piano (1100 pounds [599 kg]) that was being lifted, yet many construction workers think nothing of standing under several tons of unsupported soil walls.

2402.8 kg/m^3), which in most situations can be averaged to 120 pcf (1922.2 kg/m^3).

To illustrate the seriousness of the problem, lie down on the floor and have someone sit on your chest. Depending on the size and weight of your friend, that mass would amount to approximately 150 to 200 pounds (68.4–90.8 kg). Now imagine what just 10 cubic feet

(1200 pounds [544.3 kg]) of soil on your chest would feel like. Clearly, it would be very hard to breathe.

Horizontal Force

Lateral earth pressure, or **lateral soil pressure**, results in the force that soil exerts in the horizontal direction. The amount (magnitude) of lateral pressure that certain soils exert is a result of vertical **overburden pressure** (the weight of soil pushing down) and the internal strength of the soil (resistance). Civil engineers use the term **active earth pressure coefficient** to describe how much (percentage) of the vertical pressure translates to horizontal pressure pushing outward. So, if you have soil that has minimal internal strength (weak soil), the active earth pressure coefficient can be as high as 0.71. That means that the horizontal pressure is 71% of the vertical pressure (weight). On the other end of the spectrum, soils with very high levels of internal strength (strong soil) can have an active earth pressure coefficient as low as 0.17, which means the horizontal pressure is only 17% of the vertical pressure.

A great analogy for this would be to compare a block of gelatin, a brownie, and a brick. If you put the same weight on all, you will get very different results. The gelatin will squish, and the sides will bulge out. This is similar action to what you would see in weak clay soils and would result in the largest amount of lateral pressure. The same load on a brownie will just compress the brownie a little without really bulging the sides much. This would be like a medium clay or a medium compact granular soil. If you put the load on a brick, you would notice no changes. This would be similar to a strong clay or a very compact granular soil and results in the lowest lateral pressure because the material is so strong it resists it internally.

The total amount of vertical force that is translated horizontally into a lateral pressure depends on the strength of the soil to resist lateral movement. If the internal strength of the soil is greater than the lateral force, visible movement will not occur, and the trench wall will remain standing for a while. The amount of time that the trench wall will remain standing is dependent on several factors, such as soil composition, grain size/shape, creep, temperature, permeability, and external forces. A determination of how long a trench wall will remain stable while unsupported is beyond the scope and training of trench rescuers, so we will simply assume that if the internal strength is greater than the lateral forces, then the trench will remain standing for a while.

The way that lateral forces are typically handled by civil engineers is by assuming a triangular or rectangular unit volume of soil and multiplying it by the active earth pressure coefficient. This is based off the wealth of case studies and testing on permanent flexible retaining structures, such as heavy concrete retaining walls. However, trench shoring is not a permanent entity with a large mass and is a braced (or supported) excavation system wherein small movement can result in a complete loss of system strength. In these braced excavations, the lateral earth pressures can exceed those of the ductile systems by as much as 10 to 15 percent.

Rescuers need to be able to rapidly assess the amount of lateral soil force that their shoring system will need to support. In Chapter 7, *Operations Level Trench Rescue Shoring*, we will present a recently developed soil method designed by geotechnical engineers for use by first responders. It is called the T-L method and when accompanied by collapse specific tabulated data (also found in Chapter 7), it allows rescuers to rapidly determine lateral soil forces and safely select shoring equipment and systems.

Soil Strength

The soil's strength includes soil friction, cohesion, and moisture content. Gravity (vertical pressure) and the soil's strength each create a direction and magnitude called a **vector**. The combined effect of those vectors, or resultant, represents the lateral force on a trench wall. When the lateral force is greater than the soil's internal strength, the soil will move horizontally and create a lateral force.

If the internal strength of the soil is greater than the lateral driving force, movement will not occur, and the trench wall will remain standing for a while. That soil condition is considered to be at rest. Soil that has been at rest may begin to move as changes (moisture content, surcharged loads, etc.) occur. When the lateral force overcomes the resistant forces, the soil will become active and the wall will collapse.

To gain a better understanding, let us expand upon the major factors that influence soil strength (soil friction, cohesion, moisture content, and the resulting unconfined compressive strength).

Soil Friction

Soil friction contributes to a soil's internal strength. Friction adds resistance. Angular shapes have more friction than rounded or flat shapes. Soils with angular-shaped particles are stronger than soils with rounded-shaped particles because of the amount of friction caused by angular interlock. That is because the angular shape of the grains locks them together, preventing them from rolling freely past each other.

FIGURE 2-11 Try to stack marbles on a flat tabletop.
© Denis Botarev/Shutterstock.

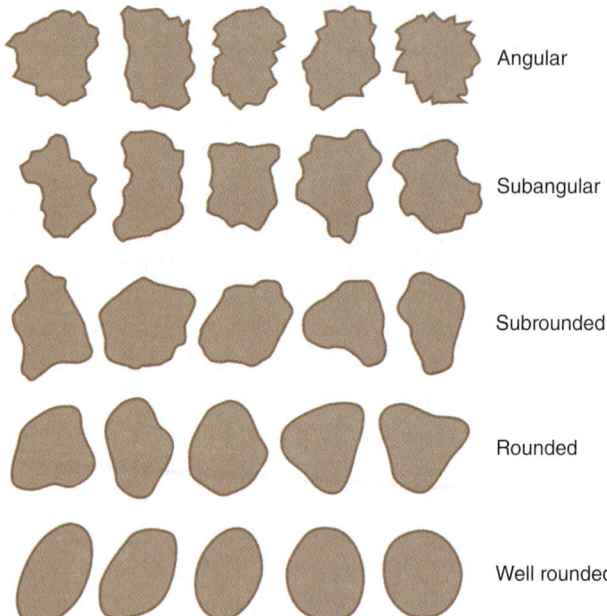

FIGURE 2-12 Angular particles have more friction allowing them to resist more lateral forces.

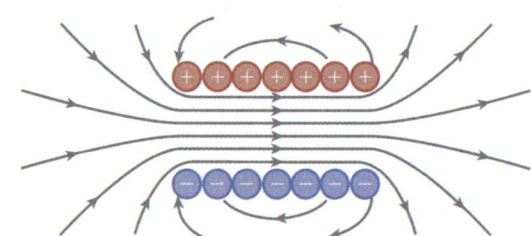

FIGURE 2-13 Electromagnetism, or cohesion.

A good way to understand soil friction is to consider round, hard spheres, like marbles or ball bearings. Perfectly round ball bearings will not stand in a pile. Ball bearings have a zero **internal angle of friction**. Due to their shape and hardness, they are nearly frictionless. This means that they are very poor at supporting forces unless they are confined. Dry sand, on the other hand, will form a cone-shaped pile when poured on a table. This is the way that granular soil (sand) acts **FIGURE 2-11**.

When attempting to stack marbles, we will find that we may be able to arrange a single layer, but if we try to put one on top for a second layer, it will push the others apart. Why? The vertical weight of a sphere pushing down between first layer spheres acts like a wedge, creating lateral forces, trying to force the first layer spheres apart. Because they are nearly frictionless, they have no ability to resist this lateral force and they move, preventing stacking from happening.

If we confine the spheres by putting a frame around them, we will be able to stack them. The frame around the first-level spheres confines them and provides the ability to resist the lateral forces.

Now, if we do the same experiment with irregularly shaped objects, the stacking becomes easier. The individual objects begin to lock together in a way that can internally resist more lateral force. The more angular those particles get, the more they lock together and the more they resist lateral forces internally **FIGURE 2-12**.

Eventually, if we stack soil high enough, the weight of the soil above the particles, and the wedging forces it creates, will be able to overcome the internal capacity of the soil to resist lateral force. This is when the soil collapses.

Cohesion

Cohesion is an attractive force between like molecules. It is what makes clays sticky. The shape and structure of the molecules, which makes the distribution of orbiting electrons irregular when molecules get close to one another, creates electrical attraction that can maintain a macroscopic structure. **Cohesive soils**, commonly referred to as clay soils, primarily develop internal strength from cohesion, rather than friction as the clay particles are small and flat. Clay is the only soil type capable of electrical attraction (cohesion). The strength of clay is determined by its cohesion and the amount of water within the soil. The greater the cohesion, the greater the attraction to water. In most cases, a clay particles' attraction to water is greater than a clay particles' attraction to other clay particles. We can imagine cohesion as a series of magnets; the closer they are, the stronger the attraction (as in dry clay) and the farther apart they are, the weaker they are (as in really wet clay). In a trench scenario, if we cut into a wet clay, the sides will collapse; however, if the clay is dry and dense, the sides will remain vertical and not collapse **FIGURE 2-13**.

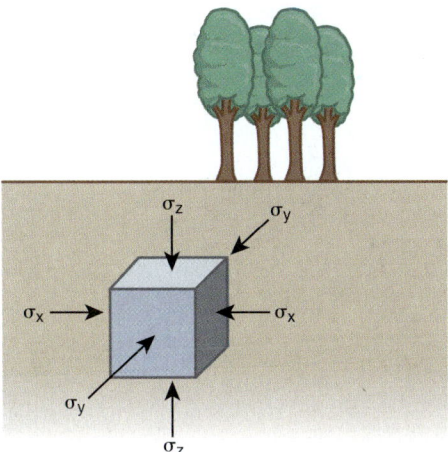

FIGURE 2-14 Soil equilibrium.

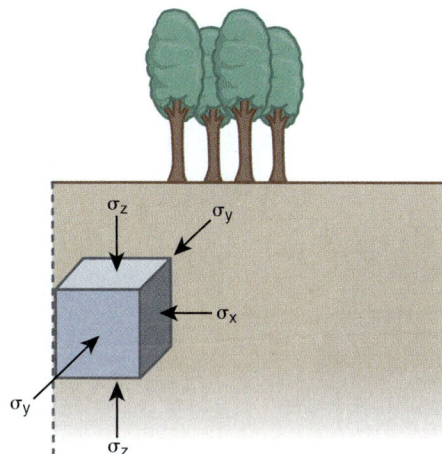

FIGURE 2-15 An open trench creates an unconfined surface area.

Moisture Content

Under certain conditions, the presence of water can add strength to particular soils; however, those conditions do not exist with open excavations. For trench walls, as the presence of water increases, the strength of the soil will decrease and eventually the soil will behave as a fluid and flow into the trench. The presence of moisture compounds the hazards of trench failure, because not only does the probability of failure increase, but the unit weight of the soil also increases. This equates to more lateral forces from the soil and a higher probability of secondary collapses.

Unconfined Compressive Strength

The most commonly used indicator of soil strength, or the amount of resistance the soil has to internal pressure, is a measurement of its **unconfined compressive strength (UCS)**. Like diving in water, as we go deeper, the pressure increases. In the ground, as we go deeper, the weight of the overlying soil, or overburden pressure, on the soil increases. As in water, the at-rest ground is pushing in every direction with equal force; the adjacent soil, in turn, is pushing back with equal force. This balancing act explains why the ground stays stable and does not move **FIGURE 2-14**.

When one side of the ground is open, however, it is unconfined—we have removed the section of earth that it was using for stability. Now the internal strength of the remaining soil must resist the force previously held by the soil that has been removed **FIGURE 2-15**.

When the soil strength generated by soil cohesion and internal friction (UCS), is lower than the driving forces (the force trying to push the trench wall into the trench), the trench wall loses its ability to stand. A higher UCS suggests a soil with more internal strength (cohesion or higher internal friction); a lower UCS indicates a soil with less internal strength (cohesion, less internal friction, or possibly a water-saturated soil). Thus, UCS will vary with the water content of the soil. Notably, it can decrease as the cut trench face dries out. But, more often the UCS can also decrease when too much water is present in the soil.

Trench Collapse

Types of Trench Collapse

Familiarity with the types of collapse will help we determine the trench's potential for collapse and the proper protective system appropriate for making it safe.

Spoil Pile Slide

A **spoil pile slide** is the result of excavated earth being placed too close to the lip of the trench. We may occasionally hear the term *overburden pressure* used to describe the effect of the weight of the spoil pile or other objects placed on the trench. Most proficient contractors are able to recognize and prevent this hazard **FIGURE 2-16**.

A spoil pile slide occurs when the soil's natural **angle of repose**—that is, the natural angle at which loose particulate products will support their own weight and can be expected not to flow from a standing position—is greater than the soil's cohesive tendency. When this occurs, the soil flows back into the trench at the rate and volume necessary to bring the situation back into relative balance for the conditions. Remember that a hole in the ground wants to fill itself back in naturally; thus, we should evaluate every pile of dirt around the trench for its potential to become active.

FIGURE 2-16 Spoil piles are frequently placed close to the trench opening and must be considered as problematic in the rescue operation.
Courtesy of Dennis Walus.

FIGURE 2-17 Rescue personnel should always stand on ground pads while moving the spoil pile.
© Jones and Bartlett Publishers. Courtesy of MIEMSS.

One factor contributing to this type of collapse is the presence of moisture in the newly excavated dirt, which may allow it to be piled at a steeper angle than would otherwise be possible. The moisture provides cohesion, holding the soil together. As the soil dries, it becomes less stable until its weight overcomes its cohesive properties. At this point, the spoil pile relieves the unbalanced pressure by flowing to a lower level.

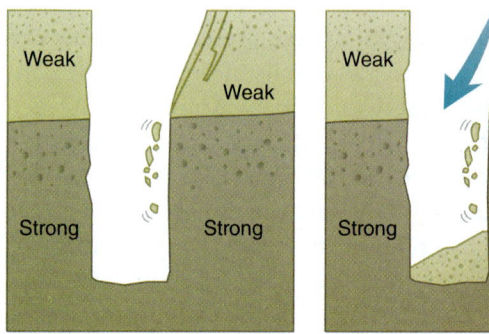

FIGURE 2-18 Weak (granular) soil over strong (cohesive) soil will result in a slide. A slide will leave an angled wall at the back of the void.

> **TIP**
>
> Angle of repose is a natural condition and should not be confused with maximum allowable slope, which is a human-made angle referenced in OSHA documents.

If the spoil pile maintains any potential to become active, it must be moved. This operation may require a few rescue personnel with shovels or, in a recovery operation, a heavy equipment operator and an excavator. The spoil pile must be moved to reduce the overburden pressure being transmitted to the open wall of the trench and to eliminate any possibility that it could flow back into the trench. While OSHA standards direct the spoil pile to be 2 feet (0.6 m) away from the lip, more distance is better, resulting in less additional lateral pressure on the trench wall **FIGURE 2-17**.

Lip Shear Failures

A lip shear is a common failure when the soil is layered. The void pattern (size, shape, and angle) is dependent on how the layered soil is stacked. Granular (sand) soils are likely to have lip shears, where a wedge of upper trench wall just slides into the trench. This type of failure usually occurs suddenly with no warning—a mass of soil comes falling down. Weak (granular) soil over strong (cohesive) soil will result in a slide **FIGURE 2-18**.

Strong, cohesive (clay) soils on the top layer will usually result in a topple-type collapse and may include the sliding of the weaker soil below it. With cohesive soils, cracks will form on the original ground surface adjacent to the trench that run approximately parallel to the trench. If the cracks are open enough, we may be able to see that the cracks are relatively vertical in nature. Strong soil (cohesive) over weak soil will usually result in a topple collapse of the lip and has a good potential for a sliding collapse of the weak portion of the wall **FIGURE 2-19**.

The soil types that are likely for lip shear failure include the following:

- Slides—sands and noncohesive silts
- Topples—cohesive silts and clays

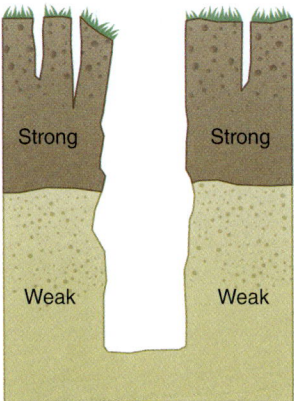 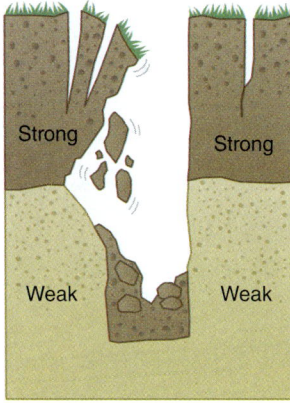 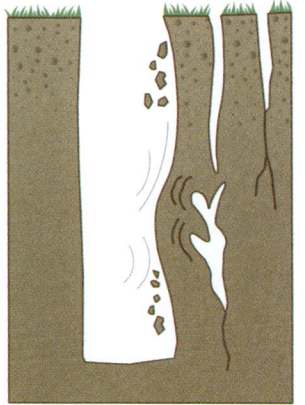 

FIGURE 2-19 A topple collapse will usually leave a near vertical wall at the back of the void.

FIGURE 2-20 A slough-in often results in an undercut (overhanging) wall section.

Conditions that are likely for lip shear failure include the following:

- Dry, loose soils
- Noncohesive silts and sands
- Soft or loose layers

Slough Failure

A **slough failure** involves the loss of part of the trench wall. It can be the result of several conditions. Frequently, the force associated with unconfined hydrostatic pressure becomes greater than the soil's ability to stand. Slough failure can also be caused by a spoil pile being placed too close to the trench lip. As extra earth is piled up, its weight is transmitted in a downward force communicated through the trench walls. When this pressure exceeds the soil's ability to support it, a failure will occur. Cracks in and around the excavated surface and multiple soil layers are key indicators that the potential for a slough collapse is present **FIGURE 2-20**.

There are several signs that sloughing may be present. Commonly, we may see the lower third of a trench wall belly or bulge into the trench. This bulging is followed by a collapse of the bulging soil. If the remaining trench wall is strong enough, we are left with an overhanging section of lip. If left long enough, the overhanging lip will eventually topple or fall into the trench. A slough failure can also be caused by a layer or pocket of soil that is different from the rest of the trench. This could be soil that is weaker, saturated with water, much drier, or of a different composition. Any of these changes could result in soil that is more likely to fail. Depending on the strength of the upper soils, this could result in a pocket in the trench wall. If the upper soils are strong enough, it may be able to stand in place with a void lower in the trench. Weaker soils are likely to have the pocket slough in, followed by the soils above falling down into the trench.

The following soil types are likely for slough failure:

- Sand
- Soils with low cohesive qualities

The conditions likely for slough failure include the following:

- Very dry soils
- Very wet soils
- Weak layer within the trench wall

Rotational Failures

A **rotational failure** may look like a slough failure to the untrained or inexperienced eye. The end result of these failures can look similar; however, the mechanics behind them are very different. A rotational failure occurs when a failure plane develops within lower-strength cohesive soils. The failure plane is circular or elliptical and can intersect the area all above the toe, at the toe, or below the toe (least likely). It can be identified by a curved failure surface and requires a deep section of weak, cohesive soils.

A rotational collapse presents as the trench wall sliding into the trench at the bottom of the failure surface. The soil slides down a circular failure plane, pushing toward the other wall of the trench. This type of failure is most likely to shove any items or workers in the trench against the opposite wall of the trench. The resulting failure plane is curved, and it extends from the original ground surface down to the trench wall.

A rotational failure area does not slide to the bottom of the trench at an angle like a wedge failure, but rather rotates downward in a circular fashion along the shear failure surface and into the bottom of the

FIGURE 2-21 Rotational failures often occur rapidly with little warning. Fissures at the surface expand moving downward and toward the trench wall face creating a bulge in the wall. The circular shape at the bottom of the fissure causes the mass of soil to rotate and create the initial force near the lower third of the failure.

FIGURE 2-23 Wet conditions or a sandy soil profile can cause toe failure at the bottom of a trench.
© Drop of Light/Shutterstock.

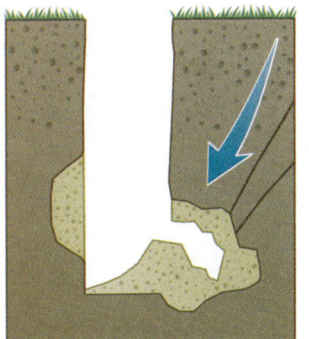
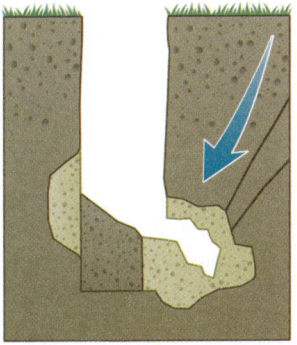

FIGURE 2-22 A toe-in may initially be small but it can result in a large collapse.

FIGURE 2-24 In a bell pier condition, the failure of both sides of the toe in a trench can create a cantilever condition; it is very difficult to backfill during rescue operations.

trench. Rotational soil failure first occurs at the bottom of the trench, which can create bulging of the trench wall **FIGURE 2-21**.

The following soil types are likely to fail in rotation:

- Cohesive silts, also known as plastic silts
- Clays

The conditions that are likely for rotational failure include the following:

- Deep layers of weak clay soils
- Shallow groundwater

Toe Failure

A **toe failure** is a slough failure that occurs at the bottom of the trench, where the floor meets the wall **FIGURE 2-22**. As the soil falls into the trench, it creates an opening at the bottom that is characteristic of a cantilever. This type of collapse can be caused by a sand pocket or the effects of water at the bottom of the trench **FIGURE 2-23**.

Toe failure is a highly dangerous type of failure for several different reasons. First, the rescuers might not notice the toe failure until they are standing on top of the cantilevered section of earth (and are therefore at risk of becoming part of a secondary collapse). Second, the situation is hard to correct until after a protective system is in place.

The effects of water accumulation can also cause a **bell pier condition**. This type of situation does not usually occur suddenly, but more often is the result of a long-term toe failure on both sides of the trench floor **FIGURE 2-24**. The bell pier condition is also dangerous for the reasons previously discussed.

Wall Shear Failures

A **wall shear collapse** occurs when a section of soil loses its ability to stand and collapses into the trench along a mostly vertical plane **FIGURE 2-25**. This condition can occur when cracks in the earth's surface

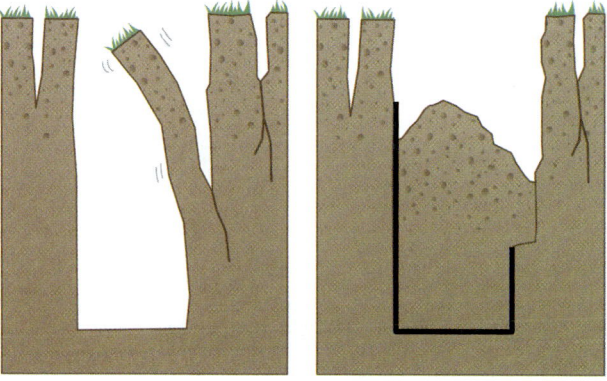
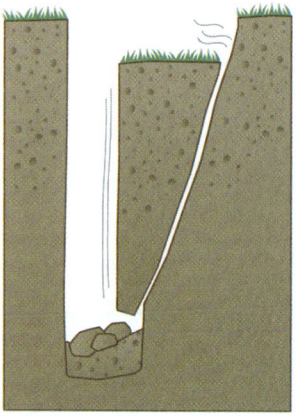
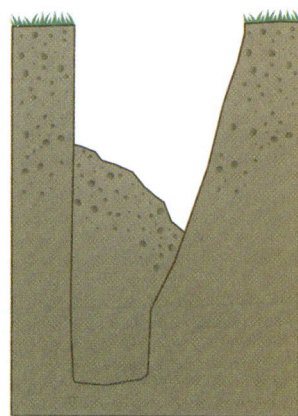

FIGURE 2-25 Shear failures typically start with a fissure at the trench lip. The cohesive soil stays together as the fissure breaks deeper into the trench. The face of the trench wall topples, often in one large piece and crashes into the wall across from it and the trench floor. A topple collapse will usually leave a near vertical wall at the back of the void. A wall shear is often a large collapse, which leaves tons of collapse debris in the trench.

FIGURE 2-27 A fissure at the surface expands, moving downward. Sliding will be followed by the collapse of the wedge-shaped section of the trench wall.

The soil types that are likely for wall shear failure include the following:

- Cohesive silts
- Clays

The following conditions are likely for wall shear failure:

- Drying wet soils

> **TIP**
>
> Collapse potential and severity are proportional to how deep and wide the trench is and how long and difficult the rescue operation will be.

FIGURE 2-26 A wall shear usually results in large collapse.
Courtesy of Ron Zawlocki.

are exposed to the weather over time. As water runs into these openings, it washes out dirt and then dries. Over time, this washing and drying action causes a hole to become deeper and deeper, until it is not supported on two sides and a wall of dirt falls into the trench **FIGURE 2-26**. This can also occur if a trench is excavated parallel to an older trench, leaving a narrow width of original soil between the old trench and the new excavation. Wall shear collapses are normally associated with fairly cohesive soils, a factor that makes them appear safe. This is a big problem, because this type of failure can create a large collapse situation.

Wedge Failures

A **wedge failure** begins with fissures along the surface of the ground. Gravity then acts to create a fracture that continues from the surface down to the area of the void, angling toward the trench wall and eventually causing the wedge of soil to collapse into the trench in an angular fashion. The presence of surface water that can enter the cracks will hasten this type of failure.

Wedge failures are often seen on the inside corners of intersecting trenches. A large section of the trench wall will collapse by sliding down the shear failure surface and into the trench **FIGURE 2-27**. A wedge failure can be sudden and catastrophic **FIGURE 2-28**.

The soil types likely for wedge failure include the following:

- Cohesive silts
- Well-graded compact sands

FIGURE 2-28 Wedge failure almost always occurs in intersecting trenches. An inside corner (wedge failure) collapse is shown here.
Courtesy of Ron Zawlocki.

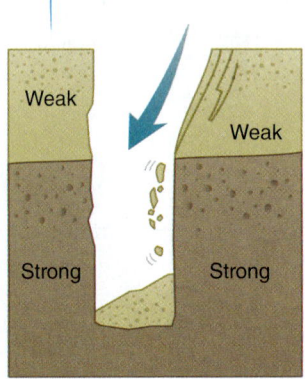

FIGURE 2-29 Lip shear (slide) pattern.

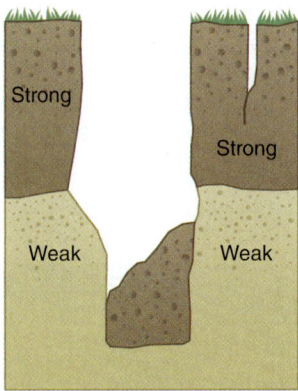

FIGURE 2-30 Lip shear (topple) pattern.

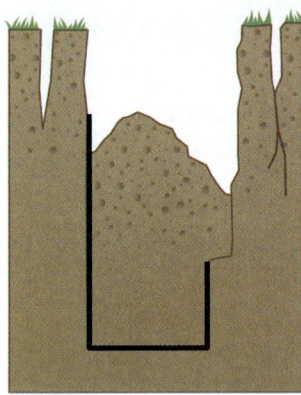

FIGURE 2-31 Wall shear (topple) pattern.

The conditions likely for wedge failure include the following:
- Intersecting trench walls
- Weak lower layer
- Wetted soils
- Sand pockets

Although it may be relatively simple to determine which type of collapse has occurred, it is often quite difficult to understand why the collapse has taken place. Commit each of these types of collapse to memory; as we discuss the physical forces associated with various collapse scenarios, the reasons why each type happens will become clear.

Collapse Patterns

Each of the types of collapse we just discussed typically leaves distinct collapse patterns and voids. Recognizing collapse patterns will help to determine appropriate backfill and shoring sequence options. (See Chapter 6, *Managing the Trench Incident*, for more detail.) Lip shear occurs when a weak soil is at the lip, the resulting collapse often slides and leaves a low angle on the back wall of the void **FIGURE 2-29**. When a strong soil is at the lip, the resulting collapse often topples and leaves a high angle on the back wall of the void **FIGURE 2-30**. Wall shears are common in cohesive soils and often result with a large section of the wall breaking off and toppling, leaving high angled and irregular void back walls **FIGURE 2-31**.

Wedge failures are commonly seen at intersecting trenches. In those cases, a wedge-shaped section of an inside corner breaks off and slides downward. A wedge collapse can leave either a high or low angled void back wall **FIGURE 2-32**. Rotational failures usually result in a collapse pattern with a high angled void back wall at the top half to two-thirds of the collapse and a lower angled (semi-circular) back wall near the bottom of the collapse **FIGURE 2-33**. Slough-in failures can result in an overhanging (cornice) section of trench wall directly above the void **FIGURE 2-34**. Similar to slough-in failures, a toe failure can result in an overhanging (cornice) section of trench wall directly above the void **FIGURE 2-35**.

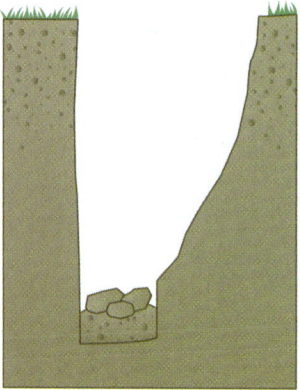

FIGURE 2-32 Wedge failure pattern.

FIGURE 2-33 Rotational failure pattern.

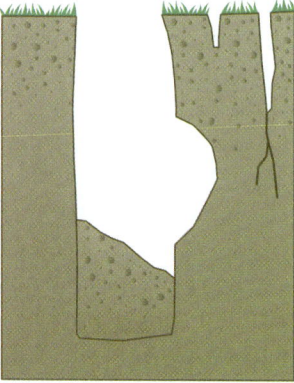

FIGURE 2-34 Slough-in (undercut) pattern.

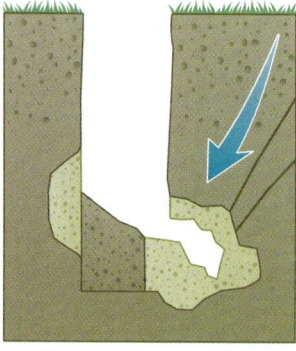

FIGURE 2-35 Toe failure pattern.

Conditions and Factors That Lead to Collapse

Many conditions and factors can ultimately lead to a trench collapse. While evaluating these factors, keep in mind that they can work synergistically to generate a serious collapse situation. None of the conditions and factors by themselves will be the ultimate straw that breaks the camel's back. Nevertheless, each can prove to be a tipping point, and therein lies part of the problem with a trench rescue. There is no definitive way to determine which single condition, or set of multiple conditions, will cause a collapse. Recognizing this complexity is the key to success when evaluating factors that could lead to trench failure.

Water

The addition of water can add tremendous weight to soil. As previously mentioned, water weighs 62.4 pcf (999.6 kg/m^3), although the effect it has on soil is influenced by many factors. For instance, the absorption rate will ultimately determine the total weight for any given volume of soil. Additionally, the effect that water has on the soil's ability to maintain its strength is critical. Some soils initially gain strength with the introduction of water, but then at some point get saturated and become weak. Watch out for soils that appear solid but are actually wet and unstable. For instance, consider clay soil. When it is dry, it can be powdery and loose. With the addition of water, it solidifies and becomes more stable and stronger. But with the addition of even more water, it becomes fluid and loses strength.

Water Table

The **water table** also dictates which type of rescue situation we may confront at a trench collapse. Near the ocean or in other low-lying areas, merely placing a shovel in the ground may create a hole for standing water. A high water table means heavier, more unpredictable soil. Even the most experienced construction workers cannot determine the amount of time the trench can remain freestanding in such an environment **FIGURE 2-36**.

Severe Environmental Conditions

The **freestanding time** of a trench or open excavation and environmental conditions can also be a factor that leads to collapse. After a trench is cut, if it is subjected to severe environmental factors such as drying, wind, water, or freezing, the internal strength of the soil can be substantially reduced. The freestanding time is also a ticking bomb with respect to the compressive forces

FIGURE 2-36 A high water table.

FIGURE 2-37 Layers of varying types of soil should be identified as having an increased collapse potential.

> **TIP**
>
> "In my own experience, I am reminded of a group of Phoenix firefighters gazing into the bottom of a 6-foot (1.8-m) deep Virginia Beach trench that contained 2 feet (0.6 m) of water. Perplexed would only begin to describe their reaction to our native soil profile in Virginia. Know the types of soil in your response area, and preplan our protective system requirements based on this determination."
>
> Courtesy of Cecil V. "Buddy" Martinette, Jr.

that the trench wall can withstand. The longer the trench is open, the closer we are to nature's attempt to fill it back in.

Varying Soil Profiles

Varying soil profiles within an excavation are a problem that rescuers face when determining the classification of the soil and its potential for collapse. Because multiple layers of different materials demonstrate different strengths and friction coefficients, it is often difficult to state with any reasonable certainty how they will react in a specific incident **FIGURE 2-37**. For example, how much of a fault is created when a layer of sand is sandwiched between two layers of clay? Certainly, the sand creates a slip potential when the earth is not supported on one side of an excavation. Additionally, we need to determine just how much sand is contained in a certain clay sample and whether that sample represents a significant portion of the excavated material.

Disturbed Soils

Previously disturbed soils, or sometimes called fill soils, are another occurrence that is easy to spot but difficult to interpret. Disturbed soils lack cohesiveness because they are broken and/or mixed with other soil types. If the trench collapse occurs in a populated area, there is a high probability that some type of utility has been placed in that location in the past. Rest assured that if a cross-section of soil in an excavation that contains bricks, bottles, or other debris is found, it has been previously disturbed.

Nevertheless, not all previously disturbed soils are dangerous. For example, when using engineered fill to build elevated roads, contractors are required to compact the soil to such a degree that it will not settle after the road is built. Engineers use mathematical calculations based on soil type and moisture content to determine maximum compaction. In these situations, previously disturbed soils are no more or less dangerous than undisturbed soils. However, in noncontrolled fills, extreme caution should be exercised, as the internal structure of the soil can be highly variable and range from very strong to very weak over a short distance.

Superimposed (Surcharge) Loads

Heavy objects and materials that are near the lip of a trench increase the vertical pressure and subsequently increase the lateral pressure on the trench wall. We call them superimposed, or surcharged loads. These objects and materials are typically construction equipment and spoil piles. Heavy equipment location (**equipment surcharge load**) represents a factor that can lead to a trench collapse. The same equipment that is digging the hole inevitably causes pressure to be exerted on the unsupported trench walls. Other heavy construction equipment includes steel or concrete pipes, front end loaders, and trench boxes, to name a few. The rule in this type of collapse is to turn off the piece of equipment and leave it in place if it is beyond the point of soil failure as measured perpendicularly from the original trench face to the farthest point of soil failure. An important point here is not to let our own heavy "rescue equipment" and apparatus become part of the problem. Keep it away from the rescue area and established hot zone **FIGURE 2-38**. Rescue vehicles should be positioned no closer than two times the depth of the trench when not running. Rescue vehicles that must continue to run in order to provide electrical or hydraulic power to tools and equipment should not be closer than three times the depth of the trench. Vacuum trucks should also be kept at a distance equal to at least three times the depth of the trench.

> **SAFETY TIP**
>
> For excavators and backhoes, the front of tracks or outriggers should be no closer than 4 feet (1.2 m) from the point of soil failure as measured perpendicular from the original trench face to the farthest point of soil failure. The center of mass of an excavator or backhoe should be no closer than 8 feet (2.4 m) from the farthest failure. If the equipment is closer than the distances given here, it is better to have a certified equipment operator move it than to keep it in place.

FIGURE 2-38 Rescue personnel need to consider the additional force that heavy equipment adds to the trench protective system requirements.
Courtesy of Chad Godfrey.

In addition to equipment, spoil piles. For contractors, their consideration for digging a hole generally revolves around the concept of getting it dug, doing the work, and filling it in again. The contractor can expedite the work process by dropping the spoils right outside the trench to minimize swing time between scoops. While this practice makes sense from the contractor's perspective—the closer the spoil pile is to the trench, the faster it can be used to fill in the trench when the work is complete—it creates two problems for rescuers. First, the spoil pile adds additional weight to the unsupported trench wall. Second, part of the spoil pile may slide back into the hole. In any case, in a trench rescue scenario, the spoil pile must be moved far enough back to alleviate the weight concern and provide an area for placing lip protection. When faced with this prospect, be sure to enlist the help of many people with shovels working off proper lip protection.

Vibration

Another factor that can cause a trench collapse is vibration, which can result from road traffic near the collapse site, the machinery digging the trench, or other machinery being operated in the area. Road traffic that is within 150 feet (45 m) of the trench must be shut down. A collapse can also be caused by directional drills and other machines used to force utilities under roadways. The key here is to shut down the drilling equipment and limit the traffic not only in the rescue area, but also around the general area of the collapse.

Although much less likely, events such as blasting, passing trains, and aircraft landings close to a trench rescue site can also cause vibrations that result in collapse. When effecting a rescue, efforts to control sources of extreme vibrations within 300 feet (91.4 m) of the rescue area must be made. Circling helicopters and their rotor wash can transmit significant vibration into the ground.

Rescuers should become familiar with all of the many factors that can lead to a collapse. Knowing what can cause a cave-in will help identify the areas of concern present when arriving at the scene. Ultimately, rescuers need to eliminate or neutralize those concerns that could lead to rescuer injury or death.

Signs of Impending Collapse

Although there are several different signs that soil will exhibit before a collapse, there is not always a warning before a collapse. The easiest thing to keep in mind is that soil fails by moving. In the case of granular (sand) soils, that movement may occur quickly and without warning. However, most cohesive (clay) soils will give warning signs. By examining the ground surface, the trench walls, the trench floor, and any previous collapse patterns that have taken place, we can begin to determine the likelihood of future failures. As we consider the signs, make sure to consider the collapse patterns described earlier. The signs mentioned in the next section, when viewed with potential collapse patterns in mind, will help envision the connection between the warning signs and collapse patterns.

Visible Cracks

A crack (fissure) is the start of a failure. Visible cracks on the ground surface or on trench walls are signs of movement that are typically related to an impending soil failure. It is most important to observe the connection of cracks to lips and other cracks. If a surface crack runs parallel to a trench, but never curves toward the trench lip, it is less likely to indicate an impending failure. When a fissure curves and intersects the trench lip, it is a guaranteed sign of impending soil failure.

Cracks on trench walls are similarly telling. A vertical crack in a wall may not be concerning if no other signs are present on the walls or ground surface. If a crack starts out as vertical and curves toward horizontal, it is telling us that a wedge, or pocket, of soil is moving and, given enough time, is going to fail.

The key to reading soil cracks is to use the cracks to determine if it outlines a mass of soil. If we start to connect the dots (cracks) and see a wedge- or block-shaped piece of earth outlined by the cracks and edges, understand that a chunk of earth is moving and at some point gravity will take over.

Visible Bulging on Walls or Floors

Bulging walls or floors are a sign of potential collapse in trenches with weak soils. This is most likely to occur in weak clay soils that are likely wet. Most commonly, these would be referred to as "toothpaste" clays. The weight above these soils is pushing down and results in the clay being squeezed, with the only outlet on the trench face. We will start to see a blob of soil that starts bulging out of the trench wall. If this fails, the next threat is the soil above the failure falling into the trench.

Bulging on the floors is typically more difficult to see because collapsed soils normally will obscure the floor. Any bulging of the floor indicates the beginnings of a rotational failure. These are the early signs that the toe of the failure is moving.

Water

Water can cause several issues that affect soil. Obviously, running water has a tendency to erode and carry away soil particles. This is something that is visible and apparent. The less visible effects are the impacts on the strength of the soil. Clay soils have maximum strength at a certain moisture content, when the soil is wetter, the strength decreases. In a wet environment, we have to be concerned about soil that is weakening over time as its moisture content increases and the internal strength decreases. In granular soils, the water soaks into the ground, filling any natural (very small) voids that exist between soil particles. The addition of water to the soil will increase the lateral pressures within the soils and make them more likely to collapse.

Another issue with water is found with fissured clay. When water fills the cracks, it creates a small lateral pressure that is trying to pry apart the soil along the fracture line. When a soil is standing at a one to one safety factor, that small difference can cause a failure. The secondary effect that water has with respect to cracks is the tendacy to lubricate the surfaces and make them slip more easily.

Undercut or Blown Out Trench Walls

Slough failures are likely to leave an overhanging mass of soil cantilevered over the edge of a trench. In urban areas, this area is likely to include road pavement, concrete (driveways, sidewalks, etc), foundation, pipes, or other built items. By definition, anything that is unsupported is currently defying gravity. It does this by using internal strength to transfer the loads to a point of support. As previously mentioned, exposed soil will start to dry and lose strength. The longer the soil overhangs, the more likely it is to fail. For most soils with overhangs, a crack will start to form at the surface before it breaks and falls. This is one of the most dangerous situations and should always be treated as an immediate hazard that needs to be addressed.

Voices of Experience

This incident occurred on private property in South Huntington Township, Pennsylvania. A professional contractor was excavating a trench with the assistance of his 18-year-old son for the purpose of installing a French drain on his own property. The trench depth ranged from 9 to 12 feet (2.7–3.6 m) in depth and was 18 to 24 inches (46–60 cm) wide. It had a straight wall and extended approximately 200 feet (61 m) from the residence at the time of the incident. The spoil from the excavation was being deposited randomly, from right at the trench edge to approximately 24 inches (60 cm) away on the right side. The soil was mostly clay and prior to the collapse could most likely have been classified as OSHA Type B soil with fissures present. The excavation was made without the use of a trench box or other protective measures. When the equipment operator (father) noticed that a terracotta pipe with flowing water had been uncovered in the excavation he summoned his son to investigate the source.

It should be noted that this was a recently acquired structure and was being extensively renovated. The 18-year-old male entered the trench and was at the location of the pipe when the left side wall of the trench collapsed trapping him. The extent of entrapment and exact sequence of events from the time of the collapse to the notification of the 911 center are unknown. However, it was evident that at some point the father (operator) made an attempt to extricate his son with the excavator (there were indications on the body that it had been contacted by the teeth on the bucket although those injuries appeared to be post-mortem and superficial).

The Westmoreland County (PA) Dispatch Center received a call for a trench collapse with entrapment and dispatched the initial alarm assignment that included the local fire companies and emergency medical services (EMS) units. My involvement in this incident began when the incident commander special called for additional rescue companies (including mine) to be dispatched to the scene. Additionally, the County Trench Rescue team was requested. These were all volunteer units (with the exception of EMS) and response times, equipment, training, and staffing levels varied.

Upon arrival, EMS workers immediately entered the trench to assess the victim. Attempts at shoring by unknown persons were made with ½-inch (2.5 cm) plywood (on-site) held in place with hastily cut polyvinyl chloride (PVC) drainpipe. This likely provided a false sense of security to the EMS workers entering the trench. After a quick assessment it was determined that the victim was deceased. The EMS workers then exited the trench and the order was given for all additional units to hold in station. This order included the special rescue companies and the trench rescue team. It is assumed that this order was given upon verification that the victim was deceased and the supposition that a rescue would no longer be conducted. After further discussion by the officers on scene, they came to the realization that the additional resources would still be needed to recover the victim's remains. At that point the resources were instructed to resume their response.

When I arrived on scene the trench had yet to be secured and responders as well as bystanders were still in close proximity to the trench edges. I advised the incident commander that this was a serious safety concern given the demonstrated instability of the ground and that precautions needed to be taken to ensure a safe recovery. At this point I was placed in charge of the recovery aspect of the incident and updated the responding units as to the conditions on scene and the resources that would be needed.

Prior to beginning the recovery operation, we were instructed to wait for representatives from the County Coroner's Office and U.S. OSHA to

Voices of Experience

arrive and complete their investigation and documentation. (Because this was private property OSHA representatives simply documented the incident and remained on scene to ensure safe operations in the recovery phase.)

This downtime was used to gather additional resources and formulate a recovery plan. As soon as investigations were completed and adequate resources arrived on scene, a briefing was conducted, and a plan of action disseminated.

It was decided that based on the fissured condition of the walls at the trench end nearest the excavator and its close proximity to the work area, it would be safest to move the excavator back and have the operator slope the trench end and walls before placing shoring around the victim. This action also facilitated an extrication pathway that enabled us to more safely and efficiently remove the victim. Once this was completed, edge protection and shoring panels were placed.

We began with edge protection on both sides of the trench working in from the sloped end and removing any dirt from the spoil pile as needed to place ground pads. Once this was done, the first panels were placed. These panels were placed horizontally in the section of trench beginning at around 4 ft in depth from the end of the trench.

The initial shoring of plywood and PVC drainpipe was removed remotely with rope and pike poles to allow for placing of additional panels to secure the work area around the victim. The male victim was exposed from mid-thigh up. It was decided at this point to extend the bucket of the excavator above the victim to provide a high anchor point to assist in victim removal via a 4:1 rope system. Once the trench was secured, crews were placed in the trench to begin soil removal around the victim. The position of the victim's body made it difficult to free his legs. It was then decided to place an improvised webbing harness around the victim's upper body to keep his torso upright while the legs were freed. Once this was accomplished a body bag and Stokes basket were placed into the trench from the sloped side and the victim was removed.

Lessons Learned

- It is important for rescuers to understand soil and collapse mechanics because collapse (cave-in) poses the greatest risk to rescuers at trench rescue incidents.
- OSHA excavation standards, 29 CFR Part 1926, Subpart P contains the requirements for excavation and trenching operations for construction work. Employers must follow the trenching and excavation requirements in 29 CFR 1926.651 and 1926.652 or comparable OSHA-approved state plan requirements.
- OSHA requires construction workers to classify soil. Each type of soil represents a varying degree of danger based on the characteristics that make it a part of that class. Stable rock is the least dangerous type of soil, with Type C-80 being the most dangerous.
- Gaining a basic knowledge of soil mechanics will help rescuers understand why soil fails and why trench walls collapse. One of the keys to success in trench rescue is recognition of how the various physical elements interact and then contribute to a collapse situation.
- The T-L method accurately represents actual earth pressures and lateral forces based on the visible soil failure conditions. Rescuers should use the T-L method for all trench rescue incidents.
- There is no definitive way to determine which single condition, or set of multiple conditions, will cause a collapse.

Voices of Experience

Recognizing this complexity is the key to success when evaluating the factors that could lead to trench failure.

- The easiest thing to keep in mind is that soil fails by moving. By looking at the ground surface, the trench walls, the trench floor, and any previous collapse patterns that have taken place, we can begin to determine the likelihood of future failures.
- The signs of impending collapse (visible cracks, bulging visible on walls or floors, water, undercut or blown out trench walls), when viewed with potential collapse patterns in mind, will help envision the connection between the warning signs and collapse patterns.
- In the case of a body recovery, it is wise to eliminate a hazard rather than try to work around it. In this case removing unstable ground was much easier than trying to shore it. Risks were minimized as the victim had already expired and responder safety was paramount.

Mark Ghion, Fire Chief
Sutersville Fire Department
Sutersville, Pennsylvania

After-Action

IN SUMMARY

- A trench is a narrow underground excavation that is deeper than it is wide and is no wider than 15 feet (4.5 m).
- The basic soil types are sand, silt, clay, or a combination.
- OSHA provides soil classifications (types A, B, and C) which is a means of describing the most stable and strongest (A) to least stable and weakest (C).
- Granular soils derive their strength primarily from friction between the grains.
- Cohesive (clay) soils derive their strength from electrical attraction (cohesion), and their strength varies with moisture content.
- Collapse patterns include the following:
  - Lip shear—when the upper portion of the trench, including the lip, slides down into the trench
  - Slough—when weak soils lower in the trench bulge out from the wall and eventually fail, sometimes followed by the soil above it collapsing
  - Rotational—when a large mass of soil fails by sliding along a circular plane in the soil, which results in the bottom kicking out and the original ground surface sloped away from the trench
  - Toe—a type of slough that occurs at the bottom of the trench
  - Wall shear—when fractured elements of the trench wall topple into the trench
  - Wedge—similar to a slide, but deeper failure plane in the trench wall
- Water-saturated soil is more likely to fail due to increased hydrostatic pressure.
- Previously disturbed soils, sometimes called fill soils, are easy to identify but difficult to interpret.
- Signs of impending soil failure include the following:
  - Visible cracks on the ground approximately parallel with the trench, especially when a fissure curves and intersects the trench lip
  - Visible bulging on the trench walls or floor
  - Water seeping or running into the trench
  - Overhanging soil or other materials (concrete pavement or utilities)

KEY TERMS

Active earth pressure coefficient How much (percentage) of the vertical pressure translates to horizontal pressure pushing outward.

Active soil Soil containing energy as it relates to movement.

Angle of repose The natural angle at which loose particulate products will support their own weight and can be expected not to flow from a standing position.

Bell pier condition A condition where the bottom few feet of shaft has a larger diameter than the majority of the shaft. This can be caused by soil failure or an oversized cut.

Clay Soil consisting of very small and flat particles; it is the only soil type capable of electrical attraction (cohesion).

Cohesion The sticking together of particles of the same substance.

Cohesive soils Soils with very fine particals (cannot be seen with the naked eye) that derive their primary strength via electromagnetism at the particle level.

Equipment surcharge load The weight of equipment on the ground that translates into increased horizontal pressure on the trench walls.

Freestanding time The amount of time an excavation is open to the elements.

Granular soils Soils made up of individual grains of soil (can be seen with naked eye) that derive their primary strength from friction between individual grains.

Gravity The force of attraction between two bodies that have mass.

Hydrostatic pressure Increased pressure caused by the addition of water to the soil profile.

Internal angle of friction A measure of the ability of a soil to internally resist shear from an outside loading. When low, it results in a weak soil, and when high it results in a strong soil.

Lateral soil pressure The horizontal force produced by soil due to gravity or other loads on the soil.

Overburden pressure The pressure that the weight of the spoil pile or other object exerts on the trench.

Previously disturbed soils Sometimes called fill soils, these are soils that lack cohesiveness because they are broken and/or mixed with other soil types.

Rotational failure A scoop-shaped collapse that starts back from the trench lip and transmits itself to the trench wall in a half-moon shape; also called slough failure.

Sand Particles that can be seen with the naked eye; particles are larger than silt and clay.

Silt Particles sized between sand particles and clay particles that form a weak soil. It will feel slightly gritty when rubbing between fingers.

Slough failure The loss of part of the trench wall starting at an area back from the trench lip and extending down into the trench wall.

Soil density The weight of a standard unit of soil, typically pounds per 1 cubic foot of soil.

Spoil pile slide The result of excavated earth that is placed too close to the lip and subsequently falls into the trench.

Toe failure A slough failure that occurs at the bottom of the trench, where the floor meets the wall.

T-L soil A recently developed soil classification designed by geotechnical engineers for use by first responders. T-L soil has the attributes of worst case soil conditions that are commonly encountered at trench rescue incidents.

Type A soil A strong soil with an unconfined compressive strength of over 1.5 tons per square foot

Type B soil A medium strength soil with an unconfined compressive strength between 0.5 and 1.5 tons per square foot.

Type C soil A weak soil with an unconfined compressive strength less than 0.5 tons per square foot.

Unconfined compressive strength (UCS) The force, or load per unit area, as calculated with a penetrometer or other device and stated numerically in tons per square foot, that determines the point at which a soil will fail in compression.

Varying soil profiles Multiple layers of different soils found at various levels in the trench or excavation wall.

Vector A force having a specific magnitude and direction.

Wall shear collapse A collapse that occurs when a section of soil loses its ability to stand and falls into the trench along a mostly vertical plane.

Water table The top of the water surface in the saturated part of an aquifer.

Wedge failure A failure that usually occurs in intersecting trenches, in which an angled section of earth falls from the corner of two intersecting trenches.

REFERENCES

Occupational Safety and Health Administration (OSHA). 1926.651 - Specific Excavation Requirements. https://www.osha.gov/laws-regs/regulations/standardnumber/1926/1926.651. Accessed June 2, 2021.

Taylor O-D S, LaBaw SM. *2018 Emergency Trench Shoring and Rescue: A Simplified Method for Calculating Lateral Earth Pressures*. The Free Library (January, 1), https://www.thefreelibrary.com/Emergency Trench Shoring and Rescue: A Simplified Method for...-a0578045838. Accessed May 7, 2021.

On Scene

1. What are four common collapse patterns?
2. Name three signs of impending collapse.
3. What are the common causes of collapse?

CHAPTER 3

Awareness Level

Initial Actions

KNOWLEDGE OBJECTIVES
- Identify the initial scene management tasks at a trench rescue incident.
- Describe an initial incident command organizational structure for a trench rescue incident.
- Describe the process of performing a trench rescue scene size-up. (**NFPA 1006: 12.1.4**, pp. 46–51)
- Explain how scene size-up information is used to develop an incident action plan. (**NFPA 1006: 12.1.4**, pp. 45–46)
- Describe the purpose and components of an incident action plan. (**NFPA 1006: 12.1.4**, pp. 45–46)
- Describe the considerations for assessing a trench incident. (**NFPA 1006: 12.1.4**, pp. 46–47)
- Describe how to safely approach a trench. (**NFPA 1006: 12.1.2, 12.1.3, 12.1.4**, p. 47)
- Identify information that should be gathered from witnesses at a trench rescue. (**NFPA 1006: 12.1.1, 12.1.4**, p. 47)
- Describe how search parameters for a trench rescue are identified. (**NFPA 1006: 12.1.4**, p. 47)
- Describe the hazards associated with the trench environment. (**NFPA 1006: 12.1.3**, pp. 47–48)
- Explain considerations for gathering information about victims at a trench rescue. (**NFPA 1006: 12.1.1, 12.1.4**, pp. 48–49)
- Describe methods of controlling hazards. (**NFPA 1006: 12.1.7**, pp. 55–57)
- Identify defensive methods of managing hazards associated with trench rescue. (**NFPA 1006: 12.1.5, 12.1.7**, pp. 55–57)
- Explain the process of determining the scope of a trench rescue. (**NFPA 1006: 12.1.4**, p. 51)
- Explain the considerations for determining what additional resources may be needed at an incident. (**NFPA 1006: 12.1.4, 12.1.6**, pp. 51–53)
- Identify the resources associated with the two response tiers for trench rescue. (**NFPA 1006: 12.1.6**, pp. 52–53)
- Describe initial site control actions at a trench rescue incident. (**NFPA 1006: 12.1.3**, pp. 53–55)
- Describe awareness level defensive actions to mitigate hazards. (**NFPA 1006: 12.1.2**, pp. 55–57)
- Describe the methods of effecting a non-entry rescue. (**NFPA 1006: 12.1.2**, pp. 57, 61)

SKILLS OBJECTIVES
- Establish command at a trench incident.
- Conduct a trench incident size-up and determine the resources required. (**NFPA 1006: 12.1.4, 12.1.6** pp. 46–51)
- Safely approach a trench. (**NFPA 1006: 12.1.2**, p. 47)
- Interview witnesses at a trench incident. (**NFPA 1006: 12.1.1**, p. 47)
- Communicate information concisely and accurately during a trench rescue incident. (**NFPA 1006: 12.1.4**, pp. 49–50)
- Conduct a non-entry rescue at a trench incident. (**NFPA 1006: 12.1.2**, p. 57, 61)

You Are the Rescuer

Your engine company is the first to arrive at the scene of a trench collapse. A new subdivision is being built just north of Eight Mile Road, near Main Street. A large private contractor has been installing the underground utilities for almost 3 weeks. Today was just another day for the construction workers. For almost 8 hours, they had been digging trenches and installing utilities using a trench box as they moved south toward Eight Mile Road. As the last section of pipe was ready to be installed to make the tap into the existing utility main at Eight Mile Road, the excavator blew a hydraulic line. It was almost quitting time and, although a spare hose was available, the time needed to make the repair would require overtime. Trying to avoid overtime payments, the crew foreman decided that he would make the tap without moving the trench box into the unprotected area. Eight Mile Road is a very busy highway, and it was rush hour. As the foreman left the protection of the trench box and took about eight steps into the unprotected trench, the trench wall (12 feet [0.6 meters] deep) on the east side, which had a large spoil pile on it, collapsed and buried the foreman to just below his chest. Within 2 minutes, the wall on the west side directly across the trench collapsed and completely buried the foreman.

Upon arrival, your crew finds two young men (the sons of the foremen employed as laborers) in the trench digging to locate their father. As you are trying to convince them to leave the trench, the trench collapses again, burying one man to his knees while knocking the other man across the trench but not burying him. Your engine company responds in addition to an ambulance that is still 5 minutes away from the scene. That response will not be able to resolve this incident without the help of several outside resources, which will bring personnel with specialized training and equipment. While you are waiting for those resources to arrive, describe the actions you will take to accomplish the following:

1. Manage the scene.
2. Provide protection from hazards to your crew and the victims.
3. Summon and support the resources needed.

 Access Navigate for more practice activities.

Introduction

The actions taken by the first responders, including firefighters, have a direct impact on the safety and success of the entire rescue operation. In fact, the first 20 to 30 minutes after the arrival of first responders should be a busy and productive period. The command structure begins with the arrival of the authority having jurisdiction (AHJ), which, in most cases, is the local fire department. The highest-ranking chief or company officer will initiate scene management and assign **initial actions** and duties to on-scene personnel, including the following:

- Scene management
- Hazard assessment
- Site control
- Summoning resources
- Nonentry rescue

Scene Management

Incident Command

In order to effectively manage a trench emergency, someone must take charge. In most jurisdictions, the local fire department is the authority that is responsible for rescue incidents. In those situations, the first-arriving fire department company officer or chief must establish command, conduct a size-up, summon resources, and assign initial action duties, which include site control and defensive measures.

It is essential to the success of a trench emergency incident that the scene management be based on a well-established and exercised **incident command system (ICS)**. A system that is based on the National Incident Management System (NIMS) principles will easily adapt to trench rescue incidents. (See Chapter 6, *Managing the Trench Incident*, for more details.) Initially, the ICS can be as simple as an

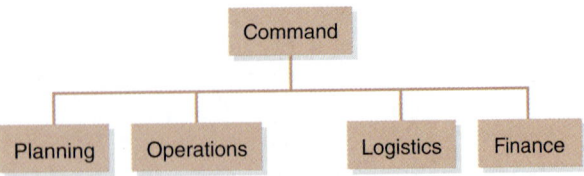

FIGURE 3-1 ICS chart.

incident commander (IC) and a couple of fire service officers supervising the initial action functions (site control and defensive measures). After the IC identifies and receives the resources necessary to resolve the emergency, the ICS will need to be expanded to manage, supervise, and support those resources. A detailed discussion of the expanded ICS can be found in Chapter 6, *Managing the Trench Incident*. To briefly summarize, the ICS comprises five main teams **FIGURE 3-1**:

- Command: Consisting of the IC and safety officer, this team determines which actions will be carried out safely.
- Planning: This team helps implement command's plans through tracking and documentation.
- Operations: This team implements command's strategic plan through tactics.
- Logistics: This team ensures that the operations team has all of the resources to carry out their tactics.
- Finance: This team is usually implemented during long-term incidents to tackle the funding of operations.

Command Post

A well-positioned and clearly marked command post provides resources (moving up from staging) with a direction and target **FIGURE 3-2**. The command post is usually the first-arriving fire apparatus or chief's car. The vehicle should keep the lights and beacons on to identify it as a command post. The command post should be placed in a position that will enable the initial action assignments to be made and facilitate the transfer of situation and status reports without being in the way of the rescue operation. Positioning the first due apparatus at about 100 feet (30 m) from the trench is a good working distance and will provide for the development of a well-placed command post.

Establishing zones will be detailed in the next section. A hot zone with a 50-foot (15-m) radius around the victim's location is needed to control the site. Having the command post spotted at 100 feet (30 m) from the trench keeps it in the warm zone (50 to 100 feet

FIGURE 3-2 Establish a clearly identifiable command post.
Courtesy of Ron Zawlocki.

[15 to 30 m] from the incident area) and places incoming first responders (fire, medical, and police) behind the command post (beyond 100 feet [30 m]) and in a good location for **forward staging**. Command vests are helpful for identifying the command staff.

Staging Area

The staging area for first-due companies will be near but not closer than the command post. If additional resources are necessary, the first-due apparatus will be moved to make room for specialized equipment. Trench and excavation incidents often require personnel and equipment from a variety of different public and private agencies. A forward staging area positioned near the command post (beyond 100 feet [30 m] from the trench) is required to get the equipment and personnel that are immediately needed close to the trench. Resources that are not directed to report to the forward staging area will need a place to stage. That location is called **primary staging** **FIGURE 3-3**. Unless a resource is specifically ordered to report to forward staging, they will report to the primary staging area. The primary staging area should be at least 300 feet (91 m) from the trench. A staging officer will help provide an organized use of acquired resources. The IC will request resources (usually achieved with radio communications), and

FIGURE 3-3 A primary staging area.
Courtesy of Ron Zawlocki.

the staging officer will direct the movement of the resources from primary staging to forward staging.

The **staging officer** is responsible for positioning and accounting for resources that are not immediately assigned. The staging officer's position is located at the primary staging area. Personnel, equipment, and supplies that are not assigned on arrival stay at the primary staging area until they are summoned by the incident commander. Once resources are requested by the IC (usually done with radio communications), the staging officer will direct the movement of the resources from primary staging to forward staging. If multiple staging areas are needed, the operations officer will assign a manager to each staging area. The staging area manages check-in of all incoming resources, dispatches resources at the operations officer's request, and requests logistics section support, as necessary, for resources located in their staging area.

Standard Operating Guidelines (SOGs) and Incident Action Plans

Written and practiced operating guidelines provide order to chaotic scenes. When members of an agency have a game plan going into the incident, levels of confusion are reduced and safety and efficiency increase. This chapter offers a SOG for initial actions. Chapter 6, *Managing the Trench Incident*, provides SOGs for incidents that go beyond initial actions.

Developing the Incident Action Plan

Incident action plans can be developed for mental use, verbal use, or command use. Although typically considered a command level task, IAPs are created for almost every task in the fire service, whether explicitly or not. This section touches on IAPs that may be required for larger incidents, and delves further into IAPs that the individual rescuer or smaller company-type rescue crews may use when operating up close and personal at rescue scenes.

Large-Scale IAPs (Written). Large-scale incidents often have written action plans for expected or potential risks, such as severe weather moving through an area. IAPs at this level are frequently broken into operational periods of hours, with details on who is working, where personnel are working, what will be done, and other parameters. These documents can be developed at a command center and disseminated through the appropriate channels to street level supervisors.

Tactical Sheets. Larger, more complex incidents can have a written worksheet or checklist—often referred to as a *tactical sheet*—to support operations. Needs such as PPE lists, radio frequencies, operational area boundaries, and individual rescue companies or team rosters may all be included in worksheets used by the incident command staff.

Medium-Scale IAPs (Verbal). Much more familiar to most rescuers are the "bread and butter" jobs we see every day on tours or during calls. ICs who oversee these operations often verbalize their IAPs over the radio, and most of these initial IAPs start with company officers reporting what they are actually seeing and what they are going to do about it.

Smaller-Scale IAPs (Mental). At the smallest scale, IAPs are developed by individual rescuers who physically perform rescue work, or by their immediate supervisors. These mental IAPs are often relayed or transmitted to the IC, who then uses them to help develop a broader IAP that may include additional resources, equipment, command needs, or other peripheral aspects of an incident. Even when developing these mental IAPs, rescuers use several sources to develop the IAP that best fits the incident at hand:

- SOGs: SOGs are requirements and directions that the AHJ develops for handling various tasks. They take into account safety, department equipment and capabilities, personnel training and capabilities, and needed results. Although SOGs provide uniformity in terms of how an incident is handled, they may also allow some variation, provided that straying from these guidelines can be justified.

- Experience: It is knowledge sometimes gained right after it was needed, but experience may be the "glue" that can hold together a plan once formulated, the "calculator" used to refigure things for the glue to work, or the "broom" that moves the plan to the brain's trashcan. Experience can also drive recognition-primed decision making, which enables seasoned rescuers to determine which action to take within a few seconds when faced with a situation.
- Training: Consider which training you, your crew, your department, and any specialty responders have. Have the skills to safely perform rescues been mastered?
- Size-up: This survey of the scene should include the following questions:
  - What will this problem take to solve (e.g., more trucks, people, higher level of expertise)?
  - How do I keep others and myself safe while we are solving it (e.g., PPE, moving farther away from the scene, heavy supervision or backups, additional resources)?
  - Will things get worse or better from actions taken or lack of action (e.g., Can I rescue someone now, or make a scene safe)?
  - Which potential hazards might I expect from this mission once I go to work (e.g., if a catastrophic failure occurs, what is in place to mitigate that outcome)?

Ideally, the ideas formed from the SOPs, experience, training, and size-up will come together to create an IAP. IAPs can be formulated within seconds, or they can require much discussion and much more time to develop. Trench rescue can be very dynamic, and rescuers and command personnel should work together to develop or change the IAP as needed. Combining all the facets of size-up, SOPs, experience, and training leads to a "system" that rescuers can use to develop, execute, and evaluate a plan for any trench rescue.

Size-Up

A size-up is the assessment and evaluation of several important conditions at the scene. The size-up involves analyzing the information and developing a plan, strategy, tactics, and resources to address the situation. It begins by gathering information.

The 911 dispatcher receiving the call for help is the first line of information. Firefighters arrive on the scene armed with the information from dispatch. Once on scene, the first responding firefighters must continue to gather information by conducting assessments. The following assessments will help turn information into intelligence:

- Situation assessment
- Hazard assessment
- Victim assessment
- Trench assessment
- Resource assessment

Situation Assessment
Dispatch Information

Fire departments must provide the 911 dispatcher with a protocol for gathering information when a call for a trench rescue is received. Dispatchers must determine the following information **FIGURE 3-4**:

- Who is the caller?
- What is the location of the emergency? This can be challenging with new construction sites, so the dispatcher may obtain nearby landmarks as well as the main address.
- What happened? Did the trench collapse and is anyone trapped?
- How many people are trapped or injured?
- When did the emergency occur and when did the digging begin?
- Why was the trench dug? Was it to install or repair a utility?
- What is the available access to the trench?

FIGURE 3-4 Dispatchers must determine the who, what, where, when, and why of the emergency.
© Jones and Bartlett Publishers. Courtesy of MIEMSS.

Initial Assessments

All first responders, including firefighters, should stop short of the trench and park about 100 feet (30 m) from the trench, as conditions permit. The trench should be approached with the wind at the back, if possible, with a four gas monitor, looking for obvious hazards. Assessments will begin during the walk to the end of the trench.

The first-arriving engine company will place lip protection (plywood on-site, medical backboard) within 6 feet [2 m] of the victim's location. While standing on the lip protection, firefighters will position a ladder within 6 feet [2 m] of the victim. After the assessments and the subsequent size-up are completed, trench rescue trucks, trailers, and equipment may be moved closer to the trench, but not closer than two times the depth of the trench or three times the distance from the trench wall to the farthest point of soil failure, whichever is greater.

Rescuers need to use both interview and visual inspection techniques to assess the situation, hazards, victims, the trench, and resources available.

- Upon arrival, the IC should identify and secure the competent person, construction workers, and other witnesses. Interviews should focus on asking questions that will provide the information needed to make the assessments listed next.
- Obtain information that only a visual inspection of the site will provide. Gather visual information for the situation assessment, the hazard assessment, the victim assessment, the trench assessment, and the resource assessment.

Interview Techniques

When interviewing, it is important to determine who (identity of interviewee and who is in trouble); what (the event that occurred and work they were doing); the number of people (in the trench and on site); where and when (location of the emergency and when they started working); and why (reason for digging the trench) **FIGURE 3-5**. In short, the interviewer needs to determine the following information:

- Who
  - Who is the person being interviewed? Is it the competent person, foreman, co-worker, or witness?
  - Who is the person in the trench? Is there more than one?

FIGURE 3-5 Interview the competent person, workers, and witnesses.
Courtesy of Ron Zawlocki.

- What
  - What happened to the person who is unable to get out of the trench?
- Where
  - Where was the trapped victim(s) last seen if no longer visible?
- When
  - When (time frame) did this emergency occur?
- Why
  - Why was this trench dug? Why did the victim enter the trench?

Classifying the Emergency

Determine if the situation is a cave-in or a noncave-in emergency **FIGURE 3-6**. If the trench did not collapse (cave-in), it will be necessary to identify the cause of the emergency. Statistically, the reasons will be that the worker fell into the trench, had a medical emergency (cardiac, seizure, etc.), or had an object fall into the trench and hit them **FIGURE 3-7**. Of course, other conditions such as a hazardous atmosphere could have made it impossible for the worker to exit the trench. If the original protective system failed, then rescuers face a bigger problem. In such a case, trench rescuers may have to systematically replace and remove what was in place and enhance the existing system. This condition should trigger a request for trench rescue technicians with advanced shoring capabilities.

Hazard Assessment

Rescuers must be aware of and look for **hazards** that are common to the trench rescue environment. Those

FIGURE 3-6 Most situations will be a cave-in emergency.
Courtesy of Ron Zawlocki.

FIGURE 3-7 Workers are sometimes incapacitated by medical conditions or injuries (noncave-in emergencies).
© Jones and Bartlett Publishers. Courtesy of MIEMSS.

hazards include: collapse (cave-in), underground utilities (natural gas, water, sewer, electrical, etc.), hazardous atmosphere (commonly carbon monoxide, hydrogen sulfide, explosives, and oxygen deficit), hazardous materials (commonly gasoline, diesel fuel, solvents, fluids), physical hazards (commonly construction and rescue equipment on the trench lip), and biological hazards that are associated with trench-related traumatic injuries **FIGURE 3-8**.

FIGURE 3-8 Assessment of the atmosphere.
Courtesy of Ron Zawlocki.

Severe environmental conditions can increase the probability of secondary collapse and can affect the victim's chances of surviving the incident. Those environmental conditions include the effects of extreme weather (rain, winds, temperatures) as well as high water tables and uncontrolled broken water lines. Excessive water cannot only quickly fill the trench and drown the trapped victim, but it can change the weight and internal strength of the soil, which can place the collapse hazard beyond the scope of the capabilities of standard rescue shoring equipment.

The hazard assessment should begin as the trench site comes into the view. Identifying the level of the hazards will help determine the hazard mitigation resources that will be necessary.

Victim Assessment

First-responding rescuers are responsible for what is often the difficult task of determining if the victims are alive or dead. The difficulty is compounded when a collapse has occurred in an unprotected trench, preventing entry and patient contact by the rescuer. Unfortunately, this is the most common situation encountered by rescuers. (Chapter 8, *Victim Care and Extrication*, covers victim assessment and care in detail.) The easy victim assessments occur when the victim is visible. With cave-in situations, the victim viability is good when they are conscious and buried below their

FIGURE 3-9 A victim assessment done from the trench lip can identify the victim location and condition and can determine the possibility of a non-entry rescue.

chest. If the victim's arms are not buried or injured, a non-entry rescue is a real possibility. Much more difficult victim assessments occur when the victims are not visible from the trench lip FIGURE 3-9. Viability of completely buried victims is statistically very low but it is possible if large objects (pipes, construction equipment, boulder, etc.) were present in the trench and near the victim at the time of the collapse. Those kinds of objects can both shelter the victim from the impact of a collapsing trench wall and can provide a void area near the object that could allow the victim to breath.

When arriving at a trench collapse where no victims are visible, the rescuers should immediately interview witnesses (competent person, foreman, laborers, excavator operator) to get information about the number of victims and their last known locations. A victim assessment includes identifying and locating all actual and potential victims to determine the number of victims involved and their probable locations.

Trench Assessment

The first trench assessment should be the depth, width, and shape of the trench. If the trench is 8 feet [2 m] or less in depth and width and is a straight trench (no intersecting trenches near the rescue area), then the trench emergency is an operations level event. While an engine company is unlikely to have a tape measure on board, it is likely to have an 8-ft (2.4-m) pike pole or another tool that the crew knows the length of and which can be used as a reference to estimate the trench dimensions. While standing on lip protection, a firefighter can lower an 8-ft (2.4-m) pike pole into the trench FIGURE 3-10. If one end of the pike pole

FIGURE 3-10 An 8-foot pike pole being used to estimate the trench depth.
Courtesy of Ron Zawlocki.

reaches the bottom (original floor) and the other end sticks up over the lip, the trench is less than 8 feet (2.4 m) deep. That technique can be used to estimate the width of the trench by holding one end at the wall where the lip protection is in place and seeing if the opposing wall is within the length of the pole.

With the trench dimension estimates, the IC can begin to determine the appropriate level of the trench rescue team needed to resolve the emergency. If the trench is straight (nonintersecting), is less than 8 feet (2.4 m) deep and less than 8 feet (2.4 m) wide, can be made safe with traditional sheeting and shoring, does not require more than 2 feet (0.6 m) of soil removal (digging) to extricate the victim, and severe environmental conditions do not exist, an operations level trench rescue team should be able to handle it.

The second trench assessment involves collapse. (See Chapter 2, *Soil and Collapse Mechanics*, for details on types of collapse and collapse patterns.) If no collapse has occurred, then it is a noncollapse situation. If a collapse has occurred, an estimate needs to be made of the size of the collapse. Communicate this and the other assessment information to the responding trench rescue team so they can prepare for the situation and inform the IC to ensure they are adequately equipped or will need additional resources. The situation, hazard, trench, and victim assessments will determine what additional (outside) resources will be needed.

Resource Assessment

Begin the resource assessment by identifying the on-site resources that can provide the needed initial actions **FIGURE 3-11**. Assess lip protection sources. Items such as fire service, emergency medical services (EMS) backboards, ladders, and construction company plywood and lumber can be used for temporary lip protection to enable initial action functions. If a spoil pile is less than 2 feet (0.6 m) from the lip, shovels will be needed to move the pile. Shovels

Initial Actions Checklists

Scene Management	Site Control

Incident Command
• Establish a clearly identifiable Command Post
• Provide a radio report that includes:
   - Situation report
   - Command Post and Staging Area locations
   - Request for resources (activate the system)
• Establish Staging Area
   - Forward Staging for First Responders and selected resources (beyond 100' from the trench)
   - Primary Staging (> 300' from the trench)
• Assign Initial Action Duties
   - Site Control
   - Hazard Management
   - Nonentry Rescue
Size-Up (Conduct Assessments-Evaluate the Situation)
• Assessments- Gather information
   Hazard Identification
     - Collapse
     - Utilities
     - Traffic
     - Atmospheric
     - Hazardous Materials
     - Physical Hazards
     - Biological
   Victim Information
     - Mechanism of injury or entrapment (cave-in or noncave-in)
     - Number of victims
     - Condition of patient
   Site Detail
     - Trench size (depth, width, and length)
     - Access to site
   On-Site Resources
     - On-site equipment/materials that can be used for rescue
     - On-site construction personnel that can be used for rescue
• Evaluation- Determine the scope of the incident based on assessments
   - Rescue or Recovery
   - Level of Risk- Based on hazards and available mitigation efforts
   - Awareness Level- Nonentry rescue or Initial Actions for higher level response
   - Operations Level- Straight trench, < 8' deep, no supplemental shoring, no severe conditions
   - Technician Level- Straight or intersecting trench, up to 20' deep, supplemental shoring, severe
   - Specialist Level- Other trenches and excavations, extreme conditions
Summon Resources (Based on assessments and evaluation)
• Rescue or Recovery
• Awareness Level (First alarm and on-site resource)
• Operations Level (Tier 1 resources)
• Technician Level (Tier 1 and selected Tier 2 resources)
• Specialist Level (Tier 1 and selected Tier 2 resources)
	• Create a hot, warm, and cold zone
	• Reroute all nonessential traffic at least 300 feet around the scene
	• Remove all nonessential civilian personnel to at least 50 feet from the incident
	• Shut down all heavy equipment operating within 300 feet of the collapse
	• Prepare the site for incoming resources
	
	Hazard Management (Defensive Actions)
	• Assure proper PPE for all first responders
	• Mitigate or isolate all hazards in the area, i.e., utilities, electric, gas, water, etc.
	• Stabilize exposed utilities, pipes, or any other obstruction in the trench
	• De-water the trench if necessary
	• Monitor the atmosphere and ventilate the trench if necessary
	• Place ingress and egress ladders in trench. There should be at least 2 ladders placed in the trench no more than 12 feet apart
	• Install appropriate Lip Protection
	
	Nonentry Rescue
	• Determine if a nonentry rescue is possible
	• If the possibility exists, assign an extrication team to implement the nonentry rescue
	• Install lip protection in the area where rescuers will need to work on the trench lip
	• For partially buried victims, tie a rope to a shovel and lower it to the victim
	• Place a ladder near the victim and support the ladder (tied off or held) as the victim climbs out
	• If the victim is unable to climb a ladder, set up a rope rescue system with an overhead anchor point (high directional), lower a simple harness (like a LSP Cinch Ring) attached to the rope, instruct the victim on how to don the harness, and raise the victim out of the trench
	• Remove the victim from the trench

FIGURE 3-11 A sample initial actions checklist.

are usually available from the construction crew and are also often part of a fire company's overhaul equipment cache. Spoil pile removal must be performed while working on lip protection. Rescuers should place a ladder in the trench; initially, the ladder is put in place as an escape ladder in the event a rescuer slips and falls into the trench. A ladder is also a possible escape route for a victim during a non-entry rescue operation.

If a non-entry rescue is not possible, a Tier 1 response will be needed (covered in detail later in this chapter). If your trench rescue response system is properly set up, all the Tier 1 response resources will be on the way before you are on the scene. The IC can then radio a simple situation and status update, such as "First engine, rescue, EMS, and police can handle. Cancel all other Tier 1 responders," or "This is a working trench rescue. Holding all Tier 1 responders."

Determine the Scope of the Rescue

The scope of the rescue is determined by the number and types of hazards, the number of victims, the extent of their entrapment and injuries, and the size and shape of the trench or excavation. More and complex hazards; multiple victims who are heavily trapped and severely injured; and trenches that are deep, wide, and have complex shapes all require more resources, such as specialized equipment and personnel with high levels of knowledge, skills, and abilities. Using the National Fire Protection Association (NFPA) 1670 rescue classification (awareness level, operations level, and technician level) criteria can help to determine the scope of the incident; however, it must be recognized that many trench and excavation incidents go beyond the scope of the NFPA's highest level (technician). The authors have added a level called specialist that addresses situations and conditions that go beyond the NFPA standards. An accurate determination of the scope of the rescue operation can be obtained by the IC's evaluation.

Evaluation

After the assessments are completed, the IC must evaluate the information gathered and determine the scope of the incident. The **evaluation** will determine the resources required to mitigate the incident and will directly impact the IAP. Whether a non-entry rescue can be completed by the awareness level rescuers (or first-responding firefighters) or will require additional resources does not change the fact that the appropriate initial actions will affect the outcome of the incident. The following guidance is presented to help determine and categorize the scope of the incident:

- Awareness level incidents
  - Hazards can be managed with defensive measures.
  - Rescues can be accomplished without entering the trench.
  - Personnel working on the trench lip can be protected from collapse.
- Operations level incidents
  - Hazards can be managed with defensive and/or offensive measures.
  - Rescues and recoveries requiring the entry of rescue personnel.
  - Trench details:
    - Depth of 8 feet (2.4 m) or less.
    - Nonintersecting trench.
    - Digging to extricate a victim will not expose more than 2 feet (0.6 m) of unshored trench wall.
  - Environmental conditions are not severe.
  - Tier 1 resources will be needed.
- Technician level incident
  - Hazards can be managed with defensive and/or offensive measures.
  - Rescues and recoveries requiring the entry of rescue personnel.
  - Trench details:
    - Depth of 20 feet (6 m) or less.
    - Nonintersecting or intersecting.
    - Digging to extricate a victim may expose more than 2 feet (0.6 m) of unshored trench.
  - Environmental conditions may be severe.
  - Tier 1 resources selected; Tier 2 resources may be needed.
- Specialist level (See Appendix A for more details.)
  - Trench depths greater than 20 feet (6 m).
  - Excavations.
  - End walls are within the victim area and have a potential to collapse.
  - Walls with angles less than 70 degrees.
  - Overhanging appurtenances (pavement, equipment, or materials) that cannot be shored or quickly removed.
  - Unstable spoil piles that are falling into the excavation/trench.
  - Excavations dug into the side of tall, steep, or unstable slopes (hills).
  - Shoring that is structurally inadequate, loosening, or moving.

- Failure of an engineered shoring system.
- Moving soil (even after the installation of shoring).
- Flowing soil (uncontrollable water and soil that are below the water table).

Summon Resources

It is a foregone conclusion that the local fire department will be dispatched to a trench/excavation emergency occurring within their jurisdiction. According to the NFPA, every individual who responds to a technical search and rescue event should have a minimum of awareness level training and capabilities. That not only applies to every firefighter, but includes all first responders, such as emergency medical and police force personnel. Most fire departments will arrive on the scene of a trench emergency within minutes of the 911 call.

All fire departments should be trained and equipped to effect awareness level tactics and non-entry rescue operations. However, when a non-entry rescue will not resolve the emergency, there will be a delay in the arrival of resources that are needed for entry rescue operations. That delay can be minimized if the AHJ has developed a comprehensive list of resources that are needed to mitigate the majority of trench incidents. The AHJ must establish agreements with each agency on the list, long before the need for them to respond occurs. In addition to the resources that are frequently needed at trench rescue incidents, the AHJ should develop a list and establish agreements with resources for trench and excavation incidents that occur but occur less frequently. The authors call the list for common trench rescue situations the Tier 1 list. Tier 1 resources should be activated upon a report of a trench rescue event. Tier 2 resources are activated upon the request from the IC.

Resources needed at a trench rescue incident include services and equipment that ultimately support the trench rescue team. The AHJ must have prearranged agreements with each resource and a resource list with 24/7 emergency contact information. A memorandum of agreement should be developed that outlines specifications for equipment and resource allocation, availability of services, procedures for procurement, and subsequent financial reimbursement for the services or equipment supplied. Agreements for trench rescue resources typically include mutual aid and "pay for" services. The authors have divided these into a two-tiered response plan. Tier 1 and Tier 2 resources need to train with first responders and trench rescue teams.

Tier 1 Response

The following services and equipment must be immediately dispatched to all reported trench emergency scenes.

- Fire Department: The highest authority at the scene from the local fire department will serve as the IC. The IC provides the initial actions, including scene management, hazard control, and site control; summons needed resources; and implements non-entry rescues. Some large fire departments also provide trench rescue teams and equipment.
- Police: Police are needed to control vehicle traffic and protect fire, rescue, and EMS personnel and equipment at the scene.
- EMS: Traumatic injuries are common at trench accident scenes. Although the responding trench rescue team should include urban search and rescue (USAR) medics, the EMS response from the AHJ needs to be at the paramedic level.
- Trench rescue team: The formations of many trench rescue teams have resulted from mutual aid agreements. A team of 20 trench rescue technicians can handle the variety of tasks that need to be conducted simultaneously during rescue operations.
- Emergency utility control: Dispatch should contact the emergency contacts for all utilities (natural gas, electric, water, and sewer) and request an immediate response to the scene of every reported trench rescue incident. One Call utility locating services provide emergency responses to assist rescue personnel as well. Response times will vary; many utility locating services suggest they can be at an emergency within 3 hours.
- Vacuum trucks: Vacuum trucks commonly come from municipal departments of public works. It is important that rescuers train on the set-up and use of vacuum trucks prior to an incident **FIGURE 3-12**.
- Recovery bags: Large vehicle recovery air bags, used by some large-scale wrecker services, can be used as back-fill options at trench collapse incidents. It is important that rescuers train on the set-up and use of recovery bags prior to an incident.

Tier 2 Resources

These services may not be automatically dispatched to a trench emergency; instead, they often are selectively requested by the IC.

FIGURE 3-12 Many municipal departments of public works have access to vacuum trucks and are willing to bring them to a trench emergency.
Courtesy of Larry Collins.

- Technical rescue advisors: The AHJ should develop a list of local trench rescue subject matter experts. This list commonly includes trench rescue instructors and experienced trench rescue specialists.
- Rescue engineers: Currently rescue engineers are few and far between. They must be developed through training with trench rescue teams and first responders. A state or local USAR team's structure specialist will help leadership find a nearby civil engineer who would be willing to train with local fire departments and trench rescue teams.
- Heavy equipment operator: In some cases, trench and excavation incidents may require the use of heavy equipment and the talents of an experienced operator. The transition from rescue to recovery operations is often best served with sloping or benching the trench walls using heavy equipment. In other situations, heavy equipment can be used to lift or stabilize heavy objects in the trench. Fire departments and trench rescue teams in several states have working agreements with the International Union of Operating Engineers (IUOE) that bring rescue-trained equipment operators to the scenes of trench rescue incidents.
- Confined space rescue team: Although a typical trench is not a permit-confined space, there can be times when a confined space rescue team is needed. Examples would include a victim who is in a pipe, tank, or vault that is in a trench or excavation. Additionally, confined space rescue teams should respond to underground shaft and tunnel rescue incidents.
- Hazardous materials team: Many trench rescue teams now have an automatic response from the local hazardous materials team to all trench and excavation rescues. Others summon hazardous materials teams as a Tier 2 response. A hazardous materials team can handle atmospheric monitoring and mitigation operations. An auto-response with the trench rescue team would be part of a Tier 1 system. In any case, a Tier 2 response from a hazardous materials team should be available for trench rescues at landfills and areas where hazardous materials are manufactured or stored.
- USAR search team: USAR search teams are helpful in the event of massive trench and excavation wall collapse. Search dogs (both live finds and human remains), search cameras, and seismic/acoustic search equipment can help pinpoint buried victims.
- Rehabilitation units: Technician- and advanced-level trench incidents can continue well beyond the golden hour. In those situations, rehabilitation units will be needed to provide food, hydration, tents, and rest areas for all rescue and support personnel on scene.

Site Control

The first-arriving responders (fire and police) must control the site. The IC must develop a plan to shut down traffic, establish and control operational zones, and prepare the site for incoming resources. The IC must then give specific assignments to control the site.

Shut Down Traffic

Road traffic must be shut down in a 300-foot (91-m) radius around the trench to help prevent cave-in. Construction equipment must also be shut down unless they are assigned to rescue operations by a technician level trench rescuer. Police officers are trained and equipped to control road traffic. Firefighters need to coordinate the movement and shut-down of construction equipment around the trench site.

Establish Operational Zones

One of the first assignments the IC will give is to establish operational zones. Firefighters assigned to this task can use caution tape and other barriers (fences, structures, etc.) to clearly mark off the zones. Initially establishing a hot zone with a radius of about 25 feet (7 m) around the victim in the trench will be easy to control and will provide enough room for rescue operations. Remove anyone who is not part of the response team from the trench and hot zone. The hot zone must have a controlled entry point that provides easy access to assigned personnel and equipment. After a hot zone has been established (barrier tape and other barricades), a warm zone should be developed. The warm zone needs to be large enough to provide for the command post, forward staging (personnel and equipment), and support functions. Everything beyond the warm zone is considered the cold zone.

As a quick reminder, the purpose of each zone is as follows **FIGURE 3-13**:

- Hot zone: This area, also known as an action zone, is for rescue teams only. It immediately surrounds the dangers of the incident, and entry into this zone is restricted to protect personnel outside the zone.
- Warm zone: This area is for properly trained and equipped personnel only.
- Cold zone: This area is for staging vehicles and equipment and contains the command post.

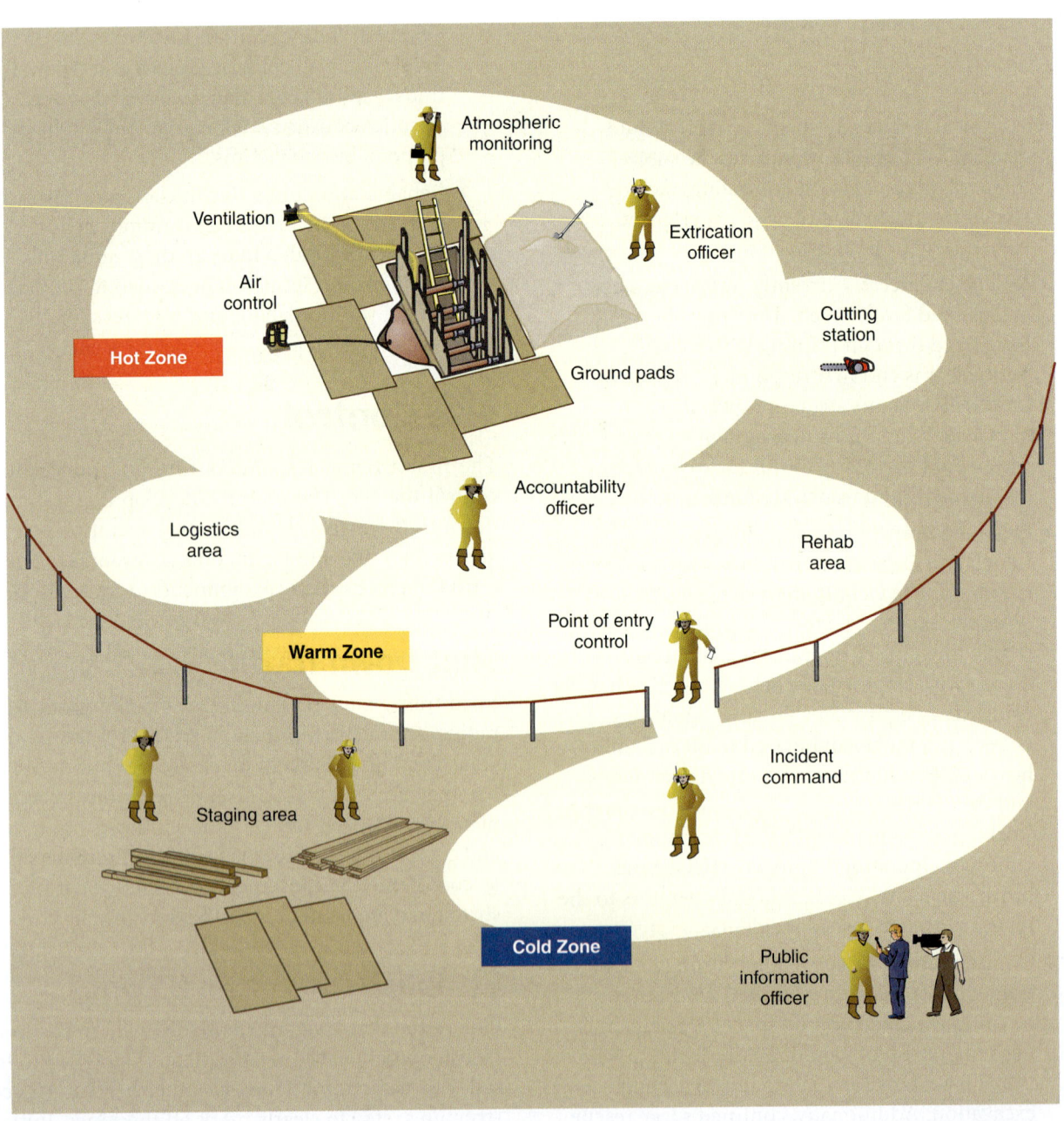

FIGURE 3-13 Hot, warm, and cold zones.

Prepare the Site for Incoming Resources

Firefighters and construction workers need to prepare the site so that when the trench rescue team and other resources arrive, they will be able to seamlessly access the trench and begin shoring and rescue efforts. Trench and excavation sites are often congested with vehicles, construction equipment, and spoil piles. Initial actions need to provide a clear access to the trench for rescue vehicles and a clear pathway for rescuers to carry equipment from their trucks and trailers to the trench. Once that path has been established, the trench lip (8 feet [2 m] on both sides of the trapped victim) must be cleared and leveled. Construction tools and equipment scattered around the trench lip are potential trip and fall hazards and must be removed. Ground pads (sheets of plywood, backboards, or ladders) should be placed on the lip near the victim to distribute the weight of rescuers and rescue equipment near the trench lip. A safe path to the trench should be identified, and an escape ladder should be placed into the trench in the event that a rescuer or assigned construction worker slips or trips and falls into the trench.

Hazard Management

Perhaps the most important function performed during the initial actions phase of the incident is recognizing existing and potential hazards. In Chapter 1, *Introduction to Trench Rescue*, we detailed the types of hazards that are common to trench and excavation sites. The most common and most deadly hazard is collapse (cave-in). In Chapter 2, *Soil and Collapse Mechanics*, we discussed soil failures and identified general collapse patterns and areas and conditions that are at risk of collapse. Firefighters typically arrive on the scene of a trench emergency with limited trench-related hazard mitigation equipment. In order to minimize risks to rescuers, bystanders, and victims, all hazards must by promptly identified and controlled. The time needed to identify hazards and initiate control (defensive measures) should be 10 to 15 minutes after arrival.

Firefighters trained at the trench rescue awareness level are needed to implement defensive measures during the initial actions phase of the response. Defensive measures include the following:

- Avoid the hazard: Isolate and keep people at a safe distance from the hazard.
- Control the hazard: Stop the hazard from entering the rescue area.
- Remove the hazard: Move the hazard away from the rescue area.
- Personal protective equipment (PPE): Don PPE that will provide protection for a specific hazard. At most trench incidents, first responding firefighters should wear helmets, boots, and gloves. If a hazardous atmospheric condition exists, full firefighting PPE with self-contained breathing apparatus (SCBA) is necessary until the hazard is mitigated.

Defensive Measures

An example of **defensive measures** taken by first responders occurred at author Ron Zawlocki's first excavation incident in West Bloomfield, Michigan. Construction workers had dug around a large steel pipe at the corner of a very busy intersection that remained open to traffic. The pipe (sewer line) was broken and leaking. Three workers were in the 13-foot-deep by 20-foot-wide (4-m by 6-m) unprotected excavation when a large lip shear collapsed. An oxygen/acetylene torch was being used to cut out the damaged section of pipe when the collapse occurred. The worker who was operating the torch was hit by a massive section of clay/sand soil. Two other workers were trapped with their lower extremities buried.

When the first engine company arrived, they were faced with several hazards. Number one was the unprotected and unstable walls of the excavation. Number two was the rush hour traffic passing by just feet away from the trench wall. Number three was the toxic and explosive acetylene that was filling up the hole after the collapse snuffed out the flame of the torch. The defensive measures taken by the West Bloomfield Firefighters (WBFD) were exemplary. Here are details of how they implemented defensive hazard control methods (avoid, remove, control, and PPE).

- Avoid: The first responders avoided the secondary collapse hazard by staying out of the trench and by working from the concrete pavement–covered lip. The decision to avoid entering the trench also kept them out of the area of toxic and explosive gas concentration.
- Control: By controlling (rerouting) the flow of vehicle traffic, dangerous vibrations that can cause secondary collapse were eliminated. Additionally, by controlling traffic, the **risk** of first responders being hit by a vehicle was eliminated. Firefighters were also able to control and stop the flow of acetylene into the trench by shutting off the valve on the acetylene tank that was located on the trench lip.

FIGURE 3-14 Defensive measures include identifying and isolating utility hazards.
Courtesy of Ron Zawlocki.

- Remove: Firefighters were able to pull the torch and hoses out of the hole after the tank valves were closed. They then wheeled the torch (removed), hoses, and tanks out of the hot zone. Additionally, part of the spoil pile was removed by firefighters using shovels. Removing the spoil pile minimized the surcharged load that could contribute to a secondary collapse.
- PPE: Firefighters dressed in turnout gear with SCBA are best suited to operate near a potentially flammable/explosive environment **FIGURE 3-14**. In this case, that was the acetylene gas component of the cutting torch. A charged hose line stretched to the edge of the lip added another level of safety.

Defensive Mitigation Actions for Underground Utilities

Especially in heavily populated areas, it is difficult to dig a trench without encountering previously installed utilities. Exposed utilities can be a threat, and broken utilities are real hazards. Both conditions must be resolved to minimize risks to rescuers, bystanders, and victims. The defensive measures listed here take that into account and provide methods for managing hazards by securing the scene, isolating hazards, and summoning hazard mitigation resources.

- Underground electrical wires: First, call the electric company. Control access to area by removing nonemergency personnel from the involved area, keeping rescue personnel away from exposed wires, and establishing perimeters (barrier tape, rope, etc.). Noncontact voltage testing should be performed if rescuers are equipped and trained, always following local SOGs.
- Natural gas line break: For a natural gas line break, call the gas company. Next, control access to the area and monitor the area for hazardous atmosphere. Rescuers should prepare for gas-related hazards such as fire, explosion, and combustion of surrounding flammables. Sources of ignition should be eliminated. Because of these hazards, suppression equipment and full structural firefighting PPE must be in place. Trench entry operations by operations level and technician level rescuers should be delayed until acceptable safety levels are met. Using the resources available, an attempt to determine the location of a shut-off should be made. However, attempts to operate a street valve should not be made without gas company supervision. Monitor and investigate surrounding structures for vapors and occupants who have been overcome.
- Sewer line break: Call the department of public works. Control access to area and monitor the area for hazardous atmosphere (methane, hydrogen sulfide, etc.).
- Water line break: Call the water department and control access to the area.

Unbroken Utility Lines

Regardless of what type of utility an exposed pipe is carrying, it should be supported before the shoring and rescue operation begins **FIGURE 3-15**. Supporting the exposed line is done to reduce the chance of the line breaking during the rescue/recovery operation. To support an exposed utility line that is still intact:

- Monitor the atmosphere in the area that needs to have a pipe supported (work from ground pads).
- Position a ladder or timber over the trench opening and directly above the exposed pipe.
- Apply ropes or straps around the pipe.
- If a pipe consists of more than one continuous section, support all sections of the pipe independently.
  - Tie off all ropes or straps to ladder rungs, beams, or timber beam.

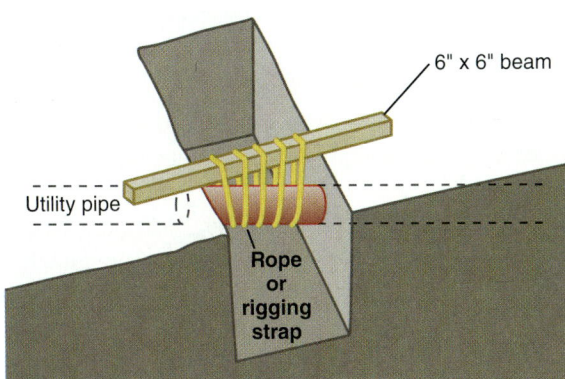

FIGURE 3-15 Exposed utilities must be supported to prevent breakage.

Nonentry Rescue and Victim Self-Rescue

Nonentry and victim self-rescues are always the preferred methods. They reduce the risk to rescue personnel and should be considered in every incident. This approach may be as simple as a worker who is only partially buried being able to dig themselves out (self-rescue) or a **non-entry rescue** where rescuers extricate the victim without ever going into the trench.

The best-case scenario is the one in which an individual just needs a ladder to get out of the trench. If the end of the trench is near the victim, the ladder can be placed into the trench while the rescuer is standing on the lip of the end wall. If the victim is partially buried and has one or both arms free, a helmet and a shovel may be lowered down to the victim and the victim can dig themselves out. Keep in mind that even individuals with broken bones will climb a ladder if they think a potential for collapse exists. If they are hurt and only partially able to help themselves, pass down a packaging device such as the chest portion of a class 3 rope rescue harness, a "screamer suit," or LSP Cinch Ring (previously called the LSP Cinch Collar) attached to a rope rescue system. The victim can then be lifted out of the trench following the release of the mechanism of entrapment **FIGURE 3-16**.

Rescuers must make sure that the weight of the rescuers assisting with the non-entry rescue operation is distributed along the trench lip **FIGURE 3-17**. Medical backboards can act as ground pads and will help with load distribution.

A non-entry rescue may require a hauling device attached to an elevated anchor (high directional) that removes the victim vertically. Access to the trench and proximity to the lip for the safe operation of an aerial ladder or a piece of heavy equipment (excavator of backhoe with a rated lifting eye) will need to be considered as a part of the operational plan in this

FIGURE 3-16 A modified packaging device like the LSP Cinch Collar can be passed down to a victim in a trench.
Courtesy of CMC Rescue, Inc.

FIGURE 3-17 A rescued worker being pulled out vertically.
Courtesy of Larry Collins.

situation. Aerial ladders and excavators used as high directional anchor points may be positioned and then completely shut down and used only as a stationary anchor point. Aerial ladders and excavators should never be extended or raised to lift a victim.

Voices of Experience

On a mild, early spring Sunday afternoon, our department was dispatched for a trench rescue. A do-it-yourself sewer line replacement project had caused a trench collapse that trapped a man working in the trench and buried him to his waist. The collapse occurred at approximately 4:30 PM after the crew had dug a trench that was about 100 feet (30 m) long. Workers had been entering the trench to replace a sewer line that ran from the home to the sewer main. The workers were in the trench placing the last section of polyvinyl chloride (PVC) pipe to the main when the collapse occurred. The trench was adjacent to a four-lane U.S. highway within the city limits. Traffic was still traveling on the highway that has a 35-mi/h (56 km/h) speed limit when first responders arrived.

To further complicate the rescue incident, the excavation had exposed a live natural gas line within the rescue area and had undermined a sidewalk and driveway. The first-due engine established command upon arrival and approached the end of the trench to perform assessments and size-up. The initial IC found several would-be rescuers in the trench, including one law enforcement officer. The IC ordered all the would-be rescuers out of the trench.

The size-up revealed unstable soil conditions consisting of a mix of clay and gravel under a few inches of topsoil, all of which were saturated from recent rain and snow melt. During the interview of the workers, it was learned that the collapse began as a semi-tractor trailer passed by on the highway. It is likely that the vibration and surcharged load of the heavy truck caused the initial shear wall collapse. The wall shear occurred under the sidewalk, leaving about a 2-foot (0.6-m) overhang of the 4-inch (10-cm)-thick concrete sidewalk. The trench was approximately 8-feet (2.4 m) deep and 4-feet (1.2 m) wide at its widest point where the cave-in had occurred. The spoil pile was immediately adjacent to the trench lip. The placement of the spoil pile not only increased the probability of a secondary collapse but made the installation of shoring equipment difficult. There was no trench box or shoring being used during the work in the trench and they were using a step ladder to enter and exit the trench.

The initial fire department response was four members responding on a fire engine and a squad truck with a technical rescue trailer. Local police and EMS were also dispatched to the scene.

After a quick size-up, initial action duties were assigned, which included installing ground pads and placing a ladder in the trench near the victim's location. A police officer was assigned to control and redirect traffic on the highway. The victim's location and utilities were marked. The first panel of the primary panel set was tied with long ropes that were thrown over the top of the spoil pile to help install it and hold it in the position that would protect the victim's head and chest. While the primary panels and struts were being installed, another crew was shoveling the spoil pile back away from the lip of the trench. As additional responders arrived on scene, secondary shoring followed. Once the panels were in place and the shores secured, two members entered the trench on a ladder placed within the safe zone (shored area) to start initial patient assessment and care and to begin the process of digging the patient out. The patient had initially been given an entrenching tool to begin aiding in his rescue. His injuries made this difficult and painful. A medic entered the trench to put oxygen on the patient and started an IV so that pain medication could be administered. The patient was uncooperative and would not wear the helmet that was placed on him or leave the blanket around his torso.

The patient was pinned in an awkward position, and his lower extremity injuries caused him to

Voices of Experience

need to support himself to try to minimize the pain resulting from his injuries. As rescuers dug nearer to the patient, the movement of the soil caused him additional pain. On the side of the trench opposite of the highway, the spoil pile had collapsed into the trench and was at approximately its natural angle of repose. As a result, the decision was made not to disturb the soil in that area in an effort to prevent the further running of the gravel into the trench.

The IC requested a vacuum truck with the thought that they would be able to remove the soil quicker and with less pain for the patient. Various shovels were used to dig around the patient. The different shovels included entrenching tools, pointed long-handled spades, transfer shovels, and a post-hole digger. The patient had numerous major lower extremity injuries that created the need to dig all the way to the soles of both the patient's boots before he was able to be lifted. It was found helpful to allow the patient to dig closest to himself and support himself to prevent as much pain as possible. Also, the patient was able to advise the rescuers of the location and orientation of his legs as they were digging. The weight of the rescuers on the cave-in pile also transmitted to the patient, causing pain, so rescuers had to be careful to keep their weight off the soil that covered the patient.

Command was transferred to the chief, which freed operations personnel for rescue work. Rescuers were frequently rotated into the digging positions. The heavy soil and digging in a bent-over position caused rescuers to fatigue quickly. Some of the clay was in large pieces that required rescuers to break them into smaller pieces to remove them. The patient was nearly freed when the vacuum truck arrived, so the decision was made not to utilize it.

A tower ladder was extended over the trench to provide a high point for a rope system removal of the patient. The rescuers found that the patient's injuries would not allow the placement of the leg straps of the harness/immobilization device so that plan was revised and the patient was manually lifted onto a backboard in an area adjacent to the trench, that was an excavation, by using webbing placed around his torso. Once secured to the backboard, he was passed hand-over-hand by rescuers out of the hazard zone to EMS personnel. The entire rescue was completed in just under 2 hours. The patient was transported to the local hospital and later airlifted to the regional trauma center. In addition to his traumatic injuries, the patient had become hypothermic. In summary, a successful rescue, albeit not textbook, was made and no rescuers were injured.

Lessons Learned

- Thorough and transparent debriefing and incident critique are vital to improving trench rescue operations.
- A dedicated and identifiable command post and zoning (especially hot zone control) are critical to trench rescue operations.
- All traffic in the area should have been stopped rather that redirected.
- Additional resources should have been called to have a strategic reserve and to enable more frequent rotation of rescuers.
- Utility companies must be requested early in the incident.
- Air quality monitoring must be performed to rule out the common atmospheric hazards (carbon monoxide, hydrogen sulfide, oxygen deficiency, and flammable).
- A forward staging area that positions fire and rescue apparatus beyond 100 feet (30 m) from the trench allows for a good working distance and will provide for the development of a well-placed command post.

Voices of Experience

- Training all first responders to the trench rescue awareness level can help direct police and EMS personnel away from trench hazards and can prevent them from parking their vehicles in areas that are either unsafe or are needed by rescue resources.
- According to the NFPA, every individual who responds to a technical search and rescue event should have a minimum of awareness level training and capabilities. That not only applies to every firefighter but includes all first responders, such as EMS and police force personnel.
- Hypothermia is a common condition for injured and trapped victims in the bottom of a trench. Hypothermia can occur in a trench even during mild and warm weather. Injuries, bleeding, and wet and cool soil will compound hypothermia effects.
- Hypothermia can be easily managed with hot packs, blankets, heated IV solutions, and electric ventilation (confined space ventilation) equipment with heating elements.
- Initial actions are performed by first-due responders trained at the trench rescue awareness level.
- Properly implemented initial actions impact the success of every trench rescue operation.
- Initial actions must include scene management, hazard control, site control, non-entry rescue (when possible), and summoning resources.
- Entry rescues and recoveries require additional resources (higher level of trained personnel and specialized rescue equipment) and require the implementation of a formal strategic plan.

Coldwater Fire Department
Coldwater, Michigan

It is unlikely that the victim has been trained in how to put on a class 3 harness or other webbing configuration, so a simple harness should be provided and the temptation to send someone down in the trench to secure the packaging system should be resisted. In all cases, never attempt to pull a partially buried victim from the trench.

Another simple but slightly more risky method of performing non-entry rescue is to have the victim place a pair of wristlets on their wrists. The best way to remove a victim with wristlets is a rope-based mechanical advantage system and a high directional anchor. A faster but more risky way to lift the victim out of the trench could involve getting several of the strongest rescuers on the rope to pull the victim up. It all comes down to the **risk versus gain analysis**, in which a decision is made based on a hazard identification and situation assessment that weighs the risks against the benefits to be gained for taking those risks.

Another rapid but somewhat risky removal method (trench width dependent) involves lowering a Stokes basket with a rope tied on each corner and having the victim roll themselves into the basket and hang on. Then, with a rescuer on each rope (at all four corners), the basket/victim is pulled up out of the trench using a hand-over-hand technique. Nonentry rescue operations are a best practice because they minimize risks to rescuers. The next section outlines steps for non-entry rescue operations.

Nonentry Rescue

First, determine if a non-entry rescue is possible. The victim must be conscious with at least upper body mobility, and the trench lip adjacent to the area where the victim is trapped must be safe enough for available lip protection techniques.

- If the possibility exists, an extrication team is assigned to implement the non-entry rescue.
  - Install lip protection in the area where the rescuers will need to work on the trench lip.
- For partially buried victims, tie a rope to a shovel and lower it to the victim.
- Place a ladder near the victim and support the ladder (tied off or held) as the victim climbs out.
- If the victim is unable to climb a ladder, set up a rope rescue system with an overhead anchor point (high directional), lower a simple harness (like a LSP Cinch Ring) attached to the rope, instruct the victim on how to don the harness, and raise the victim out of the trench.
- Remove the victim from the trench.

SAFETY TIP

Whether your victim can self-rescue or even partially assist in the removal process, mitigating the incident and not placing your personnel in harm's way is always the preferred method.

After-Action

IN SUMMARY

- The actions taken by the first-arriving firefighters have a direct impact on the safety and success of the entire rescue operation.
- The first 20 to 30 minutes after the arrival of first responders should be a busy and productive period.
- The command structure begins with the arrival of the authority having jurisdiction, which, in most cases, is the local fire department. The highest-ranking chief or company officer will initiate scene management and assign initial action duties to on-scene personnel.
- Initial actions include the following:
  - Scene management
  - Hazard control
  - Site control
  - Nonentry rescue
  - Summoning resources
- A well-positioned and clearly marked command post provides resources (moving up from staging) with a direction and target.
- When interviewing, remember to ask who, what, where, when, and why.

- The first trench assessment should be the depth, width, and shape of the trench. If the trench is 8 feet (2 m) or less in depth and width and is a straight trench, then the trench emergency is an operations level event.
- The scope of the rescue is determined by the number and types of hazards, the number of victims, the extent of their entrapment and injuries, and the size and shape of the trench or excavation.
- The evaluation will determine the resources required to mitigate the incident and will directly impact the IAP. Determine and categorize the scope of the incident to the appropriate level as follows:
  - Awareness level
  - Operations level
  - Technician level
  - Specialist level
- Tier 1 response includes the following:
  - Fire department
  - Police
  - EMS
  - Trench rescue team
  - Emergency utility control
  - Vacuum trucks
  - Recovery bags
- Tier 2 response includes the following:
  - Technical rescue advisors
  - Rescue engineers
  - Heavy equipment operators
  - Confined space rescue team
  - Hazardous materials team
  - USAR search team
  - Rehabilitation units
- Site control includes shutting down traffic, establishing and controlling operational zones, and preparing the site for incoming resources.
- The most important function performed by first responders during the initial actions phase of the incident is recognizing existing and potential hazards. The time needed to identify hazards and initiate control (defensive measures) for first responders should be 10 to 15 minutes after arrival.
- Nonentry and victim self-rescues are always the preferred methods. They reduce the risk to rescue personnel and should be considered in every incident.

KEY TERMS

Defensive measures Methods that protect people from specific hazards without directly exposing first responders and rescuers to the hazards.

Evaluation A systematic determination of scope and magnitude of the emergency using the assessment criteria. Its primary purpose is to determine the incident objectives, strategy, and tactics, and the resources that will be needed to accomplish them.

Forward staging An area located beyond 100 feet (30 m) from the trench, initially designated by the incident commander for fire, rescue, emergency medical services, and police responders and their vehicles.

Hazards Any agents that can cause harm or damage to humans, property, or the environment.

Incident commander (IC) The individual responsible for developing the strategic goals for the operation.

Incident command system (ICS) Management system by which emergency personnel are assigned specific responsibilities and areas of supervision.

Initial actions The immediate actions taken during the first 20 to 30 minutes after arrival at the site of an emergency; includes scene management, hazard control, site control, non-entry rescues, summoning resources, and establishing support functions.

Nonentry rescue A rescue that can be accomplished without rescuers entering the trench or excavation.

Primary staging An area designated by the incident commander and located at least 300 feet (91 m) from the trench, for the staging of trench rescue resources.

Risk An exposure to a hazard or hazards that will likely result in injury, health problems, or death.

Risk versus gain analysis A decision made based on a hazard identification and situation assessment that weighs the risks likely to be taken against the benefits to be gained for taking those risks.

Size-up Determination and analysis of information used to develop a rescue plan of action for an incident.

Staging officer In a rescue situation, the person responsible for positioning and accounting for resources that are not immediately assigned.

On Scene

1. List the "first due" resources that would respond to a trench rescue incident in your community.

2. A worker is partially buried in a 10 ft (3 m) deep trench that has no cave-in protection. His upper torso is not buried, but he is certain that his right leg is broken. Discuss your plan for a non-entry rescue.

CHAPTER 4

Awareness Level

Personal Protective Equipment and Equipment Basics

KNOWLEDGE OBJECTIVES

- Identify trench rescue personal protective equipment. (**NFPA 1006: 12.1.5**, pp. 66–69)
- Explain considerations for selecting personal protective equipment (PPE) for trench rescue operations. (**NFPA 1006: 12.1.5**, pp. 66–69)
- Identify task specific PPE that is used in trench rescue. (**NFPA 1006: 12.1.5**, pp. 66–69)
- Explain the considerations for selecting PPE for trench rescue operations. (**NFPA 1006: 12.1.5**, pp. 66–69)
- Identify and describe weight distribution and bridging equipment for trench rescue incidents. (**NFPA 1006: 12.1.3, 12.1.5**, pp. 70–73)
- Explain the function of trench rescue shoring equipment. (**NFPA 1006: 12.1.2, 12.1.6**, pp. 73–82)
- Explain the purpose and use of hazard mitigation equipment.

SKILLS OBJECTIVES

- Select trench rescue PPE. (**NFPA 1006: 12.1.7**, pp. 66–69)
- Rope a trench panel. (**NFPA 1006: 12.1.5**, p. 76)
- Build a cutting station table. (**NFPA 1006: 12.1.3**, pp. 87–88)

ADDITIONAL RESOURCES

- **NFPA 1670**, *Standard on Operations and Training for Technical Search and Rescue Incidents*
- **NFPA 1951**, *Standard on Protective Ensembles for Technical Rescue Incidents*

You Are the Rescuer

It has been extremely hot for the past few days. You have kicked the air conditioner up in the firehouse and are hoping for a slow day. The first call of the day comes in just after noon. The dispatcher says that a man is trapped in a hole near the corner of Wilson and Park Street. On your way home from work yesterday you remember seeing the department of public works crew digging a trench near there.

Your engine is the first to arrive at the scene. As your crew moves to the end of the trench, you see a construction worker buried to his waist. One wall of the 9-foot (2.7-m) deep trench has sloughed in, leaving an overhanging section (cornice) of soil above the trapped victim. Your crew is hustling to complete initial actions and attempt a non-entry rescue by lowering a shovel and positioning an escape ladder near the worker. You have placed a backboard on what looks like a stable section of the trench lip to help distribute your weight (load) as you assist with the self-rescue. You and your captain are trench rescue technicians, but your technical rescue personal protective equipment (PPE) and trench rescue equipment are on the special response vehicle (SRV) that is 15 to 20 minutes out, so you begin the rescue wearing your structural firefighting gear. After a few trips back to the engine to get the backboard, ladder, air monitor, and barrier tape, you are sweating profusely and starting to feel dizzy. After an all clear from the atmospheric monitor, the captain tells the crew to take off their turnout coats and pants and continue initial actions.

As the worker digs his lower torso out, you see a compound fractured femur. You realize that this incident is now going to require shoring, entry operations, patient packaging, and a patient removal system. The SRV and trench rescue technicians have arrived on the scene and your captain tells you to drink a bottle of water and put on your rescue PPE.

1. Should you have remained in your structural firefighting gear while trying to effect this type of rescue?
2. What type of PPE will you need to continue?
3. What equipment will be needed to protect the victim from the next collapse?

 Access Navigate for more practice activities.

Introduction

When trench rescues occur, the first response is typically an engine crew. At best, the crew may have one or two members that have some level of trench rescue training. They are typically outfitted with personal protective equipment (PPE) and equipment to deal with fires, vehicle accidents, and medical incidents. The PPE and equipment needed for trench rescue incidents is seldomly considered for firefighting and emergency medical companies. Generally speaking, only a small percentage of the equipment carried on a typical fire engine will be useful at trench rescue incidents.

This chapter will present the PPE and an overview of the equipment used in trench rescue. Although this information is primarily focused on trench rescue teams, it can also be used to evaluate and enhance the trench rescue–related PPE and equipment carried by firefighters at the awareness level who are responsible for conducting the initial actions.

The Importance of Proper Equipment

At every incident, the safety of personnel depends on their level of skill, as developed during both training and real-time incidents. This training and experience may give personnel an initial advantage in the rescue effort; however, if they do not have the proper PPE or operational equipment and are subsequently injured in the rescue attempt, the operation will not be considered a success overall. In other words, all training and experience can be rendered ineffective if personnel do not have the proper personal protective clothing and equipment to give them reasonable protection during the rescue. NFPA 1951, *Standard on Protective Ensembles for Technical Rescue Incidents*, sets the benchmark for technical rescue PPE. PPE for trench rescue team members should comply with that standard.

By necessity, firefighters carry what they use most and they have PPE that provides protection for the most common incidents, such as motor vehicle accidents and fires. This could lead to firefighters performing the initial actions at trench rescue incidents without a complete cache of NFPA technical rescue compliant PPE. All fire departments should consider the PPE required for trench rescue and determine if it is feasible to add to their fire apparatus.

Development of a Safety Culture

Selecting and using the proper PPE is just one component in providing a safe way for rescue personnel to operate on the scene of a trench collapse. Instilling a safety mindset into the culture of your team is also critical and does not happen overnight. Rather, development of a safety culture is the result of many hours of training and the discipline that comes from everyone on the team being accountable. Such a culture is established because individual team members demand it from each other.

When an injury brings down a team member, everyone suffers the loss. Nothing is wrong with being safe or demanding that the team maintain a positive attitude about maintaining safety. Ultimately, a safety culture is something to be proud of, and it should be emulated by other teams that deliver specialized rescue service.

Personal Protective Equipment for Trench Rescue

For the sake of simplicity, trench rescue PPE can be classified into two main categories:

- Personal-issue PPE: The garments and equipment that each member receives and keeps with them, likely in a personal bag
- Team-issue PPE: The PPE that is kept on the response vehicle and is shared (i.e., breathing apparatus)

PPE is not rocket science; it is more like common sense. The rescue personnel on the scene of a trench collapse need a variety of protective clothing to minimize the effects of the weather and any related trauma caused by working in and around machinery and tools. At a minimum, trench rescue team members need an ensemble element consisting of protection for the upper and lower torso, arms, legs, head, hands, feet, and eyes. Other items, depending on the environment expected, could include hearing protection and foul weather gear. NFPA 1951 does not address visibility concerns. In areas where visibility is important, team members should be issued reflective vests or belts to be worn.

Personal-Issue PPE

Personal-issue PPE is equipment that is given to every member of a trench rescue team. This equipment is typically stored in backpacks. Personal PPE should include coats, trousers, or coveralls, helmets, gloves, boots, and safety glasses. Although this equipment is not shared, it should be properly cleaned after each use. Proper cleaning removes contaminates, protects the equipment, ensures that it is ready for the next response and that the equipment does not fail prematurely.

Torso, Arm, and Leg Protection (Coat and Trouser or Coverall)

Garments provide protection for the upper and lower torso, arms, and legs. Garments include coats, trousers, and coveralls. Shorts or short-sleeve shirts should not be considered. If it is hot, provide for frequent rehabilitation breaks and set up misting and other ventilation fans to cool the area. If the weather is a major consideration, set up a makeshift tent (tarp or pop-up tent) over the trench to shield the rescue area from the sun, rain, or snow. Make sure that you wear the proper protection for the environment in which you are working.

The NFPA Standard 1951 garment (coat, trouser, or coverall) is a best practice for trench rescue operations **FIGURE 4-1**. One disadvantage of a coverall is that it retains heat because of its one-piece design. Also, a suit that is too large or too small, or just in the wrong proportions, can cause discomfort and limit the ease of movement. Firefighting gear is very bulky and not recommended as standard protection for trench rescue. Ideally, rescuers should be as comfortable as possible so that they can concentrate on the efforts at hand and not their clothing. This consideration alone makes a case for providing each member of the rescue team with his or her own PPE ensemble.

Head Protection (Helmet)

The most critical piece of protective equipment in a trench rescue is the helmet. Head trauma is an injury from which a victim may not easily rebound and, in fact, could prove to be a career- or life-ending mishap. Appropriate helmets must be included in special trench rescue gear. Fire helmets are not appropriate because they are heavy, uncomfortable, and the rescuers will take them off.

Considering that the most likely sources of head injuries on a trench rescue scene are from materials falling in the trench, the best helmet to wear at a trench

FIGURE 4-1 PPE for rescue personnel should be appropriate for the type of mission and the hazards present.

FIGURE 4-2 Helmets must be NFPA Standard 1951 compliant.
Courtesy of Pigeon Mountain Industries.

emergency is an NFPA Standard 1951–compliant technical rescue helmet. It is lightweight, balanced, and provides impact protection **FIGURE 4-2**. Many varieties of this type of helmet are available commercially from rescue gear manufacturers.

FIGURE 4-3 Standard leather gloves are sturdy enough to keep abrasions and splinters to a minimum and still comfortable enough to wear while moving lumber and equipment.

SAFETY TIP

Rescuers frequently need to remind each other to wear their helmets while on the rescue scene. If you need to issue this type of warning, tell rescuers that you know a great place to store their helmets. When they ask where, tell them, "On your heads." A little embarrassment plus some humor goes a long way toward reinforcing safe practices.

Hand Protection (Gloves)

The glove is often the undoing of the rescuer. We know that gloves have to be worn, but performing any rescue-related function is difficult when wearing them. When rescue personnel are standing around talking, they may have their gloves on—but all too often, the first time they are required to do something, off come the gloves. Gloves provide zero protection to your hands when stored in your back pockets.

NFPA 1951 requires design and performance compliance for cut resistance, puncture resistance, heat/flame resistance, abrasion resistance, hand function, grip, and ease of donning. Thankfully, the days of using leather work gloves and Nomex flight gloves to provide a safe and efficient glove for technical rescue operations are gone. NFPA-compliant technical rescue and extrication gloves are now available through a number of quality manufacturers.

Firefighting gloves are designed for fighting fires. They are bulky, hot, and stiff and do not provide the dexterity needed for rescue operations. Conversely, if your sole task is to move equipment and lumber, the firefighting glove will protect your hands from slivers and abrasions. A better choice at a trench rescue incident for support personnel who are performing tasks like moving equipment and lumber is the standard leather glove. These gloves are sturdy enough to keep abrasions and splinters to a minimum and still comfortable enough that rescuers will generally keep them on while working **FIGURE 4-3**. A well-equipped trench rescue team will have a dozen pairs of leather

FIGURE 4-4 Rescue-quality boots will help protect your feet from incident hazards.

FIGURE 4-5 Standard safety glasses work well for trench rescue.
© trabachar/Shutterstock.

work gloves on their rig to distribute to their support personnel.

Foot Protection (Boots)

All trench rescue team personnel should wear boots that protect their feet from the hazards that are commonly found at a trench site **FIGURE 4-4**. Not only are there many opportunities to drop something on your foot, but nails and other sharp debris are also typically found all around the trench site. In addition, it is a good idea to have a high-top boot. This will provide a measure of support in the event you step on a piece of lumber or other equipment and turn your ankle. NFPA 1951-compliant boots require protection from the following: impact, cuts, abrasion, heat, punctures, bending, electrical conduction, and slipping.

> **TIP**
>
> NFPA sets the minimum requirements for technical rescue ensemble elements. As defined by the standard, ensemble elements provide protection to the upper and lower torso, arms, legs, head, hands, and feet. The elements that are tested and certified will have a label stating that they are certified. In addition to providing the body protection described in the standard, garments that also provide protection from bloodborne pathogens will have a label stating that they protect the wearer from bloodborne pathogens.

Eye Protection (Safety Glasses)

A standard pair of American National Standards Institute (ANSI)-approved safety glasses is more than satisfactory for trench rescue **FIGURE 4-5**. Eye injuries are most likely to occur during a trench rescue when rescuers are cutting and nailing or when soil is blowing or falling. ANSI Z87.1, referenced in Occupational Safety and Health Administration (OSHA) and NFPA 1500, is the standard that covers protection for faces and eyes. Measurable characteristics of lens thickness, impact resistance, penetration by projectiles, and optical quality are detailed in Sections 9 and 10. Eyewear used by technical rescue members should be marked with Z87, indicating that the product meets the ANSI standards. Most safety glasses can be purchased with or without sun or glare protection, and they will stay in position when the rescuer's head moves. In addition, they will rarely fog because the outside of the lenses are ventilated. Make sure that any safety glasses used for trench rescue are ANSI-approved for that purpose.

Full-face goggles, while providing the most protection, will usually fog when the rescuer gets hot. When that happens, what do you think comes next? That's right, the goggles end up hanging around the neck of the rescuer and not over the eyes.

One last point about eyewear: Rescue personnel should never wear reflective sunglasses. It is frequently important for rescue personnel to confirm movement by eye contact. Do not let looking cool get in the way of

> **TIP**
>
> Although they may not exactly qualify as piece of equipment, sunblock and insect repellent (with DEET) should be part of every trench rescuer's PPE pack.

providing the best and safest rescue for the victim and fellow rescue personnel. Leave the shades on the rig!

Firefighting Turnout Gear

For most trench rescue incidents, the use of turnout gear is not a best practice. It is bulky and the insulation can overheat a rescuer performing rescue functions. However, firefighters often will arrive on scene and begin work in turnout gear. In addition, when hazards like a broken natural gas line are present, full firefighting PPE is essential for all fire and rescue personnel in the hot zone until the hazard is controlled and removed.

Team-Issued PPE

Team-issued PPE is equipment that is stored on the trench rescue team's response unit and available for use by any member who needs to use it. This equipment is either shared or disposable. Shared PPE must be properly sanitized after each use. This is important because shared equipment that has been stowed without sanitation may contaminate the next user. Additionally, dirty PPE will likely not be used by the next member who needs the protection. Nobody wants to use someone else's dirty equipment.

Respiratory Protection

Respiratory protection should always be readily available at the scene of a trench rescue incident. Whenever there is any indication that an immediately dangerous to life and health (IDLH) atmospheric problem might exist, an atmospheric monitor should be utilized, in addition to **self-contained breathing apparatus (SCBA)** or **supplied air breathing apparatus (SABA)** before entry into the environment **FIGURE 4-6**. In dry and windy conditions with flying dust and dirt, a particulate filtering respirator should be used. Whereas particulate respirators can be disposable, SCBA and SABA respirators must be sanitized after each use in accordance with manufacturers' recommended procedures.

Because of the respiratory hazards posed by trenches, all members of the trench rescue team must be compliant with OSHA 29 CFR 1910.134 training and annual fit testing requirements. An adequate supply of respirator face piece sizes must be available. Remember that if rescuers are using respiratory protection, all live victims will need the same level of protection. Rescuers must evaluate the head size and shape of the victim and make their best guess on the size of the mask needed. A comprehensive look at atmospheric monitoring appears in Chapter 5, *Hazard Mitigation*.

Hearing Protection

Hearing protection is a good idea for anyone who is working around compressors or saws. Choose a level of protection that will shield the rescuer's ears from high-noise frequencies, but will not block out all communication. It is usually more dangerous for the rescuer to be unable to hear anything than it is for them to be exposed to loud noises.

Cutting Tool Protection

Chainsaw chaps provide good leg protection and are a great idea for anyone who will be cutting with a chainsaw. Although they will not keep someone from sawing a leg off, chaps might be enough to stop or deflect a chain that bounces off the wood and inadvertently hits a rescuer's lower body. Apron-style chaps offer good protection and are easy to don and doff. Additionally, protective jackets, gloves, boots, helmets, and eye protection must be used by rescuers when operating chainsaws.

Welders wear skullcaps to keep their heads cool under their helmets while protecting them from sparks when using torches. Rescue personnel wear them for the same purpose while using cutting torches. As the heat from the head causes perspiration, it will wet the skullcap, which in turn keeps the head cooler. In hot weather conditions, the skullcap can be wetted before putting it on to get a head start on the process. Welder jackets and gloves in assorted sizes must be available on the trench rescue response vehicles. Additionally, a supply of Shade 5 eye protection must be available to rescuers using cutting torches. Remember that shared (team issued) cutting PPE must be cleaned after each use.

FIGURE 4-6 SABA is compact and can be used as a constant supply of air for trapped victims if SCBA cannot be used.

Trench Rescue Shoring Equipment Overview

Trench rescue operations rely on specialized equipment to safely prevent soil from further collapse. This equipment is a combination of equipment that is built using common materials and specialized equipment that is created to serve specific purposes. The equipment includes lip protection, trench rescue panels, struts, wales, and backfill. The specialized equipment may include pneumatic struts, lip bridges, aluminum wales, and backfill consisting of air bags or struts. Lip protection is used to minimize or prevent concentrated loading on the lips of the trench. Trench walls are supported by trench rescue panels, struts, and wales, whereas backfill is used to fill voids between the panels and soil.

The following sections describe this equipment in detail. In this section, the authors share the best practices for trench rescue shoring equipment. The term best practice is not uncommon to the fire service. In many cases, a best practice is a subjective value that becomes an opinion. The term *best practice* as used in this section represents a combination of engineering evaluations, testing results, and empirical evidence from hundreds of live trench shoring experiences.

Lip Protection

Lip protection is the act of making the lip area of a trench safe, including the selection, construction, application, and installation of ground pads and lip bridges around the affected trench area. Ground pads are a safety measure placed on a trench lip for the purpose of distributing the load, or weight, of rescuers and rescue equipment in use on the lip area. A 200-pound (91 kg) rescuer will exert a downward force of 200 pounds (91 kg) on the trench lip. If that rescuer is standing directly on the soil, the downward force is distributed over the surface area of the rescuer's boot soles, which would be less than 1 square foot (0.1 m^2). So that force would be more than 200 pounds per square foot (976 kg/m^2). Two rescuers standing on the lip will exert a downward force of 400 pounds (181 kg) over a surface area of less than 2 square feet (0.2 m^2). That is a highly concentrated load. A stiff enough, strong enough (3/4-inch [19-mm] thick lumber or greater), and large enough ground pad (4 × 8 foot [1.2 × 2.4 m]) can distribute that weight over a much larger area **FIGURE 4-7**. A concentrated load is much more likely to cause a collapse of the trench wall than a distributed load. Ground pads are used to improve safety for people working on a trench lip that has not collapsed or a trench lip with walls that are stable. Ground pad and lip bridge installation are covered in Chapter 7, *Operations Level Trench Rescue Shoring*.

FIGURE 4-7 If ground pads are appropriate, ¾-inch (19-mm) plywood sheets 4 × 8 feet (1.2 × 2.4 m) should be your first choice.
© Jones and Bartlett Publishers. Courtesy of MIEMSS.

Ground pads do, however, have many limitations and inherent dangers, one of which is that ground pads hide soil movement. In addition, ground pads cannot be used over large open lip voids, trench lips with unstable soil conditions, or over a closed lip void (slough-in with overhanging soil) due to the risk of collapse.

Once primary and secondary trench shoring has been placed, the need for ground pads is decreased. At this point, the trench shoring supporting the walls supports the lip, making it less susceptible to collapse. It is a best practice to continue to utilize ground pads while operations continue, but they can be removed if monitoring the growth of cracks or movement of the soil is deemed to be more important.

Lastly, 4 × 8 feet plywood sheets will bend, bow, and cup. To minimize tripping hazards, make sure that the edges, rather than the center, of the plywood contact the ground.

Ground pads typically consist of the following:

- 4 × 8 foot (1.2 × 2.4 m) 3/4-inch (19-mm) plywood (CDX or better) for the open areas of the trench lip, *or*
- 2 × 8 foot (0.6 × 2.4 m) 3/4-inch (19-mm) plywood (CDX or better) for the spoil pile side of the trench lip, *or*
- Two 2 × 12-inch boards (side by side) for the spoil pile side of the trench lip

SAFETY TIP

When covering up cracks on the trench lip with ground pads, mark the area on top of the ground pad with paint. This will remind people not to stand on that area, and you can also check for worsening conditions **FIGURE 4-8**.

FIGURE 4-8 Marking ground pads communicates to rescue personnel potential cracks or dangerous areas that should be monitored or avoided.
© Jones and Bartlett Publishers. Courtesy of MIEMSS.

FIGURE 4-9 Used on the spoil side of the trench, two 2 × 12-inch boards used as ground pads are usually acceptable.
Courtesy of Cecil V. "Buddy" Martinette, Jr.

FIGURE 4-10 This lip bridge is used to work directly over a large lip shear collapse. It keeps the rescuer's weight off the missing or damaged wall.
Courtesy of Ron Zawlocki.

In situations where the spoil pile is not at least 2 feet (0.6 m) back from a stable trench lip, two pieces of 12-foot (3.7-m) long 2 × 12-inch lumber can be used as a ground pad. The 2 × 12s can be used during the spoil pile removal process, and they can be moved forward along the lip as the spoil is progressively removed. The advantage comes from the time saved by only needing to clear a 2 foot (0.6 m) wide rather than a 4 foot (1.2 m) wide (4 × 8-foot [1.2 × 2.4 m] ground pad) section of the lip **FIGURE 4-9**. Remember that although the 2 × 12s help distribute the weight of rescuers, the weight is distributed over a smaller area and is not as safe or effective as ¾-inch (19-mm) 4 x 8-foot (1.2 × 2.4-m) ground pads. Another disadvantage of this type of ground pad is that it is small and difficult to balance on. Rescuers may feel like they are tiptoeing around the trench, which can create a trip or fall hazard.

Lip bridges provide an alternative to ground pads **FIGURE 4-10**. A lip bridge is built with girders (timbers), platforms (aluminum, wooden, or fire service ladders with lumber), and bases (sections of timbers placed away from the compromised wall and used to elevate the bridge above the lip). A properly installed lip bridge will prevent rescuers from riding the wall right into the trench in the event of a secondary collapse. Lip bridges require more material to be carried on the response apparatus. Although they add a few minutes to the set-up time, once they are properly installed, they offer a higher level of safety to rescuers. A good general rule is to use ground pads when the trench walls are stable and lip bridges when the walls are unstable or have collapsed, resulting in significant voids in the walls.

Lip bridges are usually a safer and more efficient method of lip protection for trench rescue incidents for several reasons. First, they allow rescuers to see what is happening (fissures, moving soil, etc.) under them. Second, because lip bridges transfer the rescue personnel and equipment load further away from the trench lip, decreasing collapse potential from rescue operations, they can be used over large open lip voids (lip or wall shears), closed lip voids (slough-in with overhanging soil), or on trench lips with unstable soil conditions. Lip bridges are constructed of the following:

- Bridge bases: 6 × 6 × 36-inch or 4 × 4 × 24-inch cribs
- Girders (timbers or laminated veneer lumber [LVL] beams): 6 × 6-inch minimum, 14 to 16-feet (4.3–4.9 m) long
- Platforms: Wood beams and 3/4-inch (19-mm) plywood decking or aluminum stages or fire service ladders with wood planks

An alternative to aluminum stages are wooden platforms. Platforms should be 2-feet (0.6 m) wide and at least 14 feet (4.3 m) long. Platforms that are 16 feet (4.9 m) long are better if the rescue team's apparatus will hold them.

Lip Protection Best Practices

Lip Bridge. The best use of lip bridges is whenever the walls are unstable or have collapsed, resulting in significant voids in the walls.

4-Foot Wide Ground Pads. The best use of 4-foot (2.4-m) wide ground pads is when the trench walls are stable and there is no interference from the spoil pile or other obstructions.

2-Foot Wide Ground Pads. The best use of 2-foot (1.2-m) wide ground pads is when the trench walls are stable and the spoil pile or other obstructions prevent the use of 4-foot (2.4-m) wide ground pads. After a short amount of use, ground pads will permanently bend (cup). By laying them with the cup facing down, the chance of tripping on the edges is minimized.

Trench Rescue Panels

A **trench rescue panel** consists of plywood and lumber **strongbacks** (or beams). Trench rescue panels must be stiffer and stronger (greater capacity) than the **sheeting** used by trench construction workers for shoring soil that is essentially at rest. Panels are the component in the **shoring system** that collect the load from one side of the trench and distributes the load to the other side of the trench **FIGURE 4-11**. For years, rescuers have used shoring guidance that was designed for use in freshly dug trenches when the soil is at rest. That guidance tells construction workers that in some soil conditions, sheeting (panels) is not required. This is based on the idea that some soil is cohesive, strong, and acts more like a block of concrete. However, the guidance tells them that if the soil is sloughing or raveling, sheeting should be used to prevent small soil particles from falling into the trench. Furthermore, the guidance describes the sheeting

FIGURE 4-11 An example of a panel being lowered into place.
© Jones and Bartlett Publishers. Courtesy of MIEMSS.

(plywood) that is used to accomplish this does not have to be very strong and in fact is said to have no structural value. Unfortunately, this guidance is not applicable to most trench rescue soil conditions. Trench rescue panel installation is covered in detail in Chapter 7, *Operations Level Trench Rescue Shoring*.

Active soil or soil that has failed is not a homogenous block like concrete or steel. It may look that way in a strong clay soil, but it is actually made up of small particles (grains) and is not a solid. To provide an analogy to compare trench collapse shoring and building collapse shoring, trench collapse shoring is like shoring broken unreinforced masonry (URM), and building collapse shoring is like shoring reinforced concrete. Structural collapse specialists are taught to use plywood behind their raker shores to retain the pieces of a URM wall. URM walls are not solid and should not be shored as solid. Likewise, trench walls are not solid and must be shored with trench rescue panels to retain and collect the load. Even if the soil appears to be cohesive (strong clay), remember that it can break apart into small pieces. These pieces need to be retained by a combination of strong panels and struts.

The plywood component of a trench rescue panel collects and retains the load coming from all of the individual grains of soil and transfers it to the struts. The plywood is rather flexible by itself, so a strongback is used to stiffen and strengthen the panel vertically. This helps to transfer the load to the struts.

Allowing rescuers to work in a trench with unstable soil conditions without sufficiently strong trench rescue panels is a high-risk endeavor. Conducting the kind of soil analysis needed to determine whether soil is stable is beyond the training and experience levels of trench rescuers. Fortunately, composite trench rescue panels used in a shoring design help reduce the risk of trench collapse in the shored area. The use of trench rescue panels is essential to the safety of both trapped victims and rescuers working in the trench.

Historically, panels have been used with struts spaced at 4 feet (1.2 m), leaving 2 feet (0.6 m) from the top or bottom of the panel to the strut. From an engineering standpoint, this spacing provides a balance in forces that is advantageous. The panel acts as a beam to transfer the load to the struts.

Beams work by resisting load. When you put a load on a beam, it wants to sag. The load causes a tension (pulling force) in the bottom of the beam and a compression (pushing force) in the top of the beam. You can continue to add load to the beam until the tension force reaches the breaking force of the wood.

Bending strength is a measure of the stress at which a material will break. This stress is usually represented in psi (pounds per square inch). The strength of a material is a function of its maximum stress and its size and shape. The strength is a force that is usually represented in pounds.

So how strong does a panel need to be? Well, the safe working load should be at least two times stronger than the force that the soil could exert on it. Based on current geotechnical soil force estimations (Oliver Taylor, PE, PhD, and Marie LaBaw, PE, PhD), soil forces in a typical trench (depth of 20 feet [6.1 m] or less) could reach 22,000 pounds (22 kips; 9980 kg) over a 4 × 4-foot (1.2 × 1.2-m; half panel) area. To safely shore, panels and struts with ultimate strengths of at least 44,000 pounds of force (44 kips, 20,000 kg) are needed to maintain a minimum factor of safety of 2 to 1.

The following are some capacities based on a 4 × 8-foot (1.2 × 2.4-m) area for common panel materials and 4-foot (1.2-m) strut spacing:

- ¾-inch (19-mm) CDX plywood capacity is 1100 pounds (499 kg).
- ¾-inch (19-mm) FinnForm plywood (14-ply arctic birch with an allowable bending strength of 3600 psi) capacity is 3800 pounds (1724 kg).
- A sawn 2 × 12 capacity is 2600 pounds (1179 kg).
- A 2 × 12 LVL (LVL with an allowable bending strength of 3100 psi) capacity is 7400 pounds (3357 kg).

The following are some capacities based on panel construction methods and 4-foot (1.2-m) strut spacing:

- Noncomposite construction: ¾-inch (19-mm) FinnForm and 2 × 12 sawn lumber provides a capacity of 6400 pounds (2903 kg).
- Noncomposite construction: ¾-inch (19-mm) CDX and 2 × 12 sawn lumber provides a capacity of 3700 pounds (1678 kg).
- Noncomposite construction: ¾-inch (19-mm) FinnForm and 2 × 12 LVL provides a capacity of 11,200 pounds (5080 kg).
- Composite construction: ¾-inch (19-mm) CDX and 2 × 12 sawn lumber provides a capacity of 18,100 pounds (8210 kg).
- Composite construction: ¾-inch (19-mm) FinnForm and 2 × 12 LVL provides a capacity of 48,600 pounds (22,045 kg).

How Panels Work

Using inexpensive (cheap) panel material will result in capacities (strength) that will compromise rescuer safety. Premium engineered materials can support nearly three times the load that can be supported by common materials. Second, the way strongbacks and plywood sheeting are assembled has a huge effect on rescuer safety in a trench. Composite construction is more than four times stronger than noncomposite construction.

It has been a common practice for rescuers to connect plywood sheeting to strongbacks (usually done at the scene) by connecting with three bolts, screws, or nails. When connected this way, the strengths of the members combine by simple addition. A sheet of ¾-inch (19-mm) CDX and a sawn lumber 2 × 12 will combine to produce a panel capacity of 3700 pounds (1100 + 2600). The same can be said for any combination of materials. You can even add a second sawn strongback, and the capacity will increase by another 2600 pounds to 6300 pounds. The panel becomes the sum of its individual components. This is called **noncomposite behavior**; each element acts on its own without dependence on the other elements.

Composite behavior is a way to get more strength from a combination of members. If the pieces can be rigidly connected to the point where they cannot slip (absolutely zero slip) relative to each other, they will act as a new, stronger single member. The most common example of this is a steel beam, which is three plates welded together into an I shape. The welds prevent any slip between the individual plates. The three plates, when welded together, provide more strength than a sum of the individual pieces.

With trench rescue panels, we cannot weld the strongback to the plywood. Mechanical connectors like nails, screws, and bolts hold the pieces together, but they are not sufficient to prevent the slip between the two pieces. Construction adhesives, properly designed and applied, can provide sufficient capacity against slip and make the two pieces of wood act like a single piece with much more strength than the two individual pieces.

For example, if you combine a ¾-inch (19-mm) CDX and a sawn 2 × 12 with mechanical fasteners (non-composite), the strength of the panel is 3700 pounds (1678 kg) as previously described. When we connect the two with sufficient adhesive, the strength increases to 18,100 pounds (8210 kg)—nearly 5 times as strong.

When we can prevent slip between the plywood and the strongback, we make a new beam that is thicker than the individual pieces. This pushes the tension and compression further away from the middle (we are essentially using a longer lever), which means less stress is created from a load. Less stress created from a load means more load can be supported. Depth of a beam has an exponential effect on the strength of a beam. This increase gives you much more strength out of the exact same materials. For this reason, composite panels are considered a best practice.

Panel Materials

When you consider sawn lumber, you must consider quality. Lumber contains checks, splits, warps, and knots, all of which can impact the strength. A knot in a piece of wood can decimate capacity. You can test several pieces of wood and come up with wildly different capacities, because of internal defects that are natural to *all* sawn wood. Plywood is made up of several thin layers of wood glued together. One layer can have a knot in it, but the impact is minimized because it does not go all the way through all of the layers. The force can redirect around it in the layers above and below it. The more layers you have, the more ability you have to bridge any defects. This has the added benefit of providing a much more consistent strength. Combine premium materials with the layering and you get the strongest option. This is where FinnForm and LVL (laminated veneer lumber) come in. They are both built like plywood, with many thin layers glued together. The wood used for those layers is higher quality and leads to a higher strength. Due to the number of layers, the strength is more reliable and repeatable.

This leads to the best practice in panels. Use construction adhesive (and screws) to connect a 2 × 12 LVL to a sheet of ¾-inch (19-mm) FinnForm to create a composite panel. This combination provides a capacity of 48,600 pounds (22,045 kg) and can resist the loads in nearly every trench situation down to a 20 foot (6.1 m) depth **FIGURE 4-12**.

In a study conducted in 2009 by Dr. Marie LaBaw, it was concluded that FinnForm panels attached to strongbacks provide the most beneficial distribution of strut pressure to the trench walls. The thicker the panels, the better—but a good compromise of pressure distribution and ease of installation (weight) is a ¾-inch (19-mm), 4 × 8-foot (1.2 × 2.4-m) FinnForm panel. Paying careful attention to make sure that the panels are set tight against the walls of the trench will ensure that the strongback/panel can transfer the necessary force from the shore to the trench wall.

Use panels and close shoring whenever the soil is active (sloughing, raveling, caving-in, etc.). Panels should be used in shoring systems designed to protect rescuers whenever possible. Panels are typically 4 × 8 feet (1.2 × 2.4 m) and are used with at least two struts per panel, which will result in a horizontal and vertical strut spacing of 4 feet (1.2 m). Placement of the struts on each panel should be a maximum of 2 feet (0.6 m) from the top and bottom of the panel. With the benefits of premium materials, the use of ¾-in (19-mm) FinnForm (Chudoform) with an allowable bending strength of 3600 psi is considered a best practice for panels.

Similarly, premium strongback that utilizes LVL with an allowable bending strength of 3100 psi is considered a best practice for panels. Using construction adhesive and screws to hold the materials together while the glue cures, to connect the LVL strongback to the FinnForm panel, results in composite behavior and much higher strengths than the individual materials. To accomplish this successfully, any special coatings need to be sanded off the panels before the glue is applied.

For finishing touches on panel construction, cutting the corners off the panels is recommended to avoid the panels chipping and separating from and falling on the corner of the panels **FIGURE 4-13**. A 45-degree cut taking 3 inches (8 cm) off the panel height and width has generally provided good results. The final detail is drilling holes to attach ropes for lowering and lifting the panels. These holes should be at least 7/8 inch (2.2 mm) in diameter and should be within the outside 3 inches (8 cm) of the edge of the panel. Keeping the holes and ropes within 3 inches (8 cm) of the panel edge allows for stacking the panels with offset strongbacks for transportation.

Best Practices for Panels

Use panels with safe working load capacities (2:1 factors of safety) for worst-case soil conditions in trenches up to 20 feet (6 m) deep, such as composite panels constructed of ¾-inch (19 mm) arctic birch with an allowable bending strength of 3600 psi and 1.75 × 12-inch LVL strongbacks.

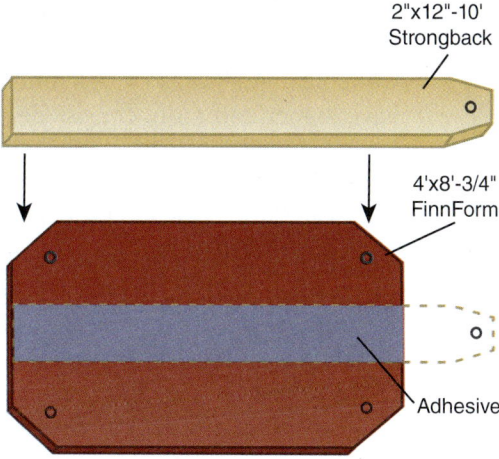

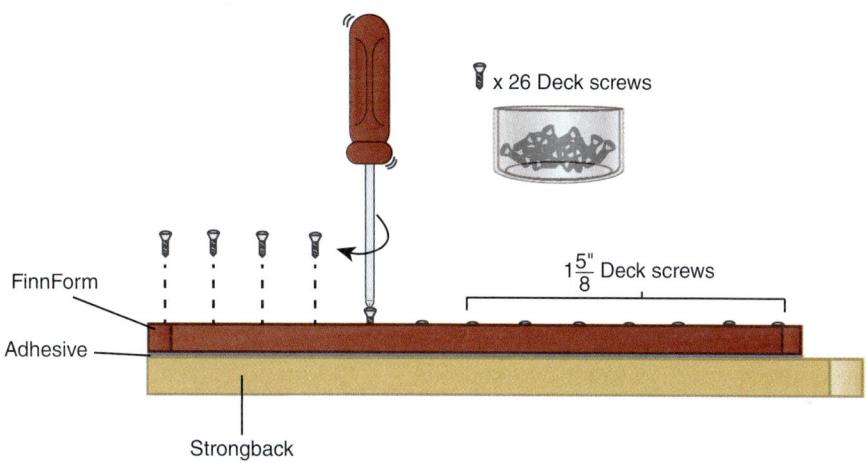

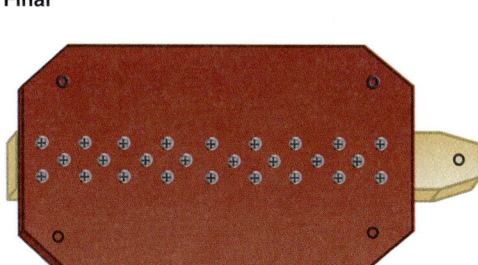

FIGURE 4-12 Screw pattern for composite panel construction.

Panel Ropes

Trench rescue panels are relatively heavy but they do not require the capacity of the recommended ½-inch (13-mm) nylon kernmantle ropes. The reason that we use ½-inch (13-mm) kernmantle rope is that size and style of rope is easy to grip with gloved hands. The panels are lowered into place with these ropes and are often repositioned using the ropes. Twenty-five foot (7.6-m) pieces of ½-inch (13-mm) kernmantle rope, usually made from retired life safety ropes, are a good choice for panel ropes. A simple overhand or stopper knot should be used to secure the rope to the panel. Both of these knots can easily be untied if there is a situation where removing the panel rope becomes necessary during placement of the shoring. Panel setting is discussed in detail in Chapter 7, *Operations Level Trench Rescue Shoring*.

FIGURE 4-13 Cutting panel corners is a method to reduce hazards to rescue personnel, it also helps reduce panel damage.
© Jones and Bartlett Publishers. Courtesy of MIEMSS.

Roping Panels. The panel team will also prepare the panels for placement. To rope the panels, follow the steps in **SKILL DRILL 4-1**.

Wales

A **wale** is used to span large areas of trench walls (without intermediate struts). The large area needing to be spanned may be a result of the voids created by a cave-in, the trench shape, or an obstruction at the wall. Wales are also used to create room for extrication work and victim removal. Timbers are commonly carried by rescue teams for use as wales. In worst-case soil conditions, 6 × 6-inch wales do not have the strength needed to give rescuers an adequate safety factor. Instead, 8 × 8-inch timbers are a better choice. Laminated wood beams, such as LVL (laminated veneer lumber) have higher and more consistent strengths than timbers. The multiple layers within the laminated beams offer internal redundancy to overcome the deficiencies (e.g., knots, cracks, checks) often found in wood. Chapter 7, *Operations Level Trench Rescue Shoring*, covers shoring skills in detail.

Aluminum wales are available from shoring manufacturers. They are not as heavy as timbers and LVL wales. Destructive testing has shown the aluminum wales to be very consistent in strength. The tests also show that the permanent deformation (yield) point of the aluminum wale is much lower than the breaking strength. For more information on inside and outside wales, see Chapter 7, *Operations Level Trench Rescue Shoring*.

Best Practices for Wales

Wales with safe working load capacities (2:1 factors of safety) for worst-case soil conditions in trenches up to 20 feet (6 m) deep have the following characteristics:

- 7 × 7-inch LVL beams have superior strength and reliability.
- Wales should extend 24 inches (60 cm) (overhang) beyond struts but not less than 12 inches (30 cm).

SKILL DRILL 4-1
Roping the Panels I (NFPA 1006: 12.1.5)

© Jones and Bartlett Publishers. Courtesy of MIEMSS.

1 Take a section of rope and place the end through a hole in the strongback side of the panel on the bottom. Tie a knot in it so that it will not pull back through, and then do the same for the other side. For deep trenches, pass the end of a 25 foot (7.6 m) section of rope from the back of the panel (at the top) to the front (the strongback side) of the panel. Tie a stopper knot at the end of the rope. Repeat this process on the other side of the of the panel.

2 Lower the panel into the trench while pulling back on the ropes—a technique that ensures the panel will get a nice vertical set against the trench wall. If you have limited ropes, you can accomplish the same objective by wrapping the closed end of a rope around the strongback and holding the terminal ends. After the panel is lowered and placed in the proper position, the rope can be pulled up and used again.

- Gaps between the wale and FinnForm (created by the strongback thickness) must be filled.
- 2-foot × 8-inch (0.6 m × 20 cm) fillers (spacers) at least 12 inches (30 cm) long should be place at the overhanging ends and at the seams of contiguous panels.
- Wales must not be bent by strut activation force during installation. Use shims and wood fillers to take up any variance **FIGURE 4-14**.

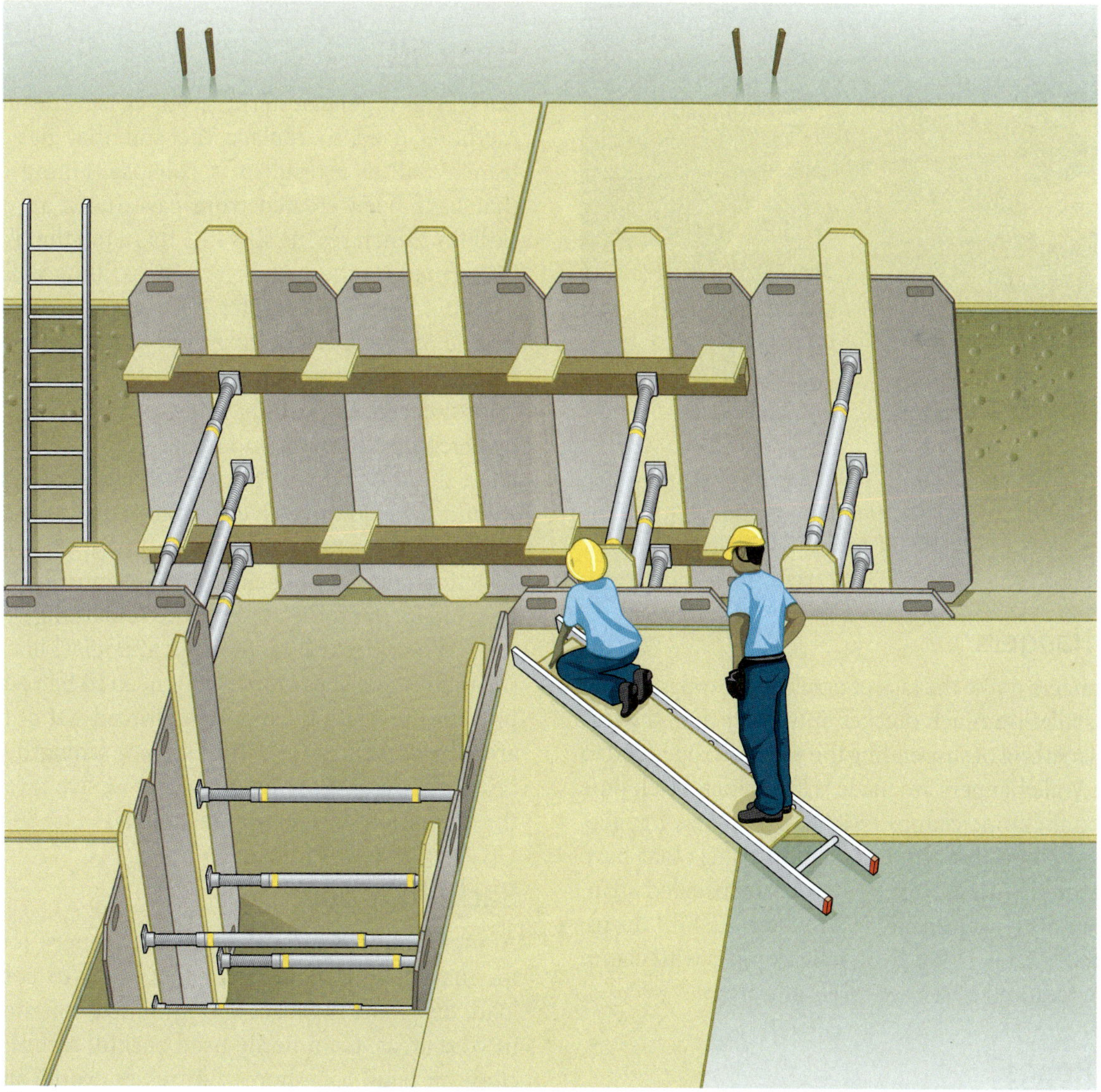

FIGURE 4-14 A. Wales made from 7 × 7-inch LVL are significantly stronger than 6 × 6-inch and 8 × 8-inch sawn lumber. **B.** An illustration showing a wale in place.

FIGURE 4-15 Wale hangers make the installation of wales faster and easier. **A.** Hangers on the frontside of panels **B.** Hanger on the backside of panel.
Courtesy of Ron Zawlocki.

Wale Hangers

Wale hangers make the task of positioning wales during strut installation much quicker and easier than the traditional method of suspending the wales on ropes tied to pickets. Wale hangers are made with ¾-inch black pipe and modified pipe clamps **FIGURE 4-15**. Wale hangers work best when they are installed on panels that have been secured with at least one properly installed strut. In operations level trenches (up to 8 feet [2.4 m] deep), four 8-foot (2.4 m) long ¾-inch black pipes with clamp ends are required to set two levels of wales.

Wale Ropes

Like trench rescue panels, wales are relatively heavy but they do not require the capacity of the recommended ½-inch (13 mm) nylon kernmantle ropes. Wales are commonly lowered into place with these ropes and are often repositioned using the ropes. The ropes can be secured to the wale by drilling a 1-inch (2.5-cm) diameter hole 6 inches (15 cm) from the end of the wale and passing the rope through the hole and tying a stopper knot. This is preferable to just tying the rope around the wale, because it prevents the wale from wanting to roll when being lowered into the trench.

SAFETY TIP

Contrary to what has been printed in previous editions of this and other trench rescue manuals, fire service–grade ground ladders should *never* be used for makeshift wales. They are extremely weak when loaded in that type of configuration.

Backfill

Backfill is a generic term given to several common methods used to replace the soil that has left the trench wall as a result of a collapse. Filling in voids that have been created from cave-ins is an essential skill for trench rescue shoring. It is a job that is usually performed by the panel team. Backfilling will be discussed in detail in Chapter 7, *Operations Level Trench Rescue Shoring*. Backfilling void areas helps to minimize soil movement and distribute the load from the opposite wall. Backfilling voids in trench walls is done by placing equipment and materials in the voids to replace the missing soil. Such materials and equipment should have compressive strength equal to or greater than the soil pressures at the area of application. The selection and use of backfill options must be based on the void size and the angle of the remaining void back wall. When backfill is installed at angles, like at a lip shear collapse, shear forces will need to be resolved to prevent the backfill from being forced out of the void area. In those instances, the system strength will not be solely dependent on the compressive strength of the equipment or material installed.

Backfill Options

As previously mentioned, *backfill* is a generic term used for material and equipment that helps to collect the load, distribute the load, and minimize soil movement in void areas. Commonly used backfill techniques include air bags, backshores, buttresses, wood, and soil.

Air Bags. Air bags are best for medium to large voids that have high angle back walls **FIGURE 4-16**.

FIGURE 4-16 Low pressure air bag used to backfill a void behind a primary panel. Soil should be placed into the voids on both sides of this bag to stabilize the bag.
Courtesy of Ron Zawlocki.

Note: some air bag manufacturers make bags that can fill voids up to 90 inches.

Open lip voids must have back wall angles between 90 and 70 degrees for air bag use.

Backshores. Backshores are struts placed on the back side of panels or wales and extending to the panel sections on the back wall of the void **FIGURE 4-17**. They are a good choice for large voids that are accessible from the lip. Backshores require high angle (70–90 degrees) surfaces to shore to and a minimum coverage of 2 × 2-foot (0.6 × 0.6 m) panel sections on the void back wall.

Buttress. A buttress can quickly and effectively resolve large lip shear voids that have left low angled walls **FIGURE 4-18**. A buttress can be constructed using timber or aluminum (pneumatic) struts. A buttress does nothing to prevent soil movement from the wall side on which it is built. When signs of soil failure are present beyond the initial lip shear failure, the void must be filled with soil backfill after the buttress is installed.

Soil. The best use of soil is as backfill for small voids accessible from the lip with low or high angled back walls. Washed stone or other backfill soil that the construction company has onsite is appropriate. The use of soil from the spoil pile is permissible. The use of soil for back fill is only practical when a void opening exists at the lip. Soil from the spoil pile may be shoveled into voids from the lip after the panels and at least one strut is in place **FIGURE 4-19**.

FIGURE 4-17 Backshores used on a large wall shear void.
Courtesy of Ron Zawlocki.

FIGURE 4-18 Buttress used to shore a wall opposite a large lip shear.
Courtesy of Ron Zawlocki.

Wood. The use of wood is a good choice for small voids (2–12 inches [5–15 cm]) beyond the trench wall) accessible from the lip. Sections of timber, lumber, and wood shims may be used to fill voids in trench walls **FIGURE 4-20**. No more than two pieces of wood may be stacked in the same direction. The use of wood for back fill is only practical for use with open lip voids **FIGURE 4-21**. Wood should not be placed into the

FIGURE 4-19 Soil temporarily held in place with wedges until additional panels and struts can be installed.
Courtesy of Ron Zawlocki.

FIGURE 4-20 Adding wedges will improve the load distribution to the soil.
Courtesy of Ron Zawlocki.

FIGURE 4-21 Soil and wood combination.
Courtesy of Ron Zawlocki.

Struts

Struts are the components in the protective system that transfer the force from one side of the trench to the other. Struts are made of a variety of materials, and different methods are used to compress them between the trench walls. Each style of strut has its own strengths and limitations. Most struts have been designed for construction purposes and are acceptable for shoring soil that has not collapsed or is stable, as determined by a competent person. In rescue situations, only struts that can be installed and removed from outside the trench and do not require entering the trench provide the desired level of safety needed for rescue operations. For instance, the installation of timber struts, cut to fit or with screw jacks, require a rescuer to be at the level of the strut to install it. This puts the rescuer at risk. Methods that reduce that risk can be used, but there is always risk. Struts that can be placed and removed from outside the trench and on lip bridges, will eliminate nearly all that risk.

Struts are sometimes called shores or, in the case of timber, cross braces. For the sake of simplicity, we will call any component of a shoring system that transfers the force from one side of a trench to the other side of the trench a strut. Chapter 7, *Operations Level Trench Rescue Shoring*, covers strut installation in detail.

void until at least one strut has pressurized the panel set. It is very important to initially create a minimum 2 × 2-foot (0.6 × 0.6 m) contact area between the panel and the existing wall directly behind the strut. Expand the contact area to the width of the panel as soon as possible.

Best Practices for Struts

Struts with safe working load capacities (2:1 factors of safety) for worst-case soil conditions in trenches up to 20 feet (6 m) deep have the following characteristics:

- They can be installed and removed without entering the trench.
- They have adjustable lengths (expandable) and have measurable (gauges) and controllable activation pressures.
- Struts with activation forces of 1000 to 1250 pounds (454–567 kg; 2.5-inch [6.4-cm] diameter struts at 200 to 250 psi or 3-inch [7.6-cm] diameter struts at 150–175 psi). A minimum of two struts installed on each 4 × 8-foot (1.2 × 2.4-m) panel. The struts must be installed between 1 and 2 feet (0.3 and 0.6 m) from both the trench lip and bottom of the panel, and there must be no more than 4 feet (1.2 m) between struts.

Pneumatic Struts

A **pneumatic strut**, like those manufactured by Paratech, Prospan, ResQtech, and Hurst/AirShore (no longer being produced) come in a wide variety of lengths. Made from lightweight aluminum, pneumatic struts are quick, strong, and dependable **FIGURE 4-22**. In general, these materials are available in lengths from about 18 inches to 16 feet (0.5–4.9 m) and come with a multitude of extensions and attachments. For example, swivel base attachments allow the shore to be extended at a trench wall angle between 90 and 70 degrees and still be effective.

All pneumatic struts operate under the same principle. The strut is extended by using compressed air at pressures recommended by the manufacturer to extend to the needed length. The initial extension of the struts can be provided at a low force. That helps stabilize the shoring system as it is being installed. After extension, the strut either locks by itself or is manually locked to prevent a collapse under load **FIGURE 4-23**. Pneumatic struts require relatively low activation forces because their capacity is based on the mechanical strength (locked collar) of the strut rather than the pressure of the strut. Gauges in the pneumatic strut system provide measurable and controllable strut activation forces. Some pneumatic struts (Paratech and ResQtech) can be installed and removed without having a rescuer in the trench (non-entry shoring). Non-entry shoring techniques dramatically enhance rescuer safety during the shoring installation and removal phases of the trench rescue operation. The shoring installation and removal phases are the most dangerous stages of the rescue operation.

> **SAFETY TIP**
>
> Entry shoring practices expose rescuers to collapse. The risk of collapse is elevated in trenches that have active soil and have already collapsed and/or are showing signs of failure.

Timber Struts

An inexpensive and common type of strut is the **timber strut**. Timber struts usually are 4 × 4-inch, 4 × 6-inch, and 6 × 6-inch sections of Douglas fir or select pine with a bending strength of not less than 1500 psi (105.5 kg/cm). Timber struts have the advantage of low cost when compared with other shores; in addition, they can be cut to varying lengths with little difficulty **FIGURE 4-24**. A disadvantage of timber struts is that they are time consuming to cut with the precision needed for proper installation. A bigger

FIGURE 4-22 Pneumatic struts use air from a remote source (usually SCBA air bottles) to expand and then are locked manually.
Courtesy of Ron Zawlocki.

FIGURE 4-23 The manual system used to lock shores varies according to the shore manufacturer.

FIGURE 4-24 Wood can be cut to various lengths and used as shores. Wood is also relatively inexpensive and widely available in most communities.
© Jones and Bartlett Publishers. Courtesy of MIEMSS.

FIGURE 4-25 Pipe jacks are not recommended for trench rescue shoring operations.

disadvantage is that timber shores require rescuers to be in the trench to install them and the capacity of a strut is highly dependent on its quality. Knots and splits can result in a widely varying strength from member to member. In the unstable soil conditions associated with trench rescue incidents, that type of shoring places rescuers at high risk.

For construction workers, the size and length of the timber strut selected is based on the depth and width of the trench as determined by the type of soil. Appendix C of the OSHA standard CFR 1926, Subpart P, contains information that can be used when timber is selected as the type of shoring material for use in a protective system. For any trench that is greater than 10 feet (3 m) in depth and 4 feet (1.2 m) in width, with any kind of soil, the minimum timber shore size is 4 × 6 inches, and in the case of Type C soil, it is 8 × 8 inches. That is a big piece of wood, which most lumberyards need to special order. If the OSHA standard on timber shoring and using OSHA soil types is being followed, some extremely big and heavy pieces of lumber need to be on hand. A much better alternative is to find a licensed professional engineer who understands collapsed soil conditions and trench rescue shoring to design a trench rescue shoring chart (tabulated data) for timber struts. Chapter 7, *Operations Level Trench Level Shoring*, covers tabulated data and rescue shoring charts in detail.

Screw Jack Struts

A **screw jack strut** is a tool commonly used in conjunction with timber struts. This type of strut has a boot end, which fits over a piece of timber (4 × 4 inch or 6 × 6 inch). The timbers are cut to fit the width of the trench and then tightened by a thread and yoke assembly. Another type of screw jack is used with pipe. These are referred to as pipe jacks, and they are used in conjunction with varying lengths of pipe that are cut to fit the trench width on scene and then tightened by a thread and yoke assembly. Pipe jacks are relatively inexpensive; however, they are not very strong when compared to pneumatic struts. Tightening either type of screw jack creates a strut activation force. Because there is no gauge on the screw jacks, there is no way to measure and accurately regulate the amount of strut activation force that is being applied. Screw jacks can create uneven and excessive strut activation forces, which can cause a weak and unstable trench wall to collapse **FIGURE 4-25**.

Hydraulic Struts

Hydraulic struts, or **hydraulic shores** are a type of protective system that combines the struts and the upright into a single unit. A system of two struts and two rails is lowered in the trench from the top and then pressurized and expanded with a hydraulic pump and hose. After the struts are expanded, the fluid is cut off at the cylinder, and the hose is removed. The advantage of a hydraulic system is that it can be set entirely from above the trench and the strongback portion of the strut is already attached. The disadvantage for rescue use is that it does not work well if the walls of the trench are not vertical or near vertical, as when a

SAFETY TIP

Most struts are rated for only 400 pounds (181.4 kg) of shear force when installed. The bottom line is: Do not stand on them and do not shore to them!

FIGURE 4-26 Hydraulic shores are a popular technique used by construction workers. It is limited in use to walls that are mostly vertical and have stable soil conditions and is not recommended for rescue (collapse) operations.

FIGURE 4-27 A grippy material like Velcro on an extendable painter's stick enables rescuers to spin collars and lock struts remotely.
© Jones and Bartlett Publishers. Courtesy of MIEMSS.

FIGURE 4-28 A digital or smartphone angle finder app will give the panel team an accurate measurement (in degrees) of the void back wall.
Courtesy of Johnson Level & Tool Mfg. Co, Inc.

trench has caved in **FIGURE 4-26**. Hydraulic struts are capable of very high pressures. That can result in excessive strut activation forces, which can cause a weak and unstable trench wall to collapse.

Tools and Appliances

The variations in tools and appliances required to complete a trench rescue successfully are as all-encompassing as "Give me whatever you have." If it would normally be on a construction site, then it will likely be needed on a trench rescue.

Strut Collar Locking Tools

This tool enables the rescuers to install and remove struts without being in the trench during the process. An adjustable paint stick with a hook on one end and 24 to 30 inches (61–72 cm) of grippy material (Velcro, rubber, or two-sided tape) does a great job of locking the collars on some styles of pneumatic struts **FIGURE 4-27**. Some teams have become proficient in remotely locking (spinning) collars with sections of webbing.

Tape Measure and Angle Finder

A 25-foot contractor style, or wide blade, tape measure is a good choice for trench rescue work. The panel team measures the height, width, and breadth of each void within the shoring area. A digital or smartphone angle finder app will give the panel team an accurate measurement (in degrees) of the void back wall **FIGURE 4-28**.

FIGURE 4-29 Pike poles are helpful with positioning panels and wales from lip protection.
Courtesy of Ron Zawlocki.

FIGURE 4-30 Shovels are an essential tool.

Pike Poles

Fire service pike poles are helpful for positioning panels and wales FIGURE 4-29. An 8-foot and a 12-foot pike pole will be helpful at most trenches.

Pickets and Sledge Hammers

Occasionally, pickets are needed to hold panels in place (panel ropes tied back to pickets) before the struts are installed. Pickets can also be used to suspend wales until struts are installed. However, if a trench rescue team is operating properly, the shoring team will be ready with the struts as soon as panels or wales are placed in the trench. That will eliminate the need to tie panel and wale ropes to pickets. Pickets, however, are an integral component of a buttress system. In any case, sledgehammers and sometimes air hammers are used to drive the pickets into the ground beyond the trench lip. The impact force of the sledgehammer hitting the picket combined with the wedge effect of the picket as it enters the ground can cause the soil to fail. The risk of causing a significant failure with pickets can be reduced by distancing the pickets no less than 4 feet (1.2 m) beyond the farthest visible point of soil failure.

Shovels

Ultimately, when you are dealing with the movement of dirt, the shovel becomes one of the most important tools employed during a trench rescue. In the initial stages of a collapse operation, shovels will be needed to move the spoil pile and clear the area around the trench lip. This work allows the ground pads to lay flat so that they do not create an additional trip hazard at the scene. In addition, if a worker is partially buried, the shovel can be given to the worker to begin self-rescue efforts. In the author's experience, most victims who are conscious and trapped in a trench will readily engage in self-rescue.

Although the shovel might work well at the top of a trench, it has little value at the bottom of the trench. The entrenching tool is a small, collapsible version of the larger shovel that is designed to be used in situations where room is limited and a shovel is too long. It also gives rescuers a better feel if they are digging in or around a victim.

Digging operations to remove trapped victims should begin as soon as possible after the protective system is in place—this means you will be working around shoring systems and in other tight, congested places. Here the entrenching tool earns its keep. It may not carry a lot of dirt but it is a great option when you cannot use a larger shovel and still have to move dirt FIGURE 4-30.

Hammers and Nails

Another important item to have on the trench collapse site is the hammer. The hammers found in a discount store are not appropriate; instead, 20-, 22-, and 24-ounce framing hammers that will drive a 16-penny duplex nail in three hits are required, such as a palm nailer. Because trench rescue operations typically involve temporary protective systems, duplex nails, which are designed to be easily removed, are used to connect wood components. A duplex nail has two shoulders, which are not supposed to be driven completely flush with the wood and thus can be removed FIGURE 4-31. The double shoulder affords the claw end of the hammer a place to get a bite for removing the nail after the operation is over. Wood is not

FIGURE 4-31 Duplex nails are two-shoulder nails designed for easy removal after use.
© Jones and Bartlett Publishers. Courtesy of MIEMSS.

FIGURE 4-32 Extreme caution must be employed when chainsaws are in use.
Courtesy of Cecil V. "Buddy" Martinette, Jr.

inexpensive, and you should try to reuse it as much as you can.

Chainsaw

The chainsaw is a versatile saw for rescue operations involving timber shoring. Stay alert when a chainsaw is in use, regardless of whether you are the one using it. Big chainsaws will cut off a leg in half a second. There are not too many second chances with this type of cutting tool **FIGURE 4-32**.

Ventilation Equipment

The ventilation equipment used in trench rescue is normally the electric-powered fire department smoke ejector. Used on the windward side, blowing into the trench, it will afford an adequate flow of fresh air into the trench. When it is hot, this equipment also provides some relief to the rescuers in the hole. Remember that ventilation is not always called for; it should only be used when an atmospheric problem is present. If it is cold outside, ventilation may simply make the rescuers and victim even colder. It would be disappointing to spend 6 hours rescuing a trapped worker, only to have him suffer from rescuer-induced hypothermia.

Ladders

Ladders are another piece of equipment that can be used for a multitude of purposes during a trench collapse. A ladder can be placed in the trench and rescuers can ask the victim to crawl out in a self-rescue. Secondary to victim escape is the requirement for ladder egress in trenches more than 4 feet (1.2 m) deep. Egress ladders need to be placed so that workers do not have to travel more than 25 feet (7.6 m) to get out of the trench, regardless of their location. More important, there needs to be two points of egress and ingress in the trench for rescuer safety. Ladders also can also be used to span the trench opening (A-frame or ladder gin) and to provide a base for lifting operations over the trench.

Scene Lighting

Proper incident management includes determining the logistical needs of the current operation and also forecasting the needs of the operation should it extend over many operational periods.

In trench rescue, it is not unusual for operational periods to extend over many hours, and rescue events that start in the daylight may not be resolved until night. Dusk is not the time to realize lights are needed. If external lighting, either separately carried or apparatus mounted, is not a part of the equipment cache, then it will need to be secured. Also keep in mind that external lighting may need to be powered by an external source, meaning that in addition to the lights, an apparatus-mounted or portable generator will be needed.

Dewatering Devices

Dewatering devices are necessary for the control of water from both ground seepage and rainwater runoff. Excess water in the trench not only creates an uncomfortable environment in which to work, but also deteriorates the trench if it is allowed to stand. Large diaphragm pumps, affectionately called mud pumps, are great low-volume dewatering devices that will hold up to the rigors of even the worst trench

FIGURE 4-33 Dewatering device.
Courtesy of Bob Schilp.

scene **FIGURE 4-33**. These pumps can be gas, air, or battery operated, with each model having its own advantages and disadvantages. Additional dewatering devices are the municipal vacuum truck and the Rescue Vac system. These devices are discussed in detail in Chapter 5, *Hazard Mitigation*.

Utility Control

Underground utilities include natural gas, water, sewer, and electrical lines. They can be damaged by excavators during the excavation process and they can be broken by the impact of a collapsing trench wall. Utility control equipment and its use is detailed in Chapter 5, *Hazard Mitigation*.

Victim Removal Equipment

The equipment needed to treat an injured victim in the trench (medical equipment), release them from entrapment (entrenching shovels, probes, vacuum tools, etc.), package them (harnesses, spinal motion restriction devices, litter baskets, etc.), and remove them from the trench (rope systems, high directional, ladders, etc.) are the essence of trench rescue. The equipment needed to take care of a trapped or injured victim in a trench victim is detailed in Chapter 8, *Victim Care and Extrication*.

Lifting and Stabilizing Heavy Objects

Heavy objects and equipment are common at every construction site. Accidents involving an excavator such as a front end loader, involving pipes, slabs, or steel plates rolling or falling into trenches have been documented. Additionally, construction sites that are near roadways have been encroached by out-of-control vehicles, causing injuries to drivers, passengers, and construction workers alike. Rescuers responding to construction site accidents must be prepared to stabilize and move heavy objects to make the rescue work zone safe. Lifting equipment (levers, pulleys, air bags, spreaders, high directional anchor points, etc.) and stabilizing equipment (cribbings, shores, wire rope, chains, shackles, etc.) are detailed in Chapter 9, *Lifting and Load Stabilization*.

Rapid Intervention Team Equipment

Rapid intervention team (RIT) duties include assessing the most likely hazards that team members will encounter, developing a plan to rescue team members from those hazards, staging the needed RIT tools and equipment near the trench, and requesting the number of RIT members to complete the anticipated rescue assignment. RIT is responsible for rescue team members and not the initial victims in the trench. RIT will have separate extrication equipment, victim treatment, packaging, and removal equipment.

Cutting Station

When a trench rescue incident requires several pieces of lumber to be cut, a cutting station table may be built. The location of the cutting station should be just outside of the hot zone, in the warm zone. The rescuer in charge of the cutting station should have radio communication with the shoring team officer.

The cutting table itself is designed with a standard 4 × 4-foot (1.2 × 1.2-m) or 4 × 8-foot (1.2 × 2.4-m) sheet of ¾-inch (19-mm) plywood. Using a standard piece of plywood means the tabletop does not need to be cut and can be used to square up the table when it is being nailed to the frame. Once the table is put together, the top can be railed with 2 × 4-inch runners that can be spaced and marked at designated positions so standard cuts do not need to be measured each time. Many teams have prebuilt a tabletop that is stored in their equipment trailer and is quickly assembled onsite using saw horses for the legs. The cutting station usually turns out to be a work station where everyone comes to fix things. To build a cutting station table, follow the steps in **SKILL DRILL 4-2**.

SKILL DRILL 4-2
Setting Up a Cutting Station Table (NFPA 1006: 12.1.3)

1. Cut four 4 × 4-inch pieces of wood to the same length, between 32 and 36 inches (0.8 and 0.9 m) long. Lay out two 96-inch 2 × 4 side rails.

2. Cut seven 45-inch 2 × 4 cross braces.

3. Starting at one end, measure and mark the rails on 16-inch centers. Line up the seven cross braces with the marked rails and nail them together.

4. Place the plywood top on the assembled frame and nail one corner. Adjust the plywood, using it to square up the frame. Nail the top to all sides and cross braces.

(continues)

SKILL DRILL 4-2
Setting Up a Cutting Station Table (NFPA 1006: 12.1.3)

5 Turn over the table and install the legs on the corners of each side of the table. Use a framing square to make sure the legs are square with the table frame when nailed to it. Pull the chalk line across the tabletop that lines up with the 2 × 4-inch pieces that make up the table frame. Nail the tabletop to the frame using the chalk lines as a guide.

6 Stand the table on its legs and install 6-foot 2 × 4-inch runners on the top that are spaced at widths of 1¾, 3¾, and 5¾ inches. Note: Each measurement is intentionally ¼ inch larger than standard lumber widths to account for swollen or slightly warped wood. Mark the 2 × 4-inch runners at 12-, 18-, 24-, 30-, and 39-inch intervals.

7 For long-term operations, cut and install braces that are cut at opposite 45-degree angles from the table legs to the bottom of the table frame. The cutting station table is now complete.

© Jones and Bartlett Publishers. Courtesy of MIEMSS.

Voices of Experience

A mutual aid request for a trench rescue on a farm in Rising Sun, Maryland, was received just after noon. Other resources responded immediately, but my department, along with other Harford County Fire/Rescue companies, was not dispatched until 2:00 PM. We responded with a total of eight trench rescue personnel—six on the rescue truck and two on the utility truck.

Upon my department's arrival (2:25 PM), the officer and crew reported to the command post and were told to stand by in the primary staging area, which was about 200 feet (61 m) from the rescue area. I was assigned the position of rescue group leader. After completing my assessments, it became very clear that this call was much more complicated than just a person trapped in a drainage pond. The person was trapped while laying a drainage pipe from the new sewage pond to a drain field. The opening in the ground was about 2 to 3 feet (0.6–0.9 m) wide and appeared to be at least 20 feet (6.1 m) deep. The ground sloped into the pond and a skid steer loader was in the pond. From a safe position, it was difficult to determine exactly how high the bank was and how wide the opening was where the pipe was being laid. Recognizing that this would be a manpower-intensive operation, I asked command to contact Harford County and find out how many staff members would be coming to the scene. A few minutes later, I asked command to request technical rescue teams from Baltimore County ATR and Claymont.

While the resources were en route, our crew tried to pinpoint the location of the person who was trapped. One of the construction workers explained that, after the collapse, he had never moved from the spot that was directly above the trapped victim. Rescue personnel marked the spot and then escorted the construction worker away from the site. The rest of the construction workers did not want to leave the site, but followed our directions after we assured them that we had marked the spot and would start there to gain access. Because of the unstable soil conditions, we decided to shore in steps of 4-foot (1.2-m) vertical increments. Rescuers were directed to start hand digging 4 feet (1.2 m) on each side of the spot and dig down 4 feet (1.2 m). Crews worked diligently to dig and move the soil.

Around 6:00 PM, the other teams began arriving. By then it was starting to get dark, and resources were called for lighting. The weather report predicted rain to start around 10:00 PM and become heavy around midnight. As the technical rescue teams arrived, they were briefed on the site, operations, and our initial tactical action plan. The teams made several important suggestions for additional resources that would support the plan in place. As a result of those suggestions, command requested a vacuum truck, more lighting, and an additional technical rescue team.

By approximately 7:30 PM, the crews had dug a hole that was 4 feet (1.2 m) down and 8 feet (2.4 m) long and panels were placed into the freshly dug hole. During the digging process, the unstable soil continued to collapse and fall to the bottom of the step being dug. In order to minimize the falling soil, low-pressure airbags were placed between FinnForm panels until we could get the struts in place.

The MOSH (Maryland Occupational Safety and Health) representative arrived and was a great help to us. While he would not give us shoring guidance, he would warn us if he saw something that he did not like, such as when common lumberyard plywood was being installed as panels. The MOSH representative stopped the operation and informed us that plywood panels were not safe for entry into the trench (see *Lessons Learned* for details). When we approached the team that brought the plywood panels and told them to stop the installation, they became upset but conceded to the orders. FinnForm panels were available from other teams and they were used to continue shoring.

The rain started early, and while it was not a heavy rain, water and dirt turned to mud, which made the hand digging operation much more difficult

Voices of Experience

and time consuming. Crews continued to struggle, digging subsequent steps (deeper) into the ground safely. Around 9:30 PM, the rescue team leaders met and discussed with command that we were 9 hours into this and we felt that there was no hope for a live victim outcome, explaining that the weight of the dirt and the amount of time this person was covered just did not add up to a survivable outcome.

With the rain coming down harder, the incident command agreed that we should pull our equipment out and transition into a recovery mode operation. The transition tactical action started about 10:30 PM, which included a coordinated effort between rescue personnel and a heavy equipment (backhoe) operator. We talked to the backhoe operator and showed him where we had marked the most probable victim location. The equipment operator suggested a digging plan that would start clear of the marked victim area and move soil away from the victim in small increments. We agreed and assigned rescue personnel to coordinate with the operator and to watch for signs of the victim as the soil removal proceeded. About an hour later, rescuers noticed a sign of being near the victim's location. It turned out that he was directly under the spot initially marked by his co-worker, about 15 feet (4.6 m) below the lip. The backhoe was stopped at that point and rescue crews hand dug the person out of the dirt and removed his body from the trench.

Lessons Learned

- Training: This was one of the first real calls for a trench rescue for many of those involved on the initial response. My department and another department had been training in trench rescue for a little over 10 years. We worked well together. Training with other agencies is needed. Know the equipment of the other teams, know the people of the other teams.
- Weather: Get a weather update as soon as you arrive. Plan for bad weather. We were lucky to have an open barn to move to and stage resources.
- Site control: We had well over 100 personnel onsite and, while we had accountability, we did not have control of access into the site. Personnel were able to freely walk around the site, which should have been controlled.
- OSHA: Use OSHA to your advantage. While the MOSH representative may not give you guidance, have them readily available where the work is occurring. If something does not look right, they will tell you to stop. The MOSH representative had been on a call previously with the team trying to use plywood as shoring panels and had told them they could not use them in our state. OSHA's enforcement of regulations at the scene of a rescue varies from state to state. However, in the event of a body recovery, OSHA representatives in just about every state will require equipment to have engineered tabulated data.
- Resources: You can never have enough. This was an eye opener for many of us. Know where your resources are located. Find out what your local jurisdictions has to assist you. Equipment and operators for vacuum trucks and excavators should be on your resource list and should be included in your trench rescue training drills.
- Personnel: Rotate your personnel. At one time it felt like everyone who showed up just grabbed a shovel and started to dig. We know rescuers want to help, but you have to let your plan expand and utilize the personnel wisely and to your advantage.

Steven Hinch
Fire Chief
Aberdeen Fire Department, Maryland
Technical Rescue Crew Chief
Harford County Special Operations Team

After-Action

IN SUMMARY

- At every incident, the safety of personnel depends on their level of skill and the proper PPE and operational equipment.
- Instilling a safety mindset into the culture of your team does not happen overnight, but rather is the result of many hours of training and the discipline that comes from everyone on the team being accountable.
- Trench rescue PPE can be classified into two main categories:
  - Personal-issue PPE: The garments and equipment that each member receives and keeps with them, likely in a personal bag
  - Team-issue PPE: Equipment that is kept on the response vehicle and shared (i.e., breathing apparatus).
- At a minimum, trench rescue team members need an ensemble element consisting of protection for the upper and lower torso, arms, legs, head, hands, feet, and eyes. Other items, depending on the environment expected, could include hearing protection and foul weather gear. NFPA 1951 does not address visibility concerns.
- Proper cleaning removes contaminates, protects the equipment, ensures that it is ready for the next response and that the equipment does not fail prematurely.
- Trench rescue shoring equipment is a combination of equipment that is built using fairly common materials and specialized equipment that is to serve specific purposes. The built equipment includes lip protection, trench rescue panels, struts, wales, and backfill. The specialized equipment can include pneumatic struts, lip bridges, aluminum wales, and backfill consisting of air bags or struts.
- Trench rescue operations rely on specialized equipment in order to safely prevent soil from further collapse.
- Lip protection is the act of making the lip area safe, including the selection, construction, application, and installation of ground pads and lip bridges around the affected trench area.
- Backfilling—filling in voids that have been created from cave-ins—is an essential skill for trench rescue shoring.
- Struts are the component in the protective system that transfers the force from one side of the trench to the other.

KEY TERMS

Composite behavior When two members are rigidly connected to act as one larger member, significantly increasing strength.

Dewatering devices Devices that control water from ground seepage, broken water mains, and rainwater runoff (e.g., trash pumps).

Duplex nail Nail that has two shoulders, which allows it to be removed easily.

Entrenching tool A small, collapsible version of the larger shovel, designed to be used in situations where space is limited and a regular shovel is too big. Commonly known as a military shovel.

Ground pads Wooden material used to line the trench lip for the purpose of distributing rescuer and equipment load (weight).

Hydraulic shores Type of shore (strut and rails) that is lowered into the trench from the top and then expanded by using a connected hydraulic pump.

Lip bridges A type of lip protection built with girders (timbers), platforms (aluminum, wooden, or fire service ladders with lumber), and bases (sections of timbers placed away from the compromised wall and used to elevate the bridge above the lip).

Noncomposite behavior When two members are used that act individually.

Pneumatic strut Type of strut that is extended by using compressed air. After extension, the strut either locks itself or is manually locked to prevent a collapse under load.

Screw jack strut Type of shore or strut end that is tightened by a thread and yoke assembly. It is also sometimes referred to as a pipe jack when used in conjunction with varying lengths of pipe.

Self-contained breathing apparatus (SCBA) A respirator that supplies breathing air to the user from an air source that is independent of the environment. It is designed to be carried by the user.

Sheeting The portion of the protective system designed to hold back running debris.

Shoring system An assembly of shoring components consisting of panels, wales, struts, backfill, etc., to provide a complete protective system.

Strongbacks The 2 × 12-inch lumber components that are 10 or 12 feet (3.0 or 3.7 m) in length on the trench rescue panel that transmit forces along the vertical plane of the trench wall.

Struts The component in the protective system that transfers the force from one side of the trench to the other.

Supplied air breathing apparatus (SABA) A respirator that supplies breathing air to the user from an air source that is independent of the environment, supplied from a remote source. It is not designed to be carried by the user.

Timber strut A wood column with a compression strength of not less than 1500 psi (105.5 kg/cm) that transfers load from one side of a trench to the other side.

Trench rescue panel A 4 × 8-foot piece of plywood and a 2 × 12-inch or 2 × 10-inch piece of lumber that are attached together to be used in shoring.

Wale A beam used horizontally in a trench shoring system to provide resistance to struts when voids or other obstructions are present.

On Scene

1. What equipment would you add to your engine in order to provide primary shoring to protect a trapped victim?

2. What local contacts do you have or could you make to arrange for specialty equipment like vacuum trucks or assistance with plugging utilities?

3. How often does your fire department train with your area trench rescue teams?

SECTION 2

Operations Level

CHAPTER **5** **Hazard Mitigation**

CHAPTER **6** **Managing the Trench Incident**

CHAPTER **7** **Operations Level Trench Rescue Shoring**

CHAPTER **8** **Victim Care and Extrication**

CHAPTER 5

Operations Level

Hazard Mitigation

KNOWLEDGE OBJECTIVES

- Recognize the presence of factors that may lead to hazardous conditions. (**NFPA 1006: 12.2.1**, pp. 95–98)
- Explain the purpose of a hazard control plan. (**NFPA 1006: 12.2.1, 12.2.2**, p. 98)
- Explain the function and application of hazard control equipment for trench rescue incidents. (**NFPA 1006: 12.2.2**, pp. 99–102, 106)
- Identify a minimum equipment cache for operations level hazard control activity. (**NFPA 1006: 12.2.4**, pp. 99–102, 106)
- Describe the methods of controlling hazards. (**NFPA 1006: 12.2.1, 12.2.2**, pp. 99–110)

SKILL OBJECTIVES

- Select and operate hazard mitigation equipment. (**NFPA 1006: 12.2.5**, pp. 99–102)
- Conduct atmospheric monitoring. (**NFPA 1006: 12.2.1, 12.2.2**, p. 103)
- Conduct dewatering procedures. (**NFPA 1006: 12.2.1, 12.2.2**, pp. 104–105)
- Control a broken utility line. (**NFPA 1006: 12.2.2**, p. 101)
- Ventilate a trench. (**NFPA 1006: 12.2.2**, pp. 108–110)
- Control hazardous materials at a trench incident. (**NFPA 1006: 12.2.2**, p. 107)

You Are the Rescuer

You are the "Haz Matician" on the technical rescue team. You are not only a rescue technician, but a hazardous materials technician as well. The technical rescue team asks you to lead next month's training on hazard assessment and control at trench rescue incidents. You could spend hours on atmospheric monitors alone, but this lesson needs to be a practical review that includes the assessment of probable hazards found at a trench site. You have contacted the city's department of public utilities (DPU), and they will allow you to use a trench site on your training day.

You plan to set up the trench with mechanical, chemical, human-made, electrical, and water hazards. Your teammates will be evaluated on their ability to identify and mitigate these hazards. Your lesson on monitoring the atmosphere at trench rescue sites will include an operational review of your team's four-function monitor, identification of hazardous atmospheric indicators at trench sites, procedures for monitoring where atmospheric hazards are present, and procedures for monitoring when atmospheric hazards are not present. Before starting the class, you quiz the students by asking them the following questions:

1. At a trench site, which conditions would likely cause carbon dioxide build-up?
2. At a trench site, which conditions would likely cause oxygen deficiency?
3. At a trench site, which conditions would likely cause a flammable atmosphere?

 Access Navigate for more practice activities.

Introduction

Trench rescuers must be aware of and must look for hazards that are common to the trench rescue environment. Potential hazards at a trench rescue site include: collapse (cave-in), underground utilities (natural gas, water, sewer, electrical, etc.), hazardous atmospheres (commonly carbon monoxide, hydrogen sulfide, explosives, and oxygen deficit), traffic (cars, trucks, and heavy equipment), hazardous materials (commonly gasoline, diesel fuel, solvents, fluids), physical hazards (commonly construction and rescue equipment on the trench lip), and biological hazards that are associated with trench-related traumatic injuries (for example, blood). The most common hazard at a trench emergency is soil failure and the collapse of trench walls. Understanding soil and collapse mechanics is covered in detail in Chapter 2, *Soil and Collapse Mechanics*, and trench shoring techniques are discussed in Chapter 7, *Operations Level Trench Rescue Shoring*, and Chapter 10, *Technician Level Trench Rescue Shoring*. In this chapter, we will closely examine the hazards commonly encountered at trench and excavation sites, discuss a generic hazard control plan, detail hazard mitigation options, discuss soil evaluation techniques, and list the required specialized equipment for trench rescue teams.

Hazard Identification

A variety of hazards may be present at an underground construction site. As previously discussed, initial actions performed by awareness level rescuers should include the identification of hazards and the control of hazards in the cold zone. This area contains staging, cutting stations, logistical support, and rehabilitation. Most hazards inside of the warm and hot zones (or the rescue area) should be isolated and avoided by awareness level rescuers. Offensive hazard mitigation in the warm and hot zones is typically performed by personnel with higher levels of training. Hazards at a trench rescue can be safely mitigated (through avoidance, control, removal, or shielding) by trained personnel using proper personal protective equipment (PPE).

Types of Hazards

Trench Collapse

The collapse of soil walls is the most prevalent hazard at a trench incident. All unprotected walls of a trench must be considered as hazardous. First-arriving firefighters and first responders should avoid this hazard by never entering a trench until it is shored and only entering a trench if they have a proper level of training. As a reminder, the area around the trench (or the hot zone) should be isolated with barrier tape; bystanders and unassigned first responders must be removed from the area. Personnel assigned to work in the hot zone must be briefed on their specific tasks and all hazards present. No one will be allowed to enter an unprotected trench, and no one will stand or walk on the trench lip (within 4 feet [1.2 m] of the wall) until lip protection is in place.

FIGURE 5-1 Electrical hazards should be secured.
Courtesy of Cecil V. "Buddy" Martinette, Jr.

Red: Electric power lines, cables, conduit, and lighting cables
Orange: Communications, alarm or signal lines, cables, or conduit
Yellow: Gas, oil, steam, petroleum, or gaseous material
Green: Sewers and drain lines
Blue: Potable water
Violet: Reclaimed water, irrigation, and slurry lines
Pink: Temporary survey markings, unknown/unidentified facilities
White: Proposed excavation limits or route

FIGURE 5-2 Utility color markings.

Trench rescue shoring is an equipment and personnel intensive endeavor and although it is an important form of hazard mitigation we have devoted three chapters to this topic. Trench rescue shoring is covered in Chapter 7, *Operations Level Trench Rescue Shoring* and Chapter 10, *Technician Level Trench Rescue Shoring*. Chapter 3, *Initial Actions*, covers the first actions of awareness level rescuers in detail.

Utilities

Underground utilities include but are not limited to water, sewer, natural gas, electric, and telecommunication lines. They can be damaged by digging or from the impact of collapsing trench/excavation walls. Exposed and unsupported utility lines spanning between two trench walls can break. Damaged and broken utility lines can be serious hazards to anyone in a trench.

Control of electricity, other than shutting off breakers, is best left to the people who do it for a living **FIGURE 5-1**. Be alert when operating around exposed telephone lines: They also carry a voltage that can cause injury or even death with contact. In addition, you should be cautious with electrical items you bring to the scene, such as lighting equipment, fans, and supplemental power.

As a routine practice, determine the locations of all utilities before digging in a collapsed area. This identification process can be as simple as calling the local utility service company. If the location of utilities is ever a hazard control issue, do not hesitate to call the local utility location service. Sometimes called "Ms. Utility," "One Call," or 811, these services are paid for by the utility companies, and their purpose is to mark existing utilities before any type of digging operation begins. In a trench collapse, such a service would confirm the locations of any utilities that are not already known but that may have an effect on the rescue operation. The standard color markings for utility locations are noted in **FIGURE 5-2**.

> **TIP**
>
> Trench rescue incident scene management may also use the terms *general area* and *rescue area* to establish boundaries of operation, in addition to hot, warm, and cold zones. The general area is the surrounding area not in the immediate vicinity of the extrication effort. Hazard control in this area would entail a large overview of the scene and, under normal circumstances, should begin first. The activities that would normally take place in this area are staging, cutting, logistical support, rehabilitation, and vehicle parking. The rescue area is located immediately surrounding the rescue site. This is a small area around the rescue effort, and hazard control here would be considered only after hazard control of the general area was completed. Extrication efforts such as air supply, panel team, shoring personnel, and the safety officer are established within this area.

Traffic

Cars, trucks, and buses running near or on roads adjacent to a trench accident can cause both trench wall collapse and collisions with rescue workers **FIGURE 5-3**. Additionally, heavy equipment (excavators, front-end loaders, etc.) operating on the trench site pose the same hazards. Remove these hazards by completely blocking off (barricades, barrier tape, apparatus positioning, etc.) all roadways within 300 feet (91 m) of the trench. Pass this duty off to police as they arrive on the scene. Shut down heavy equipment operating within 300 feet (91 m) of the trench.

FIGURE 5-3 Cars, trucks, and buses running near or on roads adjacent to a trench accident can cause both trench wall collapse and collisions with rescue workers.
© Steve Hamann/Shutterstock.

Physical Hazards

Physical hazards include construction and rescue equipment in and near the trench that can trip, cut, burn, or fall on rescuers. These items must be isolated or removed from the rescue area.

Water

Excessive water can not only quickly fill the trench and drown a trapped victim, but it can change the weight and internal strength of the soil **FIGURE 5-4**. Water can greatly increase the probability of secondary collapse and can affect the victim's chance to survive the incident. The weight of the soil increases with the addition of water. That additional weight can create high lateral earth pressures that can exceed the scope of the capabilities of standard rescue shoring equipment. The source of the water must be identified to control the hazard. Common sources of water include broken water lines, broken sewer lines, high water tables, and heavy rains.

Severe Environmental Conditions

Rescuers may be subjected to a variety of severe environmental conditions while working at trench or excavation incidents. In fact, severe environmental conditions often cause accidents at a trench site. Severe environmental conditions include, but are not limited to, heavy rain, high winds, cold temperatures, hot temperatures, snow, ice, and lightning.

Biological Hazards

Trench-related injuries are often traumatic in nature. Traumatic injuries result in bleeding, vomiting, and the presence of other body fluids. Rescuers must wear appropriate PPE to protect themselves during a rescue. Chapter 8, *Victim Care and Extrication*, covers victim assessment, management, and packaging in detail.

FIGURE 5-4 Water can create extreme forces on trenches that can compromise already poor conditions.
Courtesy of Cecil V. "Buddy" Martinette, Jr.

Hazardous Materials

Hazardous materials at a trench construction site are often related to the equipment being used and the work being performed. Gasoline, diesel fuel, and hydraulic fluids are needed to run most heavy equipment. Oxygen and acetylene are commonly used for cutting with torches. Digging in previously undisturbed soil is unlikely to uncover hazardous materials, but digging in a landfill or near hazardous material storage sites is likely to unearth them. Occupational Safety and Health Administration (OSHA) guidance for construction workers working in landfill areas advises the atmospheres in a trench or excavation deeper than 4 feet (1.2 m) be tested first. Trench rescue teams should follow that guidance.

Atmospheric

Thousands of hazardous atmospheres are possible, but only a few are probable at trench and excavation sites. A trench is unlikely to contain a hazardous atmosphere unless it is dug in a landfill area or near where hazardous substances are stored, or if a gas is released into the trench due to a ruptured line. Trench rescue teams do not have the time, nor do they have the equipment to test for every possible **atmospheric contaminant** in existence. Instead, they must test for the most common atmospheric conditions that are associated with trench and excavation environments and must look for signs of other site-specific contaminant.

The most common atmospheric hazards at a trench site are carbon monoxide, hydrogen sulfide, flammables, and oxygen deficiency. It is a best practice for rescuers to test the atmosphere for oxygen, carbon monoxide, hydrogen sulfide, and flammability upon arrival at every trench and to continuously monitor the atmosphere if any sign of hazardous conditions is present.

Hazard Control Plan

A **hazard control plan** addresses the safety of the trapped victims, bystanders, and all responders. It begins with the identification of all hazards and is followed by hazard mitigation operations. Initial hazard recognition and mitigation begins with the first-arriving firefighters (awareness level rescuers). All hazards must be identified and then dealt with in accordance with the skill level of the personnel and equipment on the scene. A fire department engine company (the first responders at a trench emergency) can dress in appropriate PPE to identify and isolate many of the hazards that are common to a trench emergency site. There will also likely be hazards that need to be avoided by awareness level rescuers but will not be completely mitigated until additional resources arrive. The trench rescue team will likely be the next level resource to arrive. Operations level rescuers must make a complete hazard assessment and determine the effectiveness of the mitigation efforts taken by the awareness level rescuers. Depending on local standard operating guidelines (SOGs), higher level rescuers may be trained to implement offensive hazard mitigation techniques to control many of the hazards that are beyond the scope of awareness level rescuers. Some hazards require an even higher level of hazard mitigation resources. Those resources may include utility company emergency response teams and hazardous material response teams.

Hazard Mitigation PPE

PPE can protect rescuers from many hazards associated with trench or excavation emergencies. Trench rescue team members should arrive at the scene wearing the basic rescue PPE (helmet, body ensemble,

FIGURE 5-5 Trench rescue team members should arrive at the scene wearing the basic rescue PPE.

boots, and gloves). Chapter 4, *Personal Protective Equipment and Equipment Basics*, covers PPE in detail. When members are assigned specific duties, like hazard control, the basic rescue PPE must be enhanced to protect them from the specific hazard they will encounter. Enhanced PPE may include self-contained breathing apparatus (SCBA) and structural firefighting gear (natural gas leaks), impervious boots, gloves and ensembles (sewage and standing water), and high-voltage gloves/over boots (electrical wires).

Briefing

Other than gathering information, no action in the hot zone should take place until the pre-entry briefing is conducted. The pre-entry briefing is important because rescue personnel need to understand the hazards and their locations, the situation and the strategic and tactical steps that are part of the overall rescue plan. Items in the pre-entry briefing include the overall goal of the operation, the hazards that have been identified, the hazard control plan, PPE requirements, protective system design, position assignments, the accountability system, and emergency procedures.

Hazard Mitigation

Defensive hazard mitigation measures primarily include isolating the hazards and the hazardous area, notifying all affected people of the hazards and the isolation technique (barriers and boundary), and requesting the resources needed to remove or control the hazard. Defensive measures are detailed in Chapter 3, *Initial Actions*.

Offensive hazard mitigation measures go beyond the isolation techniques used for defensive mitigation. Offensive measures include shutting off and controlling hazards. The techniques used to accomplish offensive methods may place the operations level rescuer at much higher risks and are dependent on local SOGs. Offensive hazard mitigation techniques should only be utilized for rescue mode operations. In situations where the victim is clearly deceased, wait for the response of the utility company, hazardous materials team, or other enhanced hazard mitigation resources.

The following section lists potential tactical options for offensive mitigation of common trench rescue hazards. These options require additional training, skill maintenance, and specialized equipment. The authority having jurisdiction (fire department) over the trench rescue team will need to make a decision to provide these services. Before attempting these techniques, be sure to have the proper equipment and PPE, and training from a subject matter expert. Training may come from utility companies, equipment manufacturers, and instructors who are subject matter experts.

SAFETY TIP

Offensive techniques for hazard mitigation must only be implemented by specially trained and equipped personnel. The training and equipment selection must come from subject matter experts and must be coordinated with local utility providers.

Hazard Control Equipment

Hazard control equipment includes tools that test for and mitigate hazards. While many hazards on a trench site, like trip hazards or unstable materials or equipment, can be seen, others like electrical current and hazardous atmospheres cannot be seen. Trench rescue teams should be prepared to control hazards when lives are at stake but should only take offensive actions on hazards that they are trained and equipped to mitigate. Potential offensive hazard control equipment includes: (1) atmospheric monitor, (2) ventilation

FIGURE 5-6 A typical hazard control equipment kit.
Courtesy of Ron Zawlocki.

equipment, (3) noncontact voltage tester, (4) high voltage gloves, (5) telescoping insulated "hot stick," (6) burlap (static electricity grounding), and (7) assorted size pipe plugs **FIGURE 5-6**. The equipment carried and utilized during an emergency is dependent on local SOGs.

Atmospheric Testing Equipment

As a minimum, trench rescue teams should be equipped with an atmospheric monitor **FIGURE 5-7** and a noncontact voltage tester **FIGURE 5-8**. A trench is unlikely to contain a hazardous atmosphere unless it is dug in a landfill area or near where hazardous substances are stored. Additionally, it is unlikely that a trench or excavation will be oxygen deficient unless there has been a release of a gas (like natural gas) that displaces the oxygen in the trench. However, it is a best practice for rescuers to test the atmosphere upon arrival at every trench and to continuously monitor the atmosphere if any sign of hazardous conditions is present. Monitoring is used not only to detect the presence of IDLH (immediately dangerous to life and health) atmospheres, but also as a tactical guide to ventilation of the trench. Atmospheric monitoring, bump testing, equipment maintenance, and calibration must comply with the manufacturer's recommendations.

FIGURE 5-7 Atmospheric monitor.
© Jones and Bartlett Publishers. Courtesy of MIEMSS.

FIGURE 5-9 Mud pumps are versatile, although they sometimes require extensive priming to pull the vacuum.
Courtesy of Bob Schilp.

trash pumps for dewatering purposes. If a trench rescue rig carries a Stanley USAR power unit, there is a self-priming hydraulic trash pump available. When pumps powered by gasoline engines are used, the exhaust must be controlled and directed away from the trench for rescuer and victim safety. Additional dewatering devices include the municipal vacuum truck and the Rescue Vacystem.

> **TIP**
>
> Do not wait for the rain to begin before locating a dewatering device. In addition, have two of these devices onsite because one of them may break down.

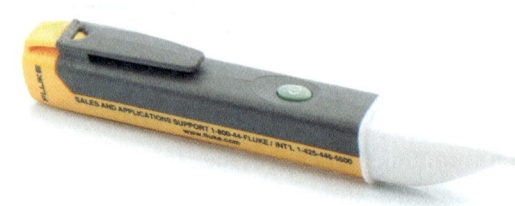

FIGURE 5-8 A noncontact voltage tester.
© Lemau Studio/Shutterstock.

Dewatering Devices

Dewatering devices are necessary for the control of water from ground seepage, broken water mains, and rainwater runoff. Excess water in the trench not only creates an uncomfortable environment in which to work, but also deteriorates the trench floor if it is allowed to stand. Large diaphragm pumps, affectionately called mud pumps, are great low-volume dewatering devices that will hold up to the rigors of even the worst trench scene **FIGURE 5-9**. Many trench rescue teams use either electric or gasoline-powered

Sewer and Water Control Equipment

Plastic, copper, lead, and steel pipes can be crimped to stop or minimize the flow of water lines up to 2 inch (5 cm) in diameter with ViseGrip style and emergency shut-off tools. These actions are dependent on local SOGs. Sewer lines are commonly larger than 2 inches (5 cm) in diameter, but the flow is low pressure and can be stopped by plugging the lines **FIGURE 5-10**. Towels, cribbing, and or shims can be used as effective plugs of sewer lines. Inflatable plugs are made for this purpose, ranging from 1-inch (2.5-cm) diameter to 96-inch (244-cm) diameter **FIGURE 5-11**. These plugs would be inserted into the pipe and inflated using an air compressor or air source **FIGURE 5-12**. Local public works departments likely have these tools and can advise trench rescue teams on the products to purchase for an equipment cache.

CHAPTER 5 Hazard Mitigation **101**

FIGURE 5-10 Small pipes can be crimped with ViseGrip devices and larger pipes (up to 2-inch [5-cm] diameter) can be crimped with emergency (hydraulic) shut-off tools.
Courtesy of Ron Zawlocki.

FIGURE 5-12 Mechanical and inflatable pipe plugs can stop fluid and gas flows.
Courtesy of Ron Zawlocki.

TIP

Many trench rescue teams are incorporating a simultaneous response from local or mutual aid hazardous materials teams to assist with hazardous atmosphere, water, and sewer mitigation operations. The hazardous materials teams will also be able to advise and assist with decontamination of the victim and rescuers if necessary.

Electric Control Equipment

Rescuers may be trained to identify primary and secondary power lines, depending on local SOGs. They may also be trained to identify high voltage and low voltage underground burial lines. Voltage (live current) can be detected with **noncontact voltage testers** like the AC HotStick (TAC Stick). Typical voltage direct burial lines (120–240 v) are common. Ultimately, the choice on how to mitigate electrical hazards will end up being a local decision.

Ventilation Equipment

In the unlikely event of a hazardous atmosphere in the trench, an **electric-powered fire department smoke ejector** can be used to ventilate the trench. Used on

FIGURE 5-11 Low pressure flows found in sewer lines can be controlled by plugging the pipe.
Courtesy of Ron Zawlocki.

FIGURE 5-13 A confined space ventilator with a heater attachment can prevent hypothermia in patients and will keep the temperature in the trench comfortable for rescuers in cold and damp conditions.
© Red_Shadow/Shutterstock.

the windward side, blowing into the trench, it affords an adequate flow of fresh air into the trench and can move the hazardous atmosphere out of the trench. Additionally, when dealing with hot weather-related temperatures, this equipment can provide relief to the rescuers and the victim in the hole. Remember that ventilation is not always called for unless an atmospheric problem is present. If it is cold outside, ventilation may simply make the rescuer and victim even colder. It would be disappointing to spend 6 hours rescuing a trapped worker, only to have them suffer from rescuer-induced hypothermia. Use ventilation only when appropriate.

An alternative method is a **confined space ventilator**, which can send fresh air directly to the victim's face by way of duct tubing **FIGURE 5-13**. Confined space ventilators can be purchased with heating elements that can keep the victim and rescuers warm during cold weather conditions. Any ventilation system should be placed so that it does not blow dirt into the faces of the victim or rescue workers.

Natural Gas Control Equipment

Underground natural gas lines can be broken during the excavation process or from the impact of a collapsing trench wall. Natural gas in a trench is explosive and can displace the oxygen in the trench, making the environment extremely dangerous for both trapped victims and rescuers. At trench rescue scenes with a broken natural gas line, the flow of gas must be stopped, and the gas must be removed (ventilated) from the trench. The safest way to do this is to have the emergency response team from the local natural gas company perform those functions. How this hazard is mitigated is dependent on local SOGs.

Potential Offensive Mitigation Techniques

Atmospheric Monitoring

The best way to detect the presence of hazardous atmospheric conditions is by using a calibrated atmospheric monitor. Many fire departments and most trench rescue teams have atmospheric monitors included in their equipment caches. Safety at trench/excavation incidents can be enhanced through the proper use of air monitors. The 4 × 4 × 4 monitoring technique (4 functions—4 locations—4-foot increments) is designed to rapidly test for the four most common hazardous atmospheric conditions found at trench and excavation sites. Additional site-specific monitoring may be required for incidents involving chemical, radiation, or other atmospheric hazards. Some hazardous atmospheric conditions like solvents and glues may not be detected by four function monitors. Identification of these types of hazards must come from interviews with workers and observations of cans, buckets, and containers at the trench site.

To check the atmosphere in a trench, you will need to locate a trench that is at least 8 feet (2.4 m) deep, have a four-function atmospheric monitor and bump gas, and follow the steps in **SKILL DRILL 5-1**.

Dewatering Devices

Dewatering the trench is necessary for the control of water from ground seepage, broken water mains, and rainwater runoff. Dewatering plans should be considered even if the trench is dry upon arrival. Pipes can break suddenly, and rainstorms can develop quickly. Building dikes around the trench should be initiated immediately if the water is entering the trench from grade level. Surface water must be directed around and downstream of the trench and not be allowed to run into the trench.

Water, sewage, or other liquids in the trench must be pumped out. Strainers on the end of the suction hose can minimize the chance of the hose getting plugged with mud. If a submersible pump is lowered into the trench, a large bucket or milk crate can be used to keep the intake of the pump from plugging with mud. Secure the pump in the bucket or crate and lower them as a unit. If a bucket is used, drill several 1-inch diameter holes around the base of the bucket to allow fluids to enter it and be drawn into the pump.

To dewater a trench, follow the steps in **SKILL DRILL 5-2**.

Sewer and Water Line Breaks. Disrupted sewer and water lines create hazardous conditions for victims and rescuers in a trench. Depending on

SKILL DRILL 5-1
Utilizing an Atmospheric Monitor (NFPA 1006: 12.2.1 and 12.2.2)

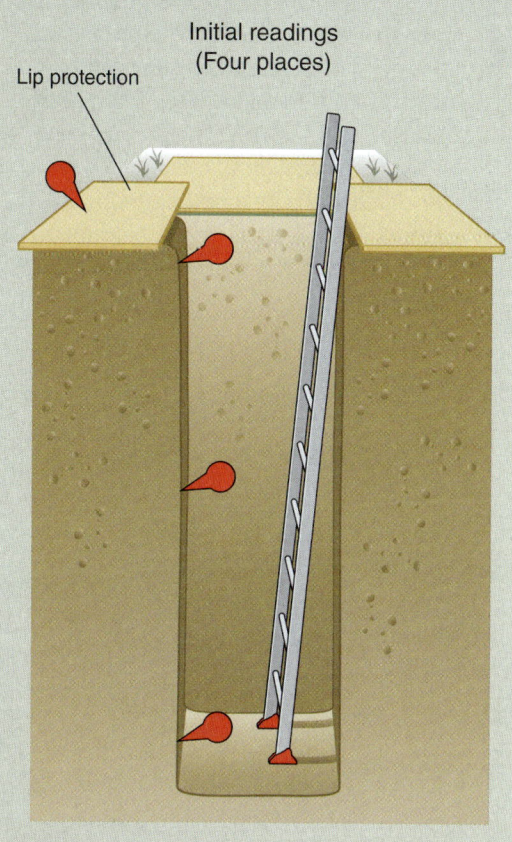

1 Review the atmospheric monitor's manufacturer's manual for proper use of the instrument. Review permissible exposure limits: Oxygen (19.5% minimum); flammables (maximum 10% of lower flammable limit); carbon monoxide (maximum 50 ppm); hydrogen sulfide (maximum 10 ppm). Check the recently calibrated monitor with bump gas. With lip protection and an escape ladder in the trench, obtain and record the readings at four locations: Outside the trench; top level: inside the trench; middle level: inside the trench; and bottom level: inside the trench.

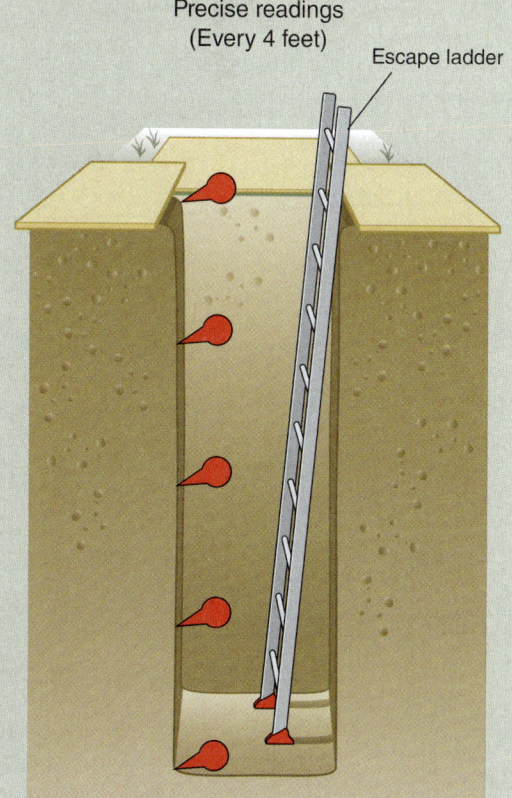

2 Following the initial readings, more precise air monitoring is performed: Obtain a reading at 4 feet (1.2 m) below the lip, then obtain readings at 4-foot (1.2-m) increments until the monitor reaches the bottom of the trench or is a safe distance from accumulated liquids.

SKILL DRILL 5-2
Dewater a Trench (NFPA 1006: 12.2.1 and 12.2.2)

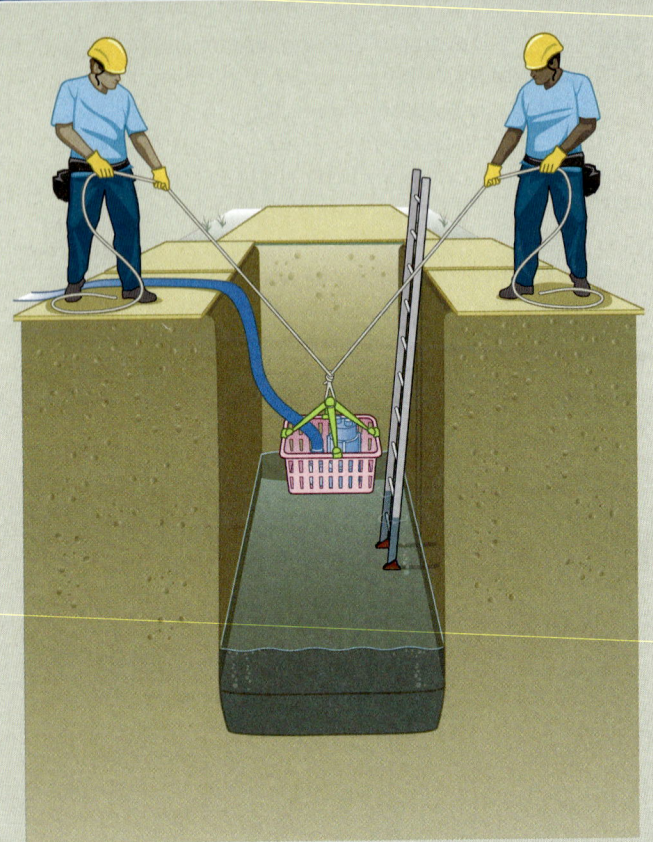

1 With an escape ladder and lip protection in place, move the dewatering pumps and hoses to the desired location. If the pump must set near the lip, it should be placed on the lip protection to minimize issues associated with vibrations. Use ropes to carefully lower and direct the placement of the pump (if it is submersible) or the suction hose.

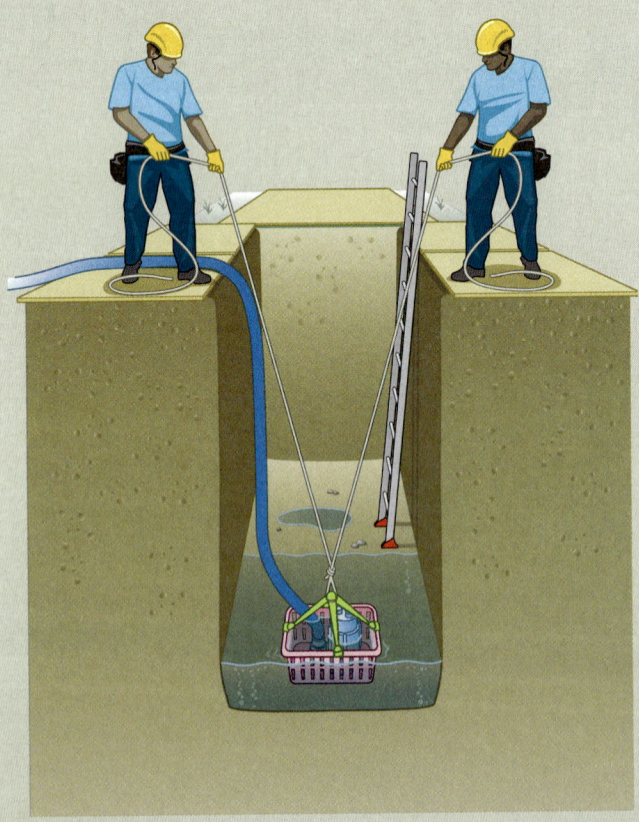

2 Position of the intake should be near the source or in the deepest portion of the trench. Position the discharge hose to direct water downhill and a minimum of 50 feet (15 m) away from the trench. Consider establishing a holding area or holding tank for the discharge if pumping sewage from the trench.

SKILL DRILL 5-2
Dewater a Trench (NFPA 1006: 12.2.1 and 12.2.2)

3 Start the pump and remove water to the lowest possible level. Once the liquid is removed to the point where the pump is no longer effective, the pump can be shut off temporarily.

4 Assign a team member to monitor the fluid level in the trench and reinitiate the dewatering process if liquids accumulate. When the trench is secured, assign two members to enter the trench to dig a sump in the floor of the trench. This will provide a low-lying place for fluids to drain to, keeping the floor of the trench drier. Reposition the pump (submersible) or intake hose and strainer into the sump (hole) and operate as needed.

local SOGs, both plastic and metal water lines may be crimped to stop or minimize the flow of water lines up to 2 inches (5 cm) in diameter with ViseGrip style and emergency shut-off tools. As stated previously, sewer lines are commonly larger than 2 inches (5 cm) in diameter and the flow is low pressure. The flow may be stopped with towels, cribbing, and or shims. These methods require coordinated instruction with a utility company, equipment manufacturer, or other instructor who is a subject matter expert on water and sewer line breaks.

To mitigate a broken natural sewer line, first pump the sewage from trench by following the dewatering procedures in Skill Drill 5-2. Once the sewage is contained in a designated holding area, the broken pipe can be plugged or crimped.

Water Line Break. Water line breaks can contribute to trench collapse and the resulting accumulation of water in a trench can drown a trapped victim. Accessing and operating a main water line shut-off valve is best left to water department personnel. Depending on local SOGs, trained rescuers may be capable of crimping broken water lines to stop or minimize water flow and pump accumulated water out of the trench.

To mitigate a broken water line, all personnel in the hot zone should don PPE. Shut off the valve above the leaking pipe whenever possible. With an escape ladder and lip protection in place, use a crimping tool to stop the water flow. Implement dewatering operations, as described in Skill Drill 5-2, and keep already operating dewatering systems running.

Underground Electrical Wires

Over the past decade, there has been an increase in the use of buried electrical wire. Existing wires are often exposed when trenches are dug. Direct burial wire is insulated and if the wire is not cut or damaged, most often it can be avoided initially. The presence of current in the wire should be determined by a noncontact voltage tester) and controlled at a disconnect that is locked out and tagged out by trained personnel before rescuers are allowed to work in close proximity to the wire.

Gas

Depending on local SOGs and the training level of rescuers, to mitigate a broken natural gas line hazard all personnel in the hot zone must be in full structural firefighting PPE and SCBA, and ignition sources must be eliminated. Please note that the shut down of a curb valve, which controls the flow of gas from the main to the pipe, should only be done by utility experts. With an escape ladder and lip protection in place, a manned and charged hose line is positioned. The broken pipe is grounded by draping a clean, wet burlap cloth over both sides of pipe, and down to the ground. The leak is controlled with a crimping tool or shut-off tool.

Ventilation

Initial ventilation, when needed, at a trench rescue incident is usually accomplished with fire service smoke ejectors that are carried on the first-arriving fire apparatus. Smoke ejectors are designed to move air through large areas, but because they do not operate with ductwork, they cannot focus the airflow to specific areas (like the victim area). This type of ventilation is directed at controlling the overall atmospheric conditions within the trench and it is called trench ventilation.

Rescue ventilation, on the other hand, is directed at improving the atmospheric conditions in the specific area where the victim is trapped. This is accomplished with confined space ventilators using duct tubes to direct the flow of air directly to the victim's area. Confined space ventilation equipment is not likely to arrive with first-due fire companies, but it is often available on trench rescue and technical rescue apparatus. When used with an approved heating device, rescue ventilation can also be used to prevent hypothermia in victims trapped in cool, wet, or cold conditions. When operating in a recovery mode, ventilation should be in place and atmospheric hazards eliminated prior to entry.

Ventilation Concerns. When ventilating a trench during a rescue, several issues may arise. Ventilation efforts in cool and damp environments can create hypothermia. Patient considerations must be evaluated during ventilation operations by the medical personnel responding to the site. When trench ventilation techniques are used in cool and damp environments, medical personnel must monitor patients for signs and symptoms of hypothermia.

Recirculation of hazardous atmospheres can occur when blowers are placed on the downwind (leeward) side of the hazard. The hazardous atmosphere exiting the trench is blown (by the wind) back into the blower intake and then circulated back into the trench. Moving the blower to the (windward side) will eliminate recirculation problems **FIGURE 5-14**.

Short circuiting occurs when fresh air moves directly from inlet to the outlet, without circulating through other areas of the space. This will be evident with atmospheric monitor measurements that are not improving the trench conditions **FIGURE 5-15**.

Trench Ventilation. The purpose of trench ventilation is to improve the atmosphere of the entire trench, rather than just the victim's area. Trench ventilation can be accomplished with electric or battery-powered fire service smoke ejectors. In trenches that have atmospheric hazards, primary ventilation needs to take

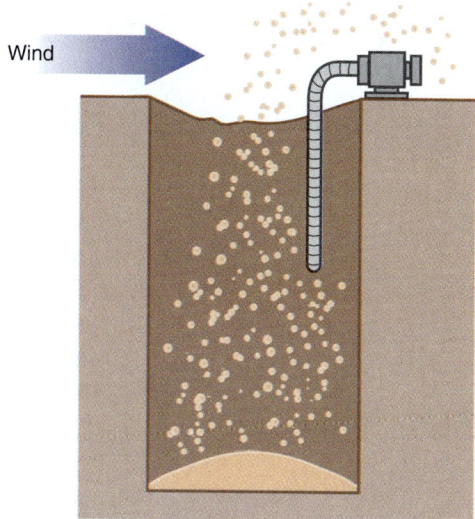

FIGURE 5-14 Recirculation.

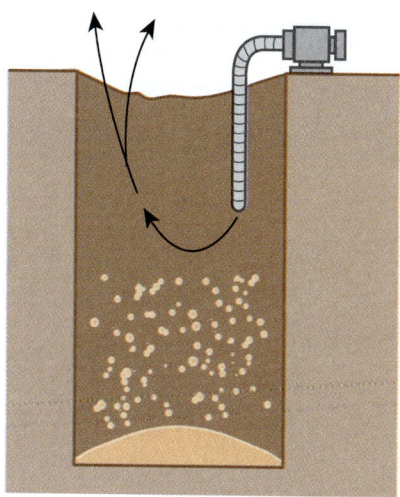

FIGURE 5-15 Short circuiting.

place before rescue entry operations begin. To supply ventilation (best for oxygen-deficient and flammable conditions), follow the steps in **SKILL DRILL 5-3**.

To exhaust ventilation (best for toxic atmospheric conditions), follow the steps in **SKILL DRILL 5-4**.

Rescue Ventilation

The purpose of rescue ventilation is to supply fresh air directly to the victim in the trench. Rescue ventilation is based on supply ventilation methods rather than exhaust ventilation methods. In this case, tubing is placed in the trench near the victims face and used to direct the supply ventilation to the victim. To provide rescue ventilation, follow the steps in **SKILL DRILL 5-5**.

Hazardous Materials

Common hazardous materials on a trench site include fuels and solvents. Rescuers trained at the hazardous materials operations level with proper PPE may control and/or remove gasoline, diesel fuel, hydraulic fluids, and solvent containers and spills per local SOGs. In the event of the presence of other hazardous materials, specially trained hazardous materials teams must be summoned.

Severe Environmental Conditions

Severe environmental conditions include heavy rain, high winds, cold or hot temperatures, snow or ice, and lightning. During heavy rainstorms, rain may be diverted away from the trench by the use of dikes and ramps. A tarp or tent may be set up over the rescue area of the trench, and water may be removed from the trench by the use of trash pumps and dewatering equipment. Rescuers should be protected with Gore-Tex type rain gear, helmets, eye protection, gloves, and water-resistant footwear.

To mitigate the interference of heavy wind, create a wind break if possible by positioning the rescue truck/trailer in a safe spot upwind of the trench. Cut off any trench panels that stick up more than 2 feet (0.6 m) above the trench lip to avoid the additional pressure created by wind hitting the large panels. Carry panels and ground pads as low to the ground as possible. For this type of weather, rescuers should be protected similarly to how they dress in heavy rain: Gore-Tex type wind breakers, helmets, eye protection, gloves, and safety footwear.

In cold temperatures, warm air from heated ventilation fans can be ducted into the trench, but rescuers will still need to be rotated out of these environments and periodically placed into warming huts. Rescuers should be protected with cold weather (insulated coverall type) ensembles, hoods, helmets, and insulated gloves and boots.

Cooling air may be circulated into the trench with ventilation equipment for rescues taking place in hot temperatures and, similarly to rescues taking place in cold temperatures, rescuers should be rotated out of hot environments and periodically placed into air-conditioned rehabilitation areas. Rescuers should be protected with light-colored cotton shirts, cotton pants, vented helmets, lightweight gloves, and breathable boots.

During snowy or icy conditions, the rescue area in the trench may be covered by tarps or tents. The snow and ice should be periodically removed (shoveled) from the trench, trench lip, and lip protection (ground pads and lip bridges). Rescuers should be protected

SKILL DRILL 5-3
Supply Ventilation (NFPA 1006: 12.2.2)

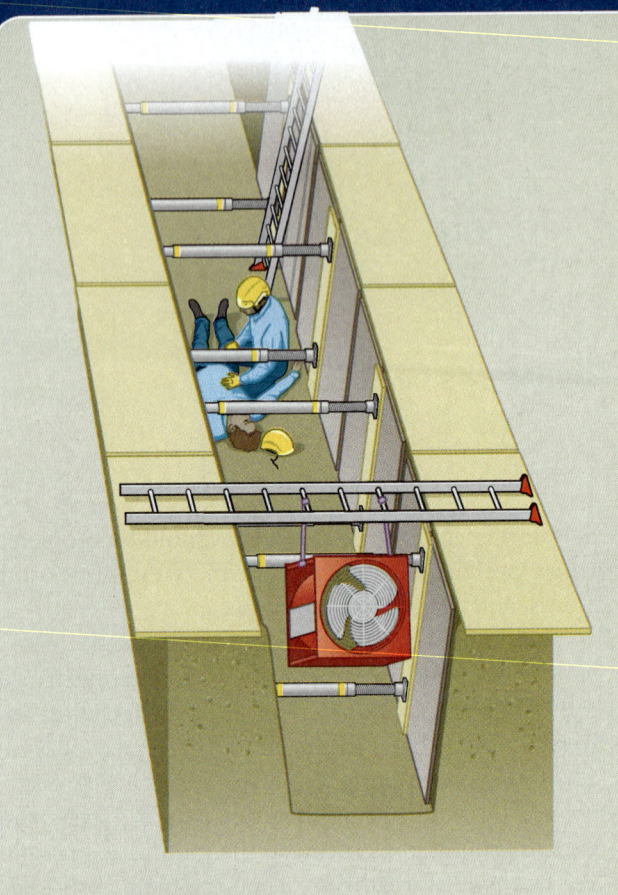

1 Use a 4 × 4 beam or ladder spanning across the trench to support the blower.

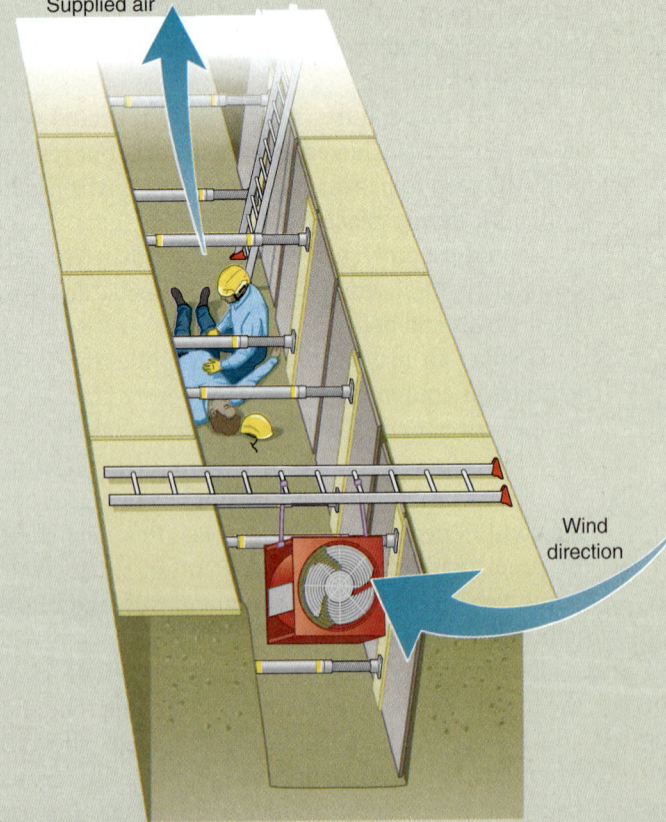

2 Attach webbing or a rope to the blower (intake side) to allow the blower to point downward. Attach heater and supply heated air at cold weather incidents.

SKILL DRILL 5-4
Exhaust Ventilation (NFPA 1006: 12.2.2)

1 Use a 4 × 4-inch beam or ladder spanning across the trench to support the blower. Attach webbing or a rope to the blower that is long enough to place the blower within 2 feet of the trench floor. Attach the webbing or rope at the exhaust side to allow the blower to point upward.

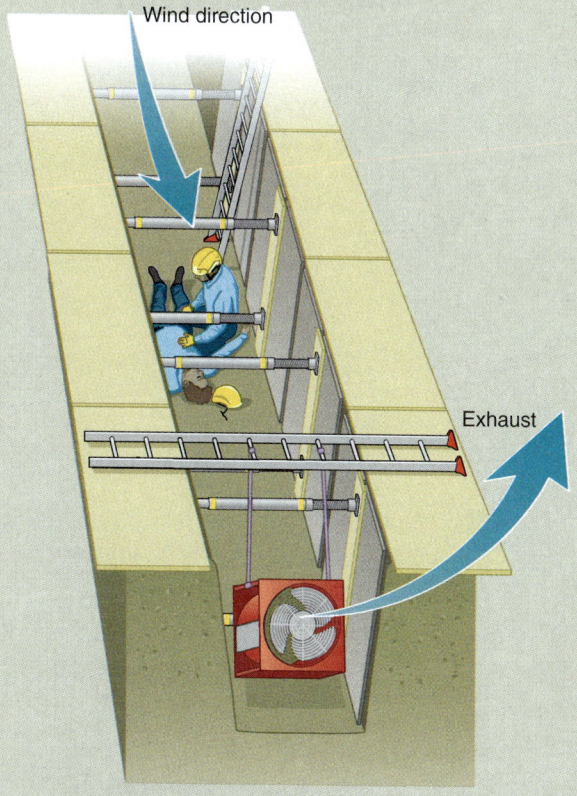

2 With an escape ladder and lip protection in place, position the blower on the downwind (leeward) side of the trench. Turn on the blower.

SKILL DRILL 5-5
Rescue Ventilation (NFPA 1006: 12.2.2)

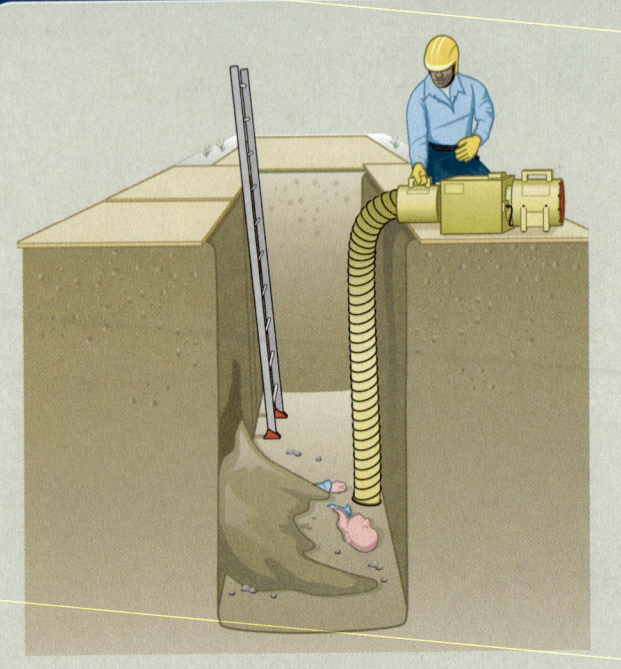

1 With an escape ladder and lip protection in place, position a blower on the trench lip (windward side).

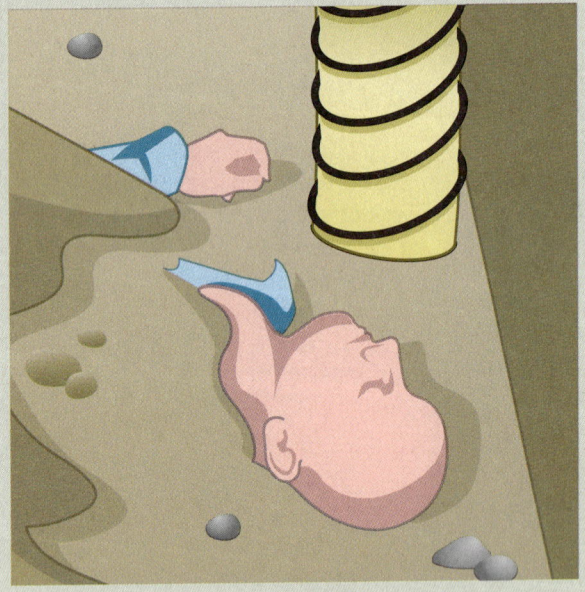

2 Use ductwork to position the flow of air as directly as possible to the victim's face.

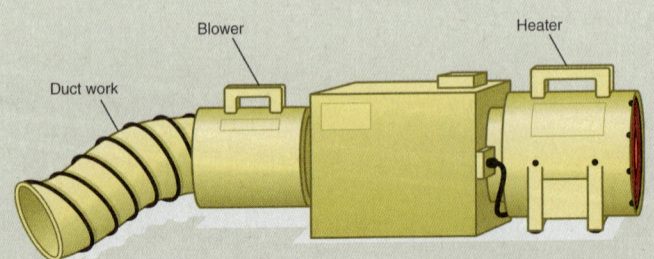

3 Ensure that a continuous flow of uncontaminated air is being delivered to the intake side of the blower. Attach heater and supply heated air.

Voices of Experience

Growing up in a plumbing and heating family exposed me to many different trenching experiences. House perimeters, parking lots, new construction, roadways, and swimming pools were among the many job sites in which I have worked. It was not until I became a firefighter, however, that I understood the dangers to which I had exposed myself. Working with lead, Portland cement, hydrochloric acid, glue, solvents, tar, gas power tools, pumps, and torches poses significant dangers to any worker in a trench. Naturally, ventilation was provided most of the time, but there was that one time—just one time—that we opted to forgo ventilation.

It was a warm June morning when we were replacing the main drain line on an in-ground pool. Polyvinyl chloride (PVC) and black poly pipe were the norm for this climate. At the bottom of the 8-foot (2.8-m) trench was the cracked pipe needing replacement. We cut out the old pipe, glued on a black poly adapter, and attempted to hook up the poly pipe. Not realizing the primer had been knocked over, I fired up the torch to soften the poly. The primer flashed and I suffered first- and second-degree burns on my hands, arms, and face. Conditions were right for this mishap, and it could have been worse. The only major injury was to my pride.

Another close call occurred when I was cutting storm sewer pipe with a gas-powered saw. The trench pit was only 7 × 7 × 7 feet (2.1 × 2.1 × 2.1 m) and it became filled with IDLH (immediately dangerous to life and health) carbon monoxide (CO) in about 5 minutes. We were using an atmospheric monitor but no one was assigned to monitor it. As a result of the noise from the power saw, the monitor's alarm could not be heard. When the saw shut down, I climbed out of the pit feeling a little light-headed and dizzy, only to find the air monitor alarming. The lessons learned in my plumbing career have been invaluable to my fire career.

Underground construction workers are required to monitor the air when hazardous substances are stored near the trench site, whenever the area may be oxygen deficient, and when digging in landfills. Absent those conditions, freshly dug trenches rarely contain hazardous atmospheres. Underground construction workers enter thousands of trenches every day without monitoring the atmospheres. It is usually humans (construction workers and rescuers) who bring hazardous atmospheric conditions into trenches. Rescuers must look for the tell-tale signs (indicators) of hazards, which include but are not limited to the following: gasoline/diesel-powered equipment/machinery, fuel cans, torches, broken utility lines, chemical storage, and geological data (radon). Construction workers down in a trench with no apparent trauma or workers feeling ill, are indicators of possible hazardous atmospheric conditions. We need to recognize and prepare for all of these possibilities. When a cave-in occurs, the chance of hazardous atmospheric conditions (e.g., broken gas/sewer lines, fuel or solvents on the lip now spilled into the trench) increases. Initial air readings should be part of every rescue safety plan. At a trench rescue, a complete hazard assessment is always necessary.

Which type of air monitoring is necessary? Most trench rescue operations will simply need a four-gas monitor to evaluate atmospheric conditions. That situation changes if a specific toxic substance is stored near the trench site, in which case a monitor for that toxin is required. In most cases, a photo ionization detector (PID) is the monitor of choice. The PID will measure O_2, flammability, CO, hydrogen sulfide, and volatile organic compounds (VOCs). Equally important are

Voices of Experience

the questions of how long and how frequently to monitor. The time and frequency will depend on the findings from your initial hazard assessment. If toxins are indicated, continuous monitoring becomes a requirement. A qualified person should be assigned to the air monitor, with results being recorded every 5 to 10 minutes. This assignment uses manpower that could be used for other needed rescue functions, but when atmospheric hazards are identified, this function is required. It is simply a function of risk versus gain analysis.

When a trench is initially monitored and nothing is noted, I look at the probability of types of toxins that might enter the trench. Are they lighter or heavier than air? Are they flammable? Then I choose a location in which to place a (manned or unmanned) monitor. If there is a lot of noise in and around the trench, make sure the monitor is where someone can hear the audible alarm.

Lessons Learned
- Always conduct a complete hazard assessment that includes use of a four-gas monitor.
- Obtain an appropriate monitor for specific hazardous substances that are stored near the trench site.
- When possible, have a support group (such as a hazardous materials team) automatically respond to trench rescue incidents to provide continuous air monitoring.
- Develop action guidelines that spell out how your team will engage when (1) hazardous substances are identified, (2) indicators of hazardous substances are present but initial monitoring does not identify current hazards, and (3) both the initial monitor reading and the assessment of hazardous substance indicators demonstrate safe conditions.

Matthew F. Ratliff
Chief of Training, Sterling Heights Fire Department
Hazmat Team Manager, MITF-1
Medical Specialist, Macomb County Technical Rescue
Planning Officer, Macomb County IMT

by water-repellent cold-weather ensembles, helmets, water-repellent insulated gloves, and boots with cleats. Any time that lightning strikes are present in the area and the incident has been declared a recovery rather than a rescue, rescuers should leave the area and only return after the lightning has stopped.

Enhanced Hazard Mitigation

Some hazards may require the direct technical assistance of a utility company emergency response team or hazardous material response teams. To minimize the delay in response time, enhanced hazard mitigation resources (or utility emergency response teams) should be dispatched on receipt of a reported trench/excavation emergency. Tier 1 resources are those that need to be immediately dispatched upon receipt of a reported trench or excavation emergency. (See Chapter 3, *Initial Actions*, for further details.) Resources that are less commonly needed at a trench emergency are listed under Tier 2 resources. Tier 2 resources should be requested immediately after the need is recognized.

Ongoing Hazard Mitigation

A technical rescue safety officer must be designated and is responsible for recognizing and mitigating hazards throughout the incident. Initial hazard and enhanced hazard mitigation efforts must be monitored and evaluated, and the development of additional hazards must be identified and mitigated throughout the incident.

After-Action

IN SUMMARY

- Hazards that are common to the trench rescue environment include: collapse (cave-in), underground utilities, hazardous atmospheres, traffic, hazardous materials, physical hazards, and biological hazards.
- The most commonly encountered hazard at a trench emergency is soil failure and the collapse of trench walls.
- A hazard control plan addresses the safety of the trapped victims, bystanders, and all responders and must begin with the identification of all hazards and is followed by hazard mitigation operations.
- At the least, hazard control equipment will include a atmospheric monitor and a noncontact voltage tester.
- Initial ventilation is accomplished using smoke ejectors, whereas rescue ventilation is performed with confined space ventilators.
- To minimize the delay in response time, enhanced hazard mitigation resources (utility emergency response teams) should be dispatched on receipt of a reported trench/excavation emergency.
- A technical rescue safety officer must be designated and is responsible for recognizing and mitigating hazards throughout the incident.

KEY TERMS

Atmospheric contaminate The accumulation of substances in the air in quantities that are large enough to produce harmful effects.

Confined space ventilator Fan that uses duct tubing to send air to a specific location; can be used with heating elements.

Dewatering devices Devices that include well points and pumping systems, that are typically installed by construction crews, and portable pumps (such as trash pumps and mud pumps); used by trench rescue teams.

Electric-powered fire department smoke ejector Fan used to provide an adequate flow of fresh air into a trench and hazardous atmosphere out of the trench.

Hazard control plan A plan developed to address the safety of the trapped victims, bystanders, and all responders by identifying and mitigating all hazards.

On Scene

1. What are the most common hazards associated with trench and excavation sites?

2. Describe a hazard mitigation technique that would be needed for a broken sanitary sewer line with contaminated (sewage) water in the trench.

3. When would trench ventilation be appropriate?

CHAPTER 6

Operations Level

Managing the Trench Incident

KNOWLEDGE OBJECTIVES

- Describe an extended command structure for trench rescue.
- Identify the common teams utilized at trench rescues.
- Describe common incident management tools.
- Identify tasks that must be completed during demobilization. (**NFPA 1006: 12.2.8**, pp. 121, 125)
- Identify considerations for terminating command. (**NFPA 1006: 12.2.8**, pp. 121, 125)
- Identify steps necessary to return all resources to a ready state. (**NFPA 1006: 12.2.8**, pp. 121, 125)
- Explain documentation considerations associated with termination an incident. (**NFPA 1006: 12.2.8**, pp. 121, 125)
- Describe how to clean and service trench rescue equipment. (**NFPA 1006: 12.2.7**, p. 125)

SKILL OBJECTIVES

- Terminate a trench rescue operation. (**NFPA 1006: 12.2.8**, pp. 121, 125)
- Conduct a post-event analysis. (**NFPA 1006: 12.2.8**, pp. 121, 125)
- Clean and service trench rescue equipment. (**NFPA 1006: 12.2.7**, p. 125)

ADDITIONAL RESOURCES

- National Incident Management System, Federal Emergency Management Agency
- **NFPA 1561**, *Standard on Emergency Services Incident Management System and Command Safety*
- **NFPA 1670**, *Standard on Operations and Training for Technical Search and Rescue Incidents*

You Are the Rescuer

You are an operations level rescuer with considerable training. As you are en route to a call, your memory takes you back to the only actual trench rescue incident you have under your belt. In that incident, a small construction company had excavated ground around a house to install waterproofing material on the basement walls. The straight trench was 2 feet (0.6 m) wide and 8 feet (2.4 m) deep, and the worker was buried to his waist.

Your fire department initially protected the victim with panels and a couple of pneumatic struts. While a second set of panels and struts was being installed, a shovel was lowered to the victim and he began digging himself out. After the second set of panels was shored, a rescuer entered the trench with another shovel and completed digging the victim out before a third set of panels was installed.

Twelve fire/rescue personnel completed the rescue in less than 50 minutes. The uninjured worker was transported (via basic life support transport) to the hospital for observation and was later released. The incident command structure at this incident was minimal and appropriate for this basic operations level incident.

Now you come back to the reality of the current incident. This intersecting trench is 16 feet (4.9 m) deep, and three workers are trapped. Two of the workers are buried to their chests but are conscious and communicating with rescue personnel. The third worker is buried with only his left hand visible. One corner of the T-trench has collapsed and presents a huge void. The other corner is still in place but shows fissures and sloughing, which are signs of impending collapse. Your fire department's trench trailer is not equipped to shore a deep intersecting trench with a large collapsed corner. Your 12-member team may be able to provide some protection (primary shoring) to the victims, but will not be able to stabilize this trench for rescuer entry and victim extrication. This scenario is clearly not the golden hour rescue that you responded to the last time. The current incident will require more equipment, more personnel, outside resources, and a much more robust command structure.

1. Which outside resources should be called in?
2. Which incident command system (ICS) positions will be needed?
3. Which additional fire, rescue, and medical resources would be appropriate for this situation?

Access Navigate for more practice activities.

Introduction

The incident command system (ICS) was developed in the 1970s to deal with interagency responses to large-scale fire incidents. The ICS was later developed as part of the National Incident Management System (NIMS) of the Federal Emergency Management Agency's (FEMA) National Response Plan. Today the incident command system (ICS) is used by most fire departments to handle local emergency incidents. The principles and practices of ICS are appropriate for trench rescue incidents. Incident command systems must address emergency situations at the strategic, tactical, and task levels. Someone needs to be in charge, and other personnel need to follow directions. Having a clearly defined approach to incident scene responsibilities and authority is critical to the safety of both victims and rescuers. Dividing trench scene duties and responsibilities helps the incident commander (IC) to implement a systematic method to handle a problem that could otherwise quickly overwhelm even the most effective and experienced officers. It also decreases the organizational **span of control** and provides a measure of needed on-scene accountability.

The ICS used for trench incidents must be flexible, expandable, and capable of managing and directing a variety of resources. Chapter 3, *Initial Actions*, provides information on the resources needed to resolve trench rescue incidents. Chapter 4, *Personal Protective Equipment and Equipment Basics*, and Chapter 5, *Hazard Mitigation*, covered the equipment needed for a trench emergency response. In this chapter, we focus on expanding the ICS and how an incident is managed and terminated.

Size-Up

A scene size-up is required to determine what *has* happened and to help predict what *will* happen. (This is discussed further in Chapter 3, *Initial Actions*.) The scene size-up provides situational awareness that allows rescuers to make informed decisions and take appropriate actions. A size-up begins with an assessment process that gathers information about the important factors of each incident: situation assessment, hazard assessment, victim assessment, trench assessment, and resource assessment. With that information in hand, the incident commander can develop a solution. The assessment process begins with information taken by the emergency dispatcher and continues with information gathered at the scene during the initial actions.

The most likely situation at a trench or exavation scene is a collapse with one or more victims trapped. Other common situations include a person who is unable to exit the trench because of a traumatic injury (i.e., hit by an object), a medical condition, a hazardous atmosphere, or electrocution. Determining the number of victims will help determine the complexity of the incident and the resources requried. An assessment of the victim's condition is essential to determining an accurate risk versus gain analysis. It is vital to determine the level of rescuers required and the resources needed as swiftly as possible.

> **TIP**
>
> To determine the level of rescuers required to resolve the incident, an assessment of the depth, width, shape, and condition (presence and degree of collapse) of the trench is critical.

Size-Up and Expanding the Incident Command Structure

The initial command structure begins with the arrival of the authority having jurisdiction (AHJ). In most cases, the AHJ is the local fire department. The highest-ranking chief or company officer will initiate the ICS. The initial command structure at a trench rescue must provide for certain functions, including command, hazard control, scene control, and if possible, non-entry rescue, and medical care. These functions are commonly managed and supervised by fire department company officers trained at the awareness or higher levels of trench rescue.

The expanded command structure enhances and increases the initial command structure. The highest-ranking chief or company officer will likely continue as the IC but will likely designate operations to a trench rescue team manager or other trench rescue subject matter expert. The expanded command structure at a trench rescue must provide for the command, operations, safety, logistics, and liaison. For complex incidents requiring specialist rescuers, a finance and planning branch is likely to be added to the command structure **FIGURE 6-1**.

Command

The IC is responsible for developing the strategic goals for the operation based on the initial scene size-up and subsequent incident developments **FIGURE 6-2**. (This is discussed further in Chapter 3: *Initial Actions*.) The IC is ultimately responsible for determining the need for and arranging the acquisition of all resources necessary to handle the incident. For example, the IC may develop a strategy specifying that the victim of a collapse in an extreme trench condition will be recovered using commercial techniques, as provided by John Doe Construction Company. The IC would then call for the resources necessary to fulfill the strategic goals. The fulfillment of those goals will rest with the **operations officer**.

The **staging officer** is responsible for positioning and accounting for resources that are not immediately assigned and is located at the primary staging area, where unassigned equipment and personnel stay. If multiple staging areas are needed, the operations officer will assign a manager to each staging area to assist

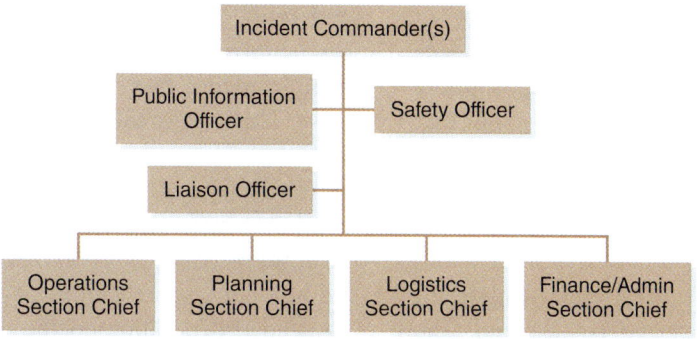

FIGURE 6-1 Incident command chart.

FIGURE 6-2 The incident commander (IC) is the head of command.
© Jones and Bartlett Publishers. Courtesy of MIEMSS.

FIGURE 6-3 The staging officer is in charge of resources that are waiting to be used.
Courtesy of Ron Zawlocki.

the staging officer. The staging area manager checks in all incoming resources, dispatches resources at the operations officer's request, and requests logistics section support, as necessary, for the resources located in their staging area **FIGURE 6-3**.

The **safety officer** monitors incident operations and advises incident command on all matters relating to operational safety, including the health and safety of emergency responder personnel. This function needs to be filled by someone who not only can spot unsafe acts, but also has the ability to anticipate activities that might lead to an accident. Although everyone on the scene is responsible for their own safety, it is necessary for someone to control the big picture. It is critical that the safety officer be familiar with the environment and its potential hazards. This is an extremely important position in a trench emergency and, consequently, the safety officer can halt the rescue effort for any safety-related reason at any time.

The ultimate responsibility for the safe conduct of incident management operations rests with the IC and supervisors at all levels of ICS. In turn, the safety officer is responsible for developing an incident safety plan. This plan includes the set of systems and procedures necessary to ensure ongoing assessment of hazardous environments, coordination of multiagency safety efforts, and implementation of measures to promote emergency management/incident personnel safety, as well as the general safety of incident operations.

At a trench rescue incident, a **technical rescue safety officer** should be appointed to work directly with the safety officer. The trench rescue team should provide a person with technical rescue safety officer qualifications.

The **liaison officer** establishes and maintains communications with governmental agencies, nongovernmental organizations, and private sector groups and provides them with organizational policies and resource needs and availability.

Trench rescues are events that draw news media attention. It is not unusual for the news media to arrive at the scene before some of the dispatched resources. Although a **public information officer (PIO)** is not going to be one of the first assignments given, it is an important function that should be filled within the first hour of the event. The PIO is responsible for disseminating accurate information about the incident's cause, size, situation, and resources being committed.

Logistics

The **logistics officer** manages all of the support needs for the incident, such as ordering resources and providing facilities, transportation, supplies, equipment maintenance and fuel, food service, communications, and medical services for incident personnel. The logistics officer position at a trench rescue incident is commonly filled by a command or company officer from the local fire department. Most trench rescue units carry only enough rescue shoring (panels, struts, wales, and associated support equipment) to stabilize an operations level trench incident. For deeper, wider, and more complex trench configurations, additional rescue shoring equipment will be required. In addition, technician level trench incidents often take hours to resolve. This creates the need for food, water, shelter, and work relief (i.e., more rescuers). Some body recoveries at trenches that are deeper, wider, and with extreme soil conditions will require the use of construction shoring techniques (e.g., trench boxes, sheet piling, sloping). Additionally, scene lighting and power sources may be needed depending on the incident conditions and duration. The logistics officer must forecast these additional requirements long before the resources are needed to ensure their timely delivery. A trench rescue team logistics officer needs to work closely with the logistics officer to keep them informed of additional rescue-related equipment needs, in addition to keeping a list of expendable and damaged rescue equipment.

The rescue team logistics officer (RTLO) is responsible for the trench rescue team equipment. Responsibilities include equipment accountability and field repairs. It is vitally important to keep all equipment not currently in use at a predetermined location. That way, the RTLO can keep track of it and determine its availability at any given time during the emergency. The RTLO must be able to predict additional equipment needs and request the needed equipment through the logistics officer.

The support team officer works at the direction of the logistics officer through the RTLO. The support team is often composed of firefighters trained at the awareness level. Assigning a member of the trench rescue team to the position of support team officer to manage and supervise the support team will pay dividends as the incident progresses. Support team duties may include transporting/carrying shoring and rescue equipment from vehicles to designated staging sites, developing a cut station, assisting with equipment maintenance, providing and maintaining power (generators) and scene lighting, and ongoing atmospheric monitoring and ventilation as needed, depending on local SOGs, and the training level of team members.

TIP

After responding to dozens of trench rescues and conducting over 100 tests on trench rescue task activities, the authors have concluded that a trench rescue team can accomplish tasks efficiently with a minimum of 18 trained personnel.

Operations

Most of the operations tasks are performed by the trench rescue team. Trench rescue teams may be broken down into sub-teams: entry, panel, and shoring. Upon arrival at a trench rescue incident, the team members may be given one of the following assignments FIGURE 6-4:

- Rescue team manager
- Technical rescue safety officer
- Entry team officer
- Entry team specialists
- Panel team officer
- Panel team specialists

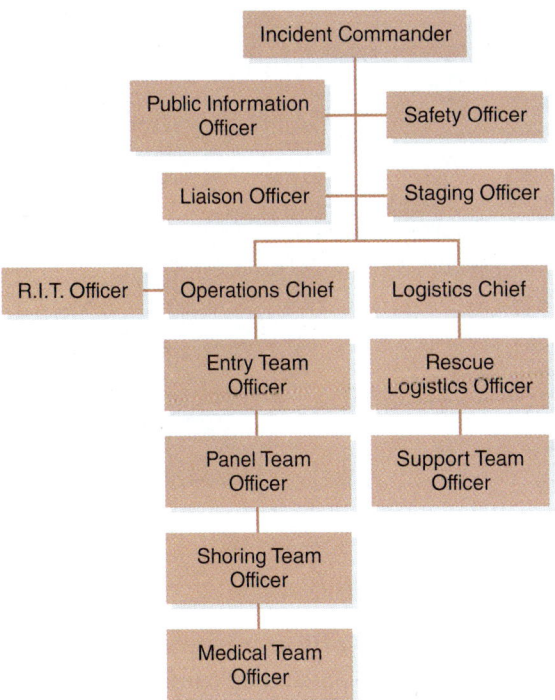

FIGURE 6-4 ICS structure with an expanded Operations section.

- Shoring team officer
- Shoring team specialists
- Rescue team logistics officer
- Support team officer

The rescue team manager manages and supervises the activities of the trench rescue team. It is common for the IC to designate the rescue team manager as the operations officer. When that is not the case, the rescue team manager works closely with the operations officer to develop and implement the incident action plan.

The entry team officer manages and supervises the entry team tasks, including non-entry and entry operations. The entry team enhances the initial actions performed by awareness level rescuers and develops and implements a hazard control plan, which is followed during all entry operations. Entry operations include victim access, treatment, packaging, extrication, and removal. The hazard control plan is covered in detail in Chapter 3, *Initial Actions*.

TIP

The authors recommend that the entry team include at least two trench rescue technicians with emergency medical technician (EMT) or advanced emergency medical technician (AEMT) certification. They recommend having entry team members trained at the Urban Search and Rescue (USAR) medical specialist level.

FIGURE 6-5 The panel team installing a lip bridge.

FIGURE 6-6 The shoring team.
Courtesy of Greg Payeur.

The **panel team officer** manages and supervises the **panel team** tasks, including conducting a shoring size-up (assessing lip and void conditions), lip protection installation, panel placement, backfill, and water placement. The panel team is required to set up, carry, and install all shields or panels **FIGURE 6-5**.

The **shoring team officer** manages and supervises the **shoring team** tasks. These tasks include a shoring size-up (trench depth, width, and soil failure L points); positioning ladders for entry and egress; shoring; and the installation of struts. The shoring team assembles and installs struts required to make the protective system safe **FIGURE 6-6**. This work may involve the installation of wood, pneumatic, or another type of strut shores.

The **rapid intervention team (RIT)** must be established and staged when rescuers enter the trench. (The RIT is discussed further in Chapter 5, *Hazard Mitigation*.) The RIT comprises members from the trench rescue team(s) on the scene and is responsible for the rescue of trench rescue team members, but not victims.

The **medical team officer** works at the direction of the operations officer. This member is responsible for establishing a medical control area to treat any on-scene rescuer injury and to provide for victim care and transportation. The medical team officer position is typically assigned to the highest-ranking local emergency medical services provider. They will know local medical protocols and have communications with the emergency room or trauma centers. The medical team officer will follow local protocols and medical direction and will supervise and manage patient care though the medical team. Additional duties of the medical team may include assisting the **rehabilitation** area with monitoring of rescuers' vital signs.

Incident Management Tools

When a trench rescue incident cannot be resolved by awareness level rescuers, an increasing number of a variety of skilled personnel and specialized equipment are required. Checklists and worksheets can be a big help to the incident command team running a trench emergency incident. Resource lists are a management tool that can be used at all levels of incidents. Ideally, resource lists would be available to all team members, from dispatch personnel to command officers. The task assignment and personnel tracking worksheets are helpful management tools at operations and technician level incidents.

The hazard control plan checklist reminds the IC of the most common hazards at a trench incident **FIGURE 6-7**. Chapter 3, *Initial Actions*, and Chapter 5, *Hazard Mitigation*, cover potential hazards in detail.

The magnitude and scope of a trench incident directly impacts the number and types of resources that will be required. Chapter 3, *Initial Actions*, details the resources that are commonly needed at trench rescue incidents. Agencies that respond to trench incidents should contact local resources that can provide the services listed in the resources list and develop

Hazard Checklist

Hazard Identification	Hazard Location	Control–Actions taken
Collapse ☐ Fissures ☐ Undercut area ☐ Bulging ☐ Raveling/sloughing ☐ Surcharge loads ☐ Other		
Traffic ☐ Road ☐ Heavy equipment		
Utilities ☐ Natural gas ☐ Electric ☐ Sewer ☐ Water ☐ Other		
Atmospheric ☐ CO >35 ppm ☐ H2S >10 ppm ☐ O2 <19.5% ☐ Flammable >10% ☐ Other		
Hazardous Materials ☐ Gasoline ☐ Diesel fuel ☐ Fluids ☐ Solvents ☐ Other		
Physical Hazards ☐ Trip hazards ☐ Sharp/Hot items ☐ Other		
Severe Environment ☐ Rain/Snow/Ice ☐ Lightning ☐ Hot temperature ☐ Cold temperature ☐ High winds ☐ Other		
Biological ☐ Bloodborne ☐ Airborne ☐ Other		

FIGURE 6-7 Hazard control plan checklist.

Trench Rescue Resource List

Tier One	Phone Number	Contact Person
Fire Department		
Police Department		
EMS		
Trench Rescue Team		
Vacuum Truck		
Recovery Bags		
Tier Two	**Phone Number**	**Contact Person**
Civil Engineer		
Excavator/Operator		
Haz Mat Team		
Rehab Unit		
Confined Space Rescue Team		
USAR Search Team		

FIGURE 6-8 Resource list.

agreements with them before adding them to their local resource list **FIGURE 6-8**. The phone numbers and contact information must be available 24/7.

Incident Termination

The incident termination phase of a trench rescue can be a dangerous time. It is at this point in the process that the adrenaline is gone, and personnel are tired. After the victim is removed, it is important to stop all activities and call everyone out of the hot zone to begin the termination process.

No part of termination procedures should begin, nor should any responders leave the scene before a postincident briefing with the IC is conducted. This is the first step in the debriefing process and summarizes the operation, answers questions, gauges the welfare of the rescuer, and offers the chance for the incident leaders to express their appreciation to the rescuers. It also is the time to tell all personnel that, from this point forward, safety is the most important consideration.

Special care needs to be taken to ensure that rescuers are rested and alert before breakdown begins. Fresh crews may be brought in to facilitate the termination and breakdown of the equipment. This phase should not be rushed. Personnel must take frequent breaks and keep in mind that the emergency is not over until overhaul of the incident is complete.

As a rule, the termination and breakdown of the protective system happen in the reverse order from how the system was built. This process includes removing unneeded equipment from the hot zone and disassembling and removing shoring equipment. In every case, while personnel are removing the protective system, they must operate from within a safe area of the protective system or remove component parts from outside the trench area, and they should continue to use any necessary PPE. Chapter 7, *Operations Level Trench Rescue Shoring*, covers the breakdown of shoring in detail.

A word of caution is necessary because this stage of the event is when rescuers need to keep an eye on each other. Rescuers should remain focused on the fact that the job is not successfully completed until the incident is terminated, and no personnel have been hurt. Clear and specific assignments should be given to each team, and supervision and accountability must be maintained through the team officers.

Voices of Experience

At about 5:15 PM, our county technical rescue team was requested on a mutual aid assignment about 40 miles (64 km) away in Morristown, New Jersey. An ongoing trench rescue operation had been in progress there since about 1:00 PM. We arrived on the scene about an hour later and were briefed about the incident.

A contractor and his crew had been working on a basement drainage project in a suburban neighborhood. The project involved installing drainage pipe around the exterior of the basement and then running the pipe out to a creek behind the home. The issue was that the backyard sloped up away from the house before it sloped back down toward the creek. Essentially, there was a small hill between the home and the creek. In order to achieve the proper pitch for the drainage, the contractor needed to dig deep through this hill to reach the creek. The contractor apparently could not dig deep enough with his rubber tire backhoe, so he dug the trench wide enough at the top to actually get the machine partially in the trench, and then used the extenda-hoe to dig further down. The sandy spoil from the trench had been piled up anywhere it could fit, including right up to the lip of the trench.

At lunchtime, the contractor (victim) sent his crew out to pick up lunch while he continued to work with the backhoe. When the crew returned, they found the machine running, but could not find their boss. They looked into the end of the trench and saw that a large portion of one of the walls and adjacent spoil pile had collapsed into the trench. Figuring they could dig the boss out, they tried using the backhoe. After several buckets of soil were removed, they accidentally struck the victim with the bucket, causing a severe head injury. The crew then called 911 to report the incident. While it was unlikely that the boss had survived the initial collapse being completely buried under about 3 feet (0.9 m) of soil, contact with the bucket ensured that it was now a recovery operation.

The local fire department and the initial mutual aid resources had focused on trying to create a safe area around the trench. A huge amount of spoil had to be moved by hand in order to create an area in which to work above the trench. Because of the collapse of the one wall, the trench in the area of the collapse was nearly 10 feet (3 m) wide at the lip. The sloughed-in area measured about 10 to 12 feet (3–3.7 m) long, with the victim buried directly beneath that area, with his head about 12 feet (3.7 m) below the original grade. At this point, the only portion of the victim visible was the top of his head. The local trench rescue team had set several panels and had begun to shore the trench, but the large void created by the collapse and width of the trench was making it difficult.

We briefed our personnel and integrated into the established command and accountability system. We began to assist in the shoring effort. Additional panels were positioned in the trench along with outside wales to bridge the collapse and support the panels. Five low pressure cushions were positioned in the void and inflated to help fill the void. Additional Paratech struts were added to supplement the shores already in place, due to the strut length and the soil conditions. Once we were comfortable with the initial shored area in the trench, personnel entered the trench to begin to assess just how badly the victim was trapped.

After a few minutes of digging, it was evident that the victim was actually in a standing position, and approximately 6 feet (1.8 m) of digging would be required. As personnel removed soil from around the victim, it was placed into buckets and removed from the trench. As soon as any progress was made, the sandy soil would slide back in and partially refill the hole. Sand was also

Voices of Experience

coming out from beneath the panels. It was clear that supplemental shoring would be needed, long before we got down to 2 feet (0.6 m) below our lowest shores. Supplemental panels (4-foot [1.2 m] wide by 2-foot [0.6 m]) high were fabricated out of double-thickness 2 × 12-inch boards. The panels were slipped in below the full-size panels as the digging progressed. This helped contain the loose soil from sliding into the trench and from potentially undermining the shoring system. The supplemental panels were used against the walls, and the rescuers doing the digging sloped the ends of the trench floor to minimize soil running in from the ends as they dug.

Two separate levels of these 2 × 4-foot (0.6 × 1.2-m) supplemental panels were needed below the original full-size panels, each with their own set of struts. The victim was now exposed to about the mid-thigh/knee level. Attempts to pull his legs free were unsuccessful. It was at that point that one of the members thought out loud that "what we really need is a post-hole digger." As luck would have it, one of the agencies had a manual post-hole digger on their trailer. It was retrieved and lowered in to the rescuers in the trench. The personnel used it to dig down directly alongside the victim's legs. Once that area was cleared, the space between the victim's legs was excavated to create a small void down to his feet. A shovel handle pinned alongside his leg needed to be cut with a reciprocating saw. Once the handle was clear and the void expanded, it was possible to wiggle the victim's feet free from the soil. Once out of the soil, the victim was packaged and removed from the trench in a basket stretcher.

It was now about 1:30 AM and our crews had been working for about 7 hours. The air temperature had dropped about 20 degrees and it had started to rain. Our team discussed the situation, and given the circumstance and conditions, we advised the IC that we would leave all of the shoring in place at that point. We recommended that a trackhoe be brought to the site in the morning and used to cut back the trench walls to a safe angle. We advised him that the equipment would not be damaged, and to just let it fall. As soon as the trench was cut back, we would retrieve the equipment, as the risk to our personnel was not justified to recover the equipment at that moment. The IC agreed, and our county units cleared from the scene around 3:00 AM.

Lessons Learned

- This incident occurred in 1999—The Paratech Long Shore System (gold struts) were not available yet. If the same conditions existed today, Paratech gold struts would be used to shore the widest areas of the trench to increase our shoring safety factors.
- As the shoring system progressed down toward the victim, we reduced the vertical spacing of the struts due to the required strut lengths and the soil conditions. We estimate that the victim was somewhere between 17 and 18 feet (5.2 and 5.5 m) deep to his feet. The technicians in the trench certainly did not mind the mental value of the additional struts, and the positioning did not hamper the digging operation.
- The use of a vacuum system and air knife would have greatly sped up the soil removal operation (we now have both items in our cache.) Removing soil and debris by the bucketful is slow and ties up personnel.
- The use of more recently developed void shoring techniques like back shores and buttress shores might have worked well for this situation. These techniques are

Voices of Experience

now part of available tactics for void management.

- The use of prefabricated 2 × 4-foot (0.6 × 1.2-m) supplemental shoring panels would have made the digging operation and supplemental shoring much easier and faster. Additionally, the prefabricated FinForm/strongback supplemental panels are stronger and easier to position than the double 2 × 12 panels we fabricated onsite. (We have since added several supplemental shoring panels to our cache.)
- When dealing with an unconscious (or deceased) victim, they need to be packaged as they are uncovered, especially if they are in a standing position; otherwise, the victim slumps down and interferes with the digging operation. In this case, an improvised webbing harness and simple rope system were used to support the victim as he was extricated.
- The decision once the victim was recovered to leave the shoring in place was critical to the safety of our personnel. After a long operation and given the environmental issues, there was no reason to risk the safety of our personnel to recover equipment.
- Cooperation between personnel from two different counties and multiple agencies was good—people need to be flexible and adapt to equipment and procedural differences as long as the operation is conducted safely. In this case, good communication between those in charge quickly smoothed over any procedural issues between the agencies.

Gary Breuer
Rescue Captain, Flemington–Raritan Rescue Squad
Hunterdon County, New Jersey
Task Force Leader, New Jersey US&R Task Force

Consideration may also need to be given to rotating crew assignments; for example, personnel who had outside-of-trench assignments during the operation might be given inside-of-trench assignments during termination and breakdown.

During the termination phase, the entry, panel, and shoring teams are responsible for gathering, inventorying, and restoring all of their equipment. Because trench rescue is set in the earth, the equipment will be dirty; it must be spotlessly clean before the incident is considered closed. Follow all manufacturers' recommendations for cleaning and inspecting equipment. Equipment that is damaged should be documented and taken out of service for repair.

TIP
All rescuer injuries must be documented.

The last part of the operation will involve after-action meetings, such as a postincident analysis and possibly a **critical incident stress debriefing (CISD)**. The postincident analysis, which involves all of the rescuers who participated in the rescue and sometimes occurs days after the incident, is important because it gives everyone on the team an opportunity to evaluate system performance and make adjustments so that the service delivery system continually improves. The key is to conduct the critique soon enough after the event so that specific aspects of the rescue are still fresh in rescuers' minds and before rescuers have moved on to other incidents.

CISD is the part of the event that allows rescuers to defuse and gear up for the next response. Just remember that the way something affects you is not always the way it affects someone else. Everyone's tolerance for critical events is different. With a significant trench collapse event, CISD should be offered to all personnel who participated in the rescue.

After-Action

IN SUMMARY

- Today, the ICS is used by most fire departments to handle local emergency incidents.
- The initial command structure at a trench rescue must provide for certain functions, including command, hazard control (identification and defensive mitigation operations), scene control (zoning and staging), and if possible, non-entry rescue, and medical.
- When a trench emergency requires resources with specialized equipment and specially trained personnel, the ICS must expand and address the incident at strategic, tactical, and task levels.
- The hazard control plan checklist and resource lists can be used to help manage an incident.
- No part of termination procedures should begin, nor should any responders leave the scene before a postincident briefing is conducted.
- The after-action meeting will include a postincident analysis to determine what worked and what needs work.

KEY TERMS

Critical incident stress debriefing (CISD) Process used to debrief rescue personnel after an emotionally charged incident.

Entry team The group of rescuers responsible for hazard identification and mitigation, initial patient care, extrication and removal from the trench.

Entry Team Officer The individual responsible for supervising and managing the entry team assignments.

Hazard control plan A plan that addresses the safety of the trapped victims, bystanders, and all responders.

Liaison officer Individual responsible for providing information and gathering information from organizations and people outside of the immediate rescue effort.

Logistics officer Individual responsible for obtaining the appropriate equipment and personnel for deployment by the operations officer.

Medical officer Individual responsible for establishing a medical control area to treat any on-scene rescuer injuries and to provide for victim care as necessary.

Operations officer Individual responsible for overall coordination of the rescue effort and the implementation of the tactical decisions that will make the incident commander's strategy successful.

Panel team The group of rescuers responsible for installing lip protection, panels, wales, and backfill.

Panel Team officer The individual responsible for supervising and managing the entry team assignments.

Public information officer (PIO) Individual who provides information to the media regarding incident activities.

Rapid intervention team (RIT) Team established and equipped to handle everything from intervention for a secondary collapse or a medical emergency involving a member of the rescue team.

Rehabilitation An area that provides for rescue personnel rotation to address medical monitoring and fluid replacement needs.

Rescue team logistics officer (RTLO) An individual responsible for tracking and determining the availability of trench rescue team equipment during the emergency.

Rescue team manager (RTM) A member of the trench rescue team who manages and supervises the activities of the trench rescue team.

Safety officer Individual responsible for all aspects of operations that deal with safety and health of the rescue personnel.

Shoring team The group of rescuers responsible for installing struts in a shoring system.

Shoring team officer The individual responsible for supervising and managing the shoring team assignments.

Span of control The number of workers whom a supervisor can manage based on the type of work being performed.

Staging officer Individual responsible for ordering and maintaining adequate resources at the scene to handle additional requests for equipment and personnel.

Technical rescue safety officer (TRSO) A trained safety officer that is a certified trench rescue technician.

On Scene

1. What is the role of an operations level rescuer during the scene size-up phase of the emergency?

2. How could the incident command structure expand? Why might this happen?

CHAPTER 7

Operations Level

Operations Level Trench Rescue Shoring

KNOWLEDGE OBJECTIVES

After studying this chapter, you should be able to:

- Explain the differences between trench rescue shoring and trench construction shoring.
- Identify the principles of trench rescue shoring and the equipment that is commonly used for this purpose.
- Explain the use of the T-L method to determine lateral soil forces and shoring requirements. (**NFPA 1006: 12.2.1, 12.2.2, 12.2.3**, pp. 133–143)
- Explain the benefits of using prescriptive shoring designs.
- Describe a safe zone. (**NFPA 1006: 12.2.2, 12.2.3, 12.2.4**, pp. 144–145)
- Explain the benefits of having a default trench rescue shoring method.
- Describe the purpose of a shoring plan. (**NFPA 1006: 12.2.2, 12.2.3**, pp. 145–147)
- Identify the steps covered in a shoring plan. (**NFPA 1006: 12.2.2, 12.2.3**, pp. 145–147)
- Identify topics to be covered in a shoring plan briefing. (**NFPA 1006: 12.2.3**, p. 145)
- Create a shoring plan for a nonintersecting trench no more than 8 feet (2.4 m) deep. (**NFPA 1006: 12.2.3**, pp. 145–147)
- Identify the requirements for the use of trench rescue shoring and trench shields as an alternative to traditional trench shoring. (**NFPA 1006: 12.2.2, 12.2.3, 12.2.4**, pp. 155–156)
- Describe procedures for installing ground pads. (**NFPA 1006: 12.2.2, 12.2.4**, pp. 157–158)
- Describe installing lip bridges. (**NFPA 1006: 12.2.4**, pp. 159–160)
- Explain the difference between entry and non-entry shoring. (**NFPA 1006: 12.2.2, 12.2.3**, pp. 161–163)
- Explain the benefits of non-entry shoring. (**NFPA 1006: 12.2.2, 12.2.3, 12.2.4**, pp. 162–163)
- Describe considerations for selecting and installing panels. (**NFPA 1006: 12.2.4**, pp. 163–167)
- Describe the procedure for installing panels. (**NFPA 1006: 12.2.4**, pp. 163–167)
- Describe the procedure for installing pneumatic struts. (**NFPA 1006: 12.2.4**, pp. 167–171)
- Describe the use of timber shores. (**NFPA 1006: 12.2.4**, pp. 171–173)
- Describe the installation of wales. (**NFPA 1006: 12.2.4**, p. 174)
- Explain methods to address voids when shoring. (**NFPA 1006: 12.2.4**, p. 176)
- Describe shoring removal processes. (**NFPA 1006: 12.2.3**, pp. 178–179)

SKILLS OBJECTIVES

After studying this chapter, you should be able to:

- Install ground pads. (**NFPA 1006: 12.2.2, 12.2.4**, pp. 157–158)
- Install lip bridges. (**NFPA 1006: 12.2.4**, pp. 159–160)
- Perform a one-side panel installation. (**NFPA 1006: 12.2.4**, pp. 163–165)
- Perform a two-side panel installation. (**NFPA 1006: 12.2.4**, p. 166)
- Install pneumatic shores from outside the trench. (**NFPA 1006: 12.2.4**, pp. 167–169)
- Install pneumatic shores with entry operations. (**NFPA 1006: 12.2.4**, pp. 169–171)
- Install timber struts. (**NFPA 1006: 12.2.4**, pp. 172–173)
- Install inside wales. (**NFPA 1006: 12.2.4**, p. 174)
- Shore voids. (**NFPA 1006: 12.2.4**, p. 177)
- Utilize a shoring plan to shore a nonintersecting trench no more than 8 feet (2.4 m) deep. (**NFPA 1006: 12.2.1, 12.2.3, 12.2.4**, pp. 145–178)
- Manually remove pneumatic shoring systems. (**NFPA 1006: 12.2.7**, pp. 178–179)
- Manually remove timber shoring systems. (**NFPA 1006: 12.2.7**, pp. 178–179)
- Complete machine removal of a pneumatic shoring system. (**NFPA 1006: 12.2.7**, pp. 178–179)

Additional Resources

NFPA 1670, *Standard on Operations and Training for Technical Search and Rescue Incidents*

You Are the Rescuer

Last year, your department went through trench training. The chief brought in a trench shoring expert from a local construction safety equipment distributor. The classroom presentation went over several ways to make a trench safe. The hands-on portion included using skip shoring along with sheeting and shoring techniques that are used by construction companies to shore stable soils immediately after the excavation process.

This afternoon, you are on your way to your first trench rescue incident in a neighboring mutual aid community. As you walk up to the end of the trench for a size-up, you notice that a large section of trench wall directly over the trapped victim is missing, leaving a void area on that wall. The wall across the trench from that has a fissure forming at 5 feet (1.5 meters) away from the lip. The victim is pinned against the wall by a pile of dirt that is 4 feet (1.2 meters) deep on the trench floor. The soil at the face of that wall is raveling down the wall. The victim is alive and talking to you. You are concerned that a secondary collapse is about to occur. This trench with a partially buried victim looks nothing like the trenches that you shored in training.

Consider the following questions:

1. Explain why the skip shoring techniques that you learned are not safe for these conditions.
2. What are the shoring strategies and tactics that will rapidly protect the trapped victim, and how can they can be easily expanded to provide a safe zone for rescuers to enter and work in?

 Access Navigate for more practice activities.

Introduction

In this chapter, you will learn about the fundamentals and installation practices used for trench rescue shoring. You also will discover the differences between rescue shoring and construction shoring and learn to recognize the components of a default trench rescue shoring system.

The shoring systems used at the majority of trench rescue incidents across the United States consist of panels, struts, wales, and support equipment. Installing that equipment at a collapsed trench where victims are trapped is called trench rescue shoring. A basic set of rescue shoring equipment can be applied in many different ways, but operations level rescuers must understand that their equipment and skills have limitations. Trench rescue shoring and underground construction shoring have similarities, but they also have several important differences that should prevent them from being used interchangeably. The Occupational Safety and Health Administration (OSHA) shoring guidelines do not address rescue operations. In fact, in some cases, the shoring practices used by construction workers are dangerous and counterproductive to rescue operations. The tactics found in this chapter must be used in conjunction with the shoring strategy found in Chapter 6, *Managing the Trench Incident*.

PRINCIPLES OF TRENCH RESCUE SHORING

The History of Trench Rescue Shoring

The sheeting and shoring techniques used by rescuers has evolved over time from those used by the underground construction industry. Methods were modified with new techniques specifically created by James Gargan for use at rescue incidents. Although these shoring disciplines share some similarities, several significant differences exist as well. In what is believed to be the only study ever published on the subject of trench rescue shoring, author Dr. Marie LaBaw concluded that, "Underground construction shoring and trench rescue shoring are significantly different. Many of the theories and field observations that have been done for construction shoring may not be applicable to trench rescue shoring."

To comprehend the fundamentals of trench rescue shoring, you will need to understand the following:

- The differences between rescue shoring and construction shoring.
- The principles of rescue shoring.
- The shoring practices that will increase the chances of a successful rescue.

Comparing Rescue Shoring to Construction Shoring

Shoring in the underground construction world (large construction projects and utility installation and repair) differs from shoring of a trench for rescue in several ways: purpose, soil conditions, time, and planning.

Purpose

The purpose of shoring a trench for construction is to provide a safe area for the installation and repair of underground utilities and structures. The purpose of rescue shoring is to protect trapped victims, enabling their rescue and removal, and to protect the rescuers who are working to remove the victim. Trench rescue shoring must begin with techniques that will provide the victim with immediate protection from additional collapse (primary shoring). That immediate need, coupled with the active soil that has collapsed and created voids in the wall(s), often requires a departure from traditional construction shoring methods and OSHA-mandated practices such as, but not limited to, always shoring from the top down.

Soil Conditions

Sheeting/shoring used for construction purposes is designed to be installed before the soil becomes active. That is accomplished by installing the shoring during, immediately after, or in some cases, even before the excavation takes place. In those cases, the OSHA shoring charts, manufacturer's tabulated data, and construction shoring techniques provide valuable guidelines **FIGURE 7-1** and **FIGURE 7-2**. Rescuers, in contrast, are often confronted with soil conditions that have already become active and usually have collapsed. As a result, OSHA shoring charts, strut manufacturers' tabulated data, and construction techniques may not be best practices for rescue

FIGURE 7-1 Static soil.
Courtesy of Speed Shore Corporation.

FIGURE 7-2 Dynamic soil.
Courtesy of Ron Zawlocki.

FIGURE 7-4 Planned excavation site.
© Robert Kneschke/Shutterstock.

FIGURE 7-3 Rescue shoring.
Courtesy of Ron Zawlocki.

FIGURE 7-5 Rescue site.
Courtesy of Marc Messier.

shoring. OSHA charts and tabulated data based on theoretical soil pressure diagrams do not consider the large lateral forces that can be developed by massive pieces of soil breaking away from the trench wall.

Time

During a rescue, time is of the essence and can be the difference between extricating a live person and recovering a dead body. Trench rescue shoring practices must be capable of rapid installation that can be performed manually **FIGURE 7-3**. Unlike rescue operations, long-term soil retention (construction shoring) requires strong equipment and techniques that are capable of sustaining their strength over extended time periods with climate (weather) and soil changes. As a result, the shoring equipment is much heavier than what is required for typical rescue operational periods (12 hours or less). OSHA charts and the tabulated data produced by manufacturers of shoring equipment are based on longer durations of shoring use and require much more robust components.

Planning

Although it is common to see sloping, benching, sheeting, soldier piling/lagging, and trench boxes used by construction workers, these techniques require a large assortment of material and heavy equipment for installation. They also require comprehensive soil analysis and a coordinated preplan to have the materials, equipment, and operators at the site when the excavation begins. These are called *planned excavations*. Planned excavations provide several options that are based on soil conditions, excavation size, time, and budget **FIGURE 7-4**. Trench rescue incidents, by comparison, are *unplanned events* **FIGURE 7-5**. Rescuers are constrained to the limited assortment of shoring equipment that is brought to the scene in trucks and trailers and that must be manually installed using rescue-specific shoring skills and techniques. We call these skills, techniques, and equipment default shoring. Default shoring provides one option for a majority of soil conditions and excavation sizes that result in collapsed trenches and trapped victims.

Dr. LaBaw's study discusses several differences between construction shoring and rescue shoring. The authors encourage trench rescuers to read that study because successful trench rescue shoring requires knowledge of those differences as well as an understanding of the proper use and application of trench rescue shoring fundamentals. The References section at the end of this chapter lists the full details of Dr. LaBaw's study.

It is extremely important that the shoring methods are designed by a rescue engineer specifically for worst-case soil conditions expected in the operating area. Rescuers should simply follow a prescription (shoring operations guide) that is designed by a professional engineer with trench rescue shoring competency. A shoring operations guide would consist of predesigned systems that can be used for common trench collapse situations, including techniques for dealing with common collapse patterns and challenges. Unfortunately, many of the shoring designs that have worked their way into trench rescue shoring practices are systems that have not been designed by professional engineers for soil conditions commonly found at collapsed trenches. Many firefighters have learned shoring techniques that were designed for at-rest soil conditions and in some cases, practices that were simply made up by firefighters alone.

Principles of Trench Rescue Shoring

Trench rescue shoring systems must be capable of collecting the load, transferring the load, distributing the load, and resisting the load. This is similar to building collapse shoring, but in this case, the load is the lateral soil pressure and the shoring is primarily horizontal rather than vertical. The best practice for constructing a shoring system with these capabilities uses components that can be installed from outside the trench. Any system that requires entry into the trench to construct shoring puts the rescuers at increased risk of injury from additional trench collapses. When systems are installed from outside the trench using ground pads and lip bridges, this risk is nearly eliminated.

High-strength composite panels are used to collect the load, struts are used to transfer the load, and composite panels are used again (on the opposite wall) to distribute the load. Additionally, trench rescue shoring equipment and individual structural components must be light enough to be carried and installed by rescuers (manually) while working above compromised trench walls.

Trench Rescue Shoring Essentials
Collect/Distribute Loads

The first step of shoring is *collecting the load* from the now unsupported soil wall. It is important to remember that while the unsupported trench wall looks solid, it really is not. It is made up of grains of soil that have the strength to maintain a vertical wall face for some amount of time. This vertical soil wall face has to be supported similar to the way an unreinforced masonry wall is supported in emergency building shoring. This is typically achieved with sheeting (plywood). Trench panels (or sheeting with strongbacks) are used to cover the soil in the trench to collect the loads from the unsupported trench wall without relying on the unknown internal strength of the soil.

Trench rescue panels are typically a combination of plywood and nominal 2-inch thick lumber (2×12 or 2×10) commonly referred to as a strongback **FIGURE 7-6**. Panels are needed to both collect the load of an unstable trench wall and to distribute the load to the resistance point (the opposing trench wall). In both cases, suitable wood panels are needed. The strength of the panels is dependent on the type of plywood and lumber used (density and thickness) and the method used to attach the plywood to the strongback.

A wale is a beam used to span large areas of trench walls without the use of intermediate struts, similar to how a beam is used in a building to minimize the use of columns or supporting. Wales are necessary when the trench rescue panel cannot be

FIGURE 7-6 Trench rescue panels.
© Jones and Bartlett Publishers. Courtesy of MIEMSS.

directly supported by a shore member bearing on the strongback. The wale also *collects the load* from the unsupported trench wall. The trench rescue panels collect the load from the soil of the trench wall, and then the wale collects the load from the trench rescue panels.

On the opposite side of the trench, the trench rescue panels are used to distribute the loads back into the soil of the trench wall. In some situations, wales are also used to distribute that load. The wales distribute the load to the individual trench rescue panels, and then the panels distribute that load to the soil of the trench wall.

Transfer Loads (Struts)

Wales are unique in that they not only collect the load, but they also transfer that load to the supporting struts. Wales use their internal bending strength to transfer the load collected from the panels to the struts. Wales are typically supported by struts near their ends, and the wale takes the loads collected between the struts' location and transfers that load through the struts to be transferred across the trench. It is important to understand that struts supporting the wales will have much higher loads on them than struts that are only supporting single trench rescue panels. Meaning that if a wale is collecting the load from four panels and is supported by two struts, each strut will need to carry the load of two panels.

Pneumatic struts are commonly used by trench rescue teams. Use of timber and hydraulic struts is less common but still exists. Struts are used to stabilize a trench wall by transferring the load from the trench wall to a suitable point of resistance (commonly the opposing trench wall) **FIGURE 7-7**. A suitable point of resistance (commonly the opposing trench wall) must exist to receive and resist the load being transferred. It is important to remember that when a suitable point of resistance does not exist to distribute the load to, a wale can be used to further transfer the load to a location where suitable resistance does exist.

Resist Loads

In a straight trench, the resistance to movement (collapse) is typically transferred through the struts to the trench wall that is directly across from the failing wall section. Filling in voids (backfill) that have been created from cave-ins is a necessary practice for shoring partially collapsed trench walls **FIGURE 7-8**. An adequately backfilled trench wall will add strength to the point of resistance and support the remaining soil, preventing further collapse.

Lateral Soil Forces and Firefighters

With the understanding of how to collect, transfer, and distribute the loads from a trench, the next important area to understand is how to determine the loading that needs to be collected, transferred, and distributed. The ability to rapidly and safely determine the maximum forces that the soil can exert on your shoring system and the ability to select and install shoring that is capable of resisting those forces are essential for safe and efficient trench rescue operations. Many rescuers have learned to use methods of making on scene estimations of lateral soil forces that were developed by OSHA for use by

FIGURE 7-7 Struts.
Courtesy of Ron Zawlocki.

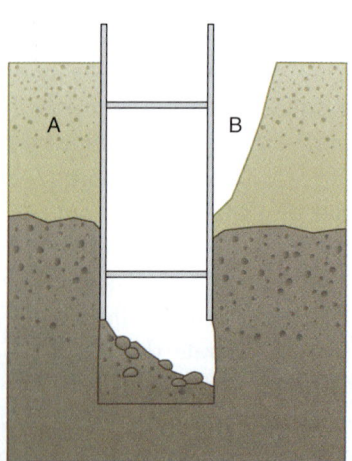

FIGURE 7-8 The void area (B) must be filled to resist the lateral forces from the opposing wall (A).

workers who are trained and experienced trench and excavation professionals. Unfortunately, those methods do not accurately represent the soil conditions and the lateral soil pressures associated with the unstable and dynamic (collapsing) soil found at the overwhelming majority of trench rescue incidents.

T-L Method

An important part of hazard mitigation and safety is determining the possibility of soil failure. In 2018, Oliver Taylor, PE, PhD, and Marie LaBaw, PE, PhD, worked as part of a Federal Emergency Management Agency (FEMA)/State Urban Search and Rescue Alliance (SUSAR)/U.S. Army Corps of Engineers (USACE) structures specialist committee. The committee was studying trench rescue shoring standards, and developed a soil assessment method that is designed for use by rescuers. The T-L method was developed using the imperial measurement system and this chapter primarily references that system of measurement during this discussion. Please see Appendix C, *The T-L Method - A Metric System* Guide for more information on utilizing the T-L method in the metric system.

The **T-L method** accurately represents actual earth pressures and lateral forces based on the visible soil failure conditions. Determining lateral soil forces and interpreting tabulated data (data organized into a table) is reduced to the use of a tape measure and charts (or pocket guides). The authors recommend that rescuers use the T-L method for all trench rescue incidents and to assist in hazard mitigation.

The T-L method relies on the failure signs observed onsite and uses those visual clues to more closely approximate the actual load that will occur in the short term of a rescue. Specifically, it uses the horizontal distance of trench failure (collapse), or signs of failure such as cracks approximately parallel to the trench lip. The "L" in the T-L method is the distance measured horizontally from the trench lip (or where it used to be) to the farthest collapse or cracking that parallels the trench, or "L distance."

TIP

Cohesive (clay) soils shrink as they dry. Many clay soils will have cracks present on the surface that have nothing to do with a trench. The presence of surface cracks that run well beyond the length of the trench are likely general shrinkage cracks that existed before the excavation. Cracks that indicate a failure will usually curve and intersect the trench lip and will be more than ½ inch (1.3 cm) wide and more than 12 inches (3.6 m) deep.

The T-L method calculates the vertical weight of the block of soil bounded by the farthest sign of failure (Trench depth × L distance × Length × Assumed soil density of 133 pounds per cubic foot [pcf]; 133 pcf is used rather than the 120 pcf average to provide a greater range of soil conditions). The method applies a coefficient of 0.5 to determine lateral earth pressure. The shoring is assumed to be 4 × 8 feet (1.2 × 2.4 m) panels with two struts per panel and the method produces a force where the only variable is the L distance. This provides easy equation that depends entirely on L for systems using 4-foot (1.2-m) wide panels.

$$\text{Lateral force} = 1063 \times L \text{ (in feet)}. \text{ For simplicity,}$$
$$\text{this is rounded to } 1100 \times L$$

This equation calculates the force that would act on a 4 × 4-foot (1.2 × 1.2-m) section of panel, which becomes a strut load or a wale load. In cases where more than two struts are being placed on a panel, one would double the result of the equation to come up with a load on the entire panel and divide by the number of supports being placed.

If signs of failure are not visible, or possibly covered by a spoil pile, a failure depth of 0.7 times the trench depth is used to determine the L value. Assigning such a high L value to a trench that has not yet collapsed might seem counterintuitive, but in the case where no failure signs are present, the soil has not given clues and a larger potential soils force must be considered (see **TABLE 7-1** for a simplified method to make this determination).

Soil Classification

Because trench rescuers lack the experience, knowledge, and equipment to accurately classify soils, they should always consider the soil to be a worst-case soil condition. That means that the soil at a trench rescue should be considered weak, unstable, and capable of creating the maximum possible physical forces on a shoring system. The attributes of T-L soil have been engineered to meet those worst-case soil conditions in soils that could stand on their own long enough to dig and enter a trench. If rescuers protect the victim and the rescuers in the trench from **T-L soil conditions**, then they have been protected from every other soil condition and classification.

Soil Forces

The T-L method utilizes soil pressure formulas and worst-case conditions (weight of 133 pcf; active earth pressure coefficient of 0.5). That means a lateral

TABLE 7-1 Depth to L Conversion Chart

Trench Depth (Ft)	SL (FT)
5	4
6	5
7	5
8	6
9	7
10	7
11	8
12	9
13	10
14	10
15	11
16	12
17	12
18	13
19	14
20	14

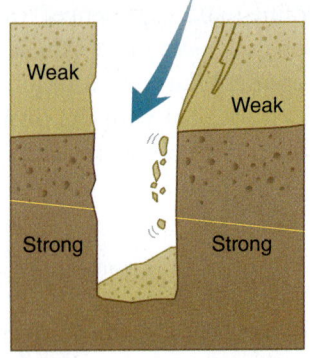

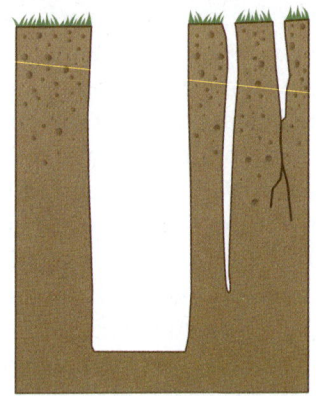

FIGURE 7-9 Common types of soil failures. **A.** Lip failures. **B.** Mid-wall failures. **C.** Fissures.

force of about 66 pounds per square foot (psf). On a 4 × 4-foot (16-square foot [ft^2]) section of shoring (half panel), 16 cubic feet (ft^3; 4 × 4 × 1-feet) of T-L soil will develop nearly 1100 pounds (499 kg) of lateral force (16 ft^2 × 66.4 psf = 1,063). Comparatively, 64 ft^3 (4 × 4 × 4-feet) or L-4 will develop nearly 4400 pounds (1996 kg) of lateral force on each half panel and the associated struts and/or wales. Surcharged loads must be added as explained next.

How To Use The T-L Method for Trench Rescue

A tape measure is used to find the simple L from the original trench face (wall) to the farthest point of soil failure **FIGURE 7-9**. Simple L (SL) is the distance (length) measured in feet from the original trench wall perpendicular to the farthest point of soil failure or cracks/fissures. When there are multiple signs of failure, always use the farthest point as the SL **FIGURE 7-10**. To use the T-L method for trench rescue, follow these steps:

1. From a safe area on the lip, measure the distance from the original trench wall (face) perpendicular to the farthest failure point.
2. Round that measurement up to the next foot and utilize that total L distance in the shoring charts. Example: In this case, we have a closed lip failure at 24 inches and fissures at both 24

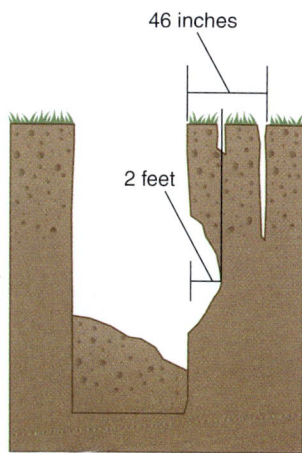

FIGURE 7-10 With multiple signs of failure, always use the farthest point as the SL.

TABLE 7-2 Surcharge Table

Surcharge (ScL)- Feet within Simple L (SL)

Spoil	Add to SL	EQUIP.	Add to SL
1	1	1	1
2	1	2	2
3	1	3	3
4	2	4	5
5	3	5	8
6	4	6	11
7	5	7	N/A
8	7	8	N/A
9	9	9	N/A
10	10	10	N/A

Note: Total L (L) = Simple L (SL) plus Surcharge L (ScL)

Charts are valid for Total L of 20 or less

and 46 inches. We measure the farthest point, which is 46 inches. Round that measurement (46 inches) up to the next foot, which, in this case, is 4 feet. The SL is 4. If no surcharged load exists, the SL is the total (L), which is used in the shoring charts.

Surcharge Loads (ScL)

Surcharge loads (ScL) are spoil piles and construction equipment that are within the area that is between the original trench faces and the farthest point of soil failure. ScL is measured in feet perpendicular to the trench wall. Note that construction equipment can include, but is not limited to, excavators, dump trucks, trench boxes, and pipes.

Surcharge loads that are within the area between the original trench face and the farthest point of soil failure can add significant lateral forces to shoring systems. Those additional forces must be added to the SL to obtain the total L (L). The **total L (L)** is the SL plus the ScL (spoil pile and equipment), if present. The total L is used to select the proper trench rescue shoring from engineered tabulated data charts approved by the authority having jurisdiction.

TABLE 7-2 allows the user to easily determine the additional value to add to the SL to account for the ScL and obtain the total L.

Determining the Spoil Pile Surcharge

The T-L method includes spoil pile surcharge by approximating the weight of a spoil pile within the SL distance. An equation was developed to determine the equivalent L value to add to the SL. To determine the SL using the table, follow these steps:

1. Measure the amount of spoil pile that is within the SL.
2. Round the measurement up to the next foot.
3. Use this measurement in the surcharge table (spoil column).
4. Table 7-2 provides the corresponding value to add to the SL.

Adding the Spoil Pile Surcharge

To determine the total L (L), when a spoil pile is present, use the following steps:

1. Determine the SL. For this example, the SL is 4. In addition, 2 feet of the spoil pile is within the SL **FIGURE 7-11**.
2. Use 2 feet as the spoil pile measurement and refer to the surcharge table (spoil column) to determine the additional L value that must be added to the SL. In this case, with a measurement of 2, the surcharge table determines an additional L value of 1, which results in a total L (L) of 5 (4 + 1).

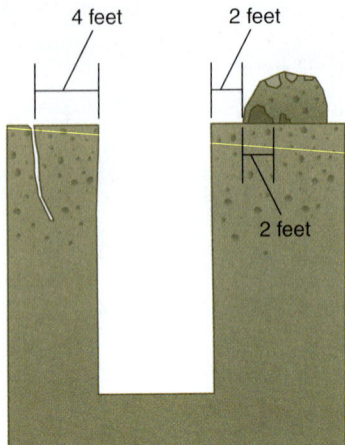

FIGURE 7-11 Determine the amount (linear feet) of spoil within the SL.

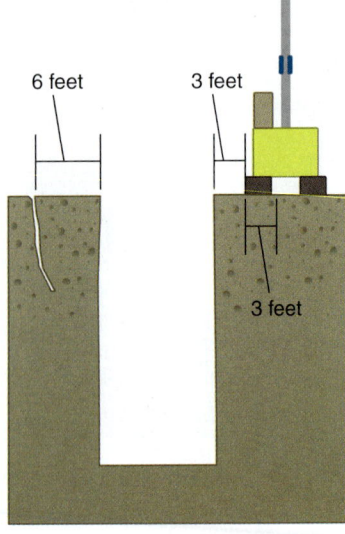

FIGURE 7-12 Determine the amount of equipment (linear feet) within the simple load (SL).

TIP

For those interested in the details of how the spoil pile surcharge is calculated in Table 7-2, the method uses a horizontal distance (that we measure), takes into account the angle of repose of the soils, and estimates the weight of soil. The weight of soil next to the trench increases the lateral earth pressure that will act on the trench rescue shoring. This increase in pressure is treated as an addition to the L value.

$$\text{Spoil pile surcharge} = 0.1 \times L^2$$

This equation is used, with results rounded up to the nearest foot, for the "Add to SL" values in Table 7-3. It is important to remember that the L in the equation includes only the amount of spoil pile within the SL distance.

For example, in trench with fissures at 6 feet from the lip of the trench, and a spoil pile that is set back 2 feet from the lip, we have 6 feet − 2 feet = 4 feet of spoil pile within the L distance. Entering this value in the equation given earlier, we get spoil pile surcharge of: $0.1 \times (4^2) = 1.6$ feet, which is rounded up to 2 feet. We then add the spoil pile surcharge, which equals 2 feet in this case, to our SL of 6 feet for a total L of 8 feet.

Determining the Equipment Surcharge

Equipment also adds additional weight on the soils adjacent to the trench, increasing lateral earth pressures. Equipment surcharge loads are handled similarly to the spoil load surcharge described previously. To determine the equipment surcharge, follow these steps:

1. Measure the amount of equipment that is within the SL.
2. Round the measurement up to the next foot.
3. Use the measurement in the surcharge table (Equipment column).
4. Use the table to locate the corresponding value to add to the SL.

TIP

For those who may be interested in the details of how the equipment surcharge is calculated, the method uses a horizontal distance (that we measure) to estimate a weight of equipment. This is done by using a 3 foot depth of soil in place of an equipment load. This weight of soil acting next to the trench increases the lateral earth pressure that acting on the trench rescue shoring. This increase in pressure is treated as an addition to the L value. The following equation is used to calculate the additional L:

$$\text{Equipment surcharge} = 0.3 \times L^2$$

This equation is used with results rounded up to the nearest foot to create the "Add to SL" values in Table 7-3. It is important to remember that the L in the equation is only the amount of equipment within the SL distance.

Adding the Equipment Surcharge

The procedure used to calculate equipment surcharge is fairly straightforward. Consider the following example: the SL is 6, and 3 feet of the equipment is within the SL. Use 3 feet as the equipment surcharge L and refer to the surcharge table (equipment column) to determine the corresponding value to add to the SL. In this case, with 3 feet of equipment within the SL, the surcharge table tells us to add 3 to the SL for a total L (L) of 9 (6 + 3) **FIGURE 7-12**.

Trench Depth to L Conversion

As mentioned previously, when absolutely no signs of failure or distress are apparent, the SL can be taken as 70 percent of the trench depth. **TABLE 7-3** provides easy-to-look-up values. It can be used by measuring

TABLE 7-3 Depth to SL Conversion Table

DEPTH TO SIMPLE L (SL) CONVERSION GUIDE

Trench Depth	SL Equivalent
4-8 feet	SL-6
9 feet	SL-7
10 feet	SL-7
11 feet	SL-8
12 feet	SL-9
13 feet	SL-10
14 feet	SL-10
15 feet	SL-11
16 feet	SL-12
17 feet	SL-12
18 feet	13
19 feet	14
20 feet	14

Note: Total L (L) = Simple L (SL) plus Surcharge L (ScL)

the trench depth, rounding up to the nearest foot, then reading the corresponding SL equivalent value from the chart.

T-L Method in Action

To get a feel for the T-L method in action, let's consider an example. We are called to a trench rescue. The trench is 9-feet deep, there is a failure in the wall that measures 3 feet from the lip with a spoil pile set back 2 feet from the lip. Part of the spoil pile slid into the trench when the first collapse occurred. The other side of the trench has fissures 2 feet from the lip, 3 feet from the lip, and 5 feet from the lip and an empty wheelbarrow right at the edge of the trench. No other equipment is within 25 feet of the trench.

The SL is the measurement from the lip to the farthest sign of failure, in this case 5 feet. With the spoil pile set back 2 feet from the trench lip, that means that 3 feet (5 feet − 2 feet = 3 feet) of the spoil pile is within the SL distance, which adds a surcharge load to the soil. Using the surcharge table, under the Spoil column, we enter the table with a value of 3 feet and it provides a corresponding value of 1 foot to add to the SL. Next, we must consider equipment surcharge. An empty wheelbarrow has little weight so it does not count toward a surcharge load. In this case, the total L is equal to 6 feet (5 feet SL + 1 foot ScL [due to the spoil pile]).

Tabulated Data for Shoring Equipment

Engineered tabulated data is information that has been developed by a licensed professional engineer and arranged in easy-to-read rows and columns. Valid engineered tabulated data is developed for specific purposes using appropriate theories and assumptions. It is very important to carefully read all the engineer's notes to determine if the tabulated data addresses the type of trench and soil conditions that you are dealing with at the scene.

Most trench shoring tabulated data is designed for use by construction workers. Construction workers may begin shoring either during or shortly after the excavation (digging) process begins and the trench walls are usually intact. With a trench collapse, the condition of the trench walls is different, the soil is unstable and dynamic, and the lateral soil forces are different than what construction-based tabulated data is designed to address. Organizations such as Michigan Urban Search and Rescue (MUSAR) have created tabulated data specifically for collapsed trench walls and rescue conditions. Their engineered tabulated data is based on worst case soil (T-L soil) and their shoring charts may be used at all rescue incidents and in most soil conditions.

To interpret and use tabulated data from organizations such as MUSAR, rescuers only need to measure the trench failure point (SL) and calculate the Total L (L) if surcharge loads are present. Then use the charts to determine the appropriate shoring component (strut, panel, or wale). If you know what struts, panels, and wales your team carries, it becomes very easy to know a maximum Total L that each component can support.

TABLES 7-4 through **7-13** provide tabulated data on shoring components ranging from struts to sole anchor pickets. Remember to follow the standard operating guidelines for your agency, including any tabulated data created for your agency.

TABLE 7-4 Paratech Strut Chart

Vertical Spacing with Maximum 4' Horizontal Spacing

Total L	Gray Struts		Gold Struts			
	Width ≤8	Width 8'-10'	Width <10	Width ≤12	Width ≤14	Width ≤16
L-1	4'	4'	4'	4'	4'	4'
L-2	4'	4'	4'	4'	4'	4'
L-3	4'	4'	4'	4'	4'	4'
L-4	4'	4'	4'	4'	4'	4'
L-5	4'	4'	4'	4'	4'	4'
L-6	4'	4'	4'	4'	4'	3'
L-7	4'	4'	4'	4'	4'	3'
L-8	4'	4'	4'	4'	4'	3'
L-9	4'	4'	4'	4'	4'	2'
L-10	4'	3'	4'	4'	4'	2'
L-11	4'	3'	4'	4'	3'	N/A
L-12	4'	3'	4'	4'	3'	N/A
L-13	4'	3'	4'	3'	3'	N/A
L-14	4'	3'	4'	3'	3'	N/A
L-15	4'	2'	4'	3'	3'	N/A
L-16	4'	2'	4'	3'	3'	N/A
L-17	4'	2'	4'	3'	2'	N/A
L-18	4'	2'	4'	3'	2'	N/A
L-19	4'	2'	4'	3'	2'	N/A
L-20	4'	N/A	4'	4'	2'	N/A

Chat data based on strut capacities provided by Paratech at 2:1 factor of safety and converted for spacings depending on Total L.
Reproduced from MUSAR Training Foundation. (2020). *Trench rescue shoring operations guide* (2nd ed.). Author. https://paratech.com/wp-content/uploads/2021/01/MUSAR-SOG-2020.pdf

TABLE 7-5 ResQTec Strut Chart

Maximum Vertical Spacing with Maximum 4' Horizontal Spacing

ResQTec Struts Total L	PxM470 PxM600 PxM880 PxM1400 Width ≤6	PxM470 PxM600 PxM880 PxM1400 PxM2300 Width ≤8'	PxM1400 PxM2300 Width ≤10	PxM1400 PxM2300 Width ≤12	PxM2300 Width ≤14	PxM2300 Width ≤16	PxM2300 Width ≤18
L-1	4'	4'	4'	4'	4'	4'	4'
L-2	4'	4'	4'	4'	4'	4'	4'
L-3	4'	4'	4'	4'	4'	4'	4'
L-4	4'	4'	4'	4'	4'	4'	4'
L-5	4'	4'	4'	4'	4'	4'	4'
L-6	4'	4'	4'	4'	4'	4'	4'
L-7	4'	4'	4'	4'	4'	4'	4'
L-8	4'	4'	4'	4'	4'	4'	3'
L-9	4'	4'	4'	4'	4'	4'	3'
L-10	4'	4'	4'	4'	4'	3'	3'
L-11	4'	4'	4'	4'	3'	3'	2'
L-12	4'	4'	4'	4'	3'	3'	2'
L-13	4'	4'	4'	4'	3'	3'	2'
L-14	4'	4'	4'	4'	3'	2'	2'
L-15	4'	4'	4'	4'	3'	2'	N/A
L-16	4'	4'	4'	4'	3'	2'	N/A
L-17	4'	4'	4'	3'	2'	2'	N/A
L-18	4'	4'	4'	3'	2'	2'	N/A
L-19	4'	4'	4'	3'	2'	N/A	N/A
L-20	4'	4'	4'	3'	2'	N/A	N/A

Chart data is based on strut capacities provided by ResQTec user manual for PROFIX Max Stabilization System (DM058k08. Capacities interpolated from graphs and multiplied by 2 to get a 2:1 factor of safety.
Reproduced from MUSAR Training Foundation. (2020). *Trench rescue shoring operations guide* (2nd ed.). Author. https://paratech.com/wp-content/uploads/2021/01/MUSAR-SOG-2020.pdf

TABLE 7-6 Holomatro Strut Chart

Vertical Spacing with Maximum 4' Horizontal Spacing

Total L	Width ≤5'	Width ≤6'	Width ≤8'	Width ≤10'	Width ≤12'	Width ≤14.75
L-1	4'	4'	4'	4'	4'	4'
L-2	4'	4'	4'	4'	4'	3'
L-3	4'	4'	4'	4'	3'	2'
L-4	4'	4'	4'	4'	3'	N/A
L-5	4'	4'	4'	3'	2'	N/A
L-6	4'	4'	4'	3'	N/A	N/A
L-7	4'	4'	3'	2'	N/A	N/A
L-8	4'	4'	3'	2'	N/A	N/A
L-9	4'	4'	3'	N/A	N/A	N/A
L-10	4'	4'	2'	N/A	N/A	N/A
L-11	4'	4'	2'	N/A	N/A	N/A
L-12	4'	4'	2'	N/A	N/A	N/A
L-13	4'	3'	N/A	N/A	N/A	N/A
L-14	4'	3'	N/A	N/A	N/A	N/A
L-15	4'	3'	N/A	N/A	N/A	N/A
L-16	4'	2'	N/A	N/A	N/A	N/A
L-17	4'	2'	N/A	N/A	N/A	N/A
L-18	4'	2'	N/A	N/A	N/A	N/A
L-19	4'	2'	N/A	N/A	N/A	N/A
L-20	3'	2'	N/A	N/A	N/A	N/A

Chart data based on strut capacities provided by Holmatro for a 2:1 factor of safety.
Strut capacities were interpolated for even foot increments and rounded down.
Reproduced from MUSAR Training Foundation. (2020). *Trench rescue shoring operations guide* (2nd ed.). Author. https://paratech.com/wp-content/uploads/2021/01/MUSAR-SOG-2020.pdf

TABLE 7-7 Hurst/Airshore Strut Chart

Vertical Spacing with Maximum 4' Horizontal Spacing

Total L	Struts with 1 Pin		Struts with 2 Pins		
	Width ≤8	Width 8'-12'	Width ≤4	Width ≤8	Width ≤12
L-1	4'	4'	4'	4'	4'
L-2	4'	4'	4'	4'	4'
L-3	4'	4'	4'	4'	4'
L-4	4'	4'	4'	4'	4'
L-5	4'	4'	4'	4'	4'
L-6	4'	4'	4'	4'	4'
L-7	4'	4'	4'	4'	4'
L-8	4'	4'	4'	4'	4'
L-9	4'	4'	4'	4'	4'
L-10	4'	4'	4'	4'	4'
L-11	4'	4'	4'	4'	4'
L-12	4'	4'	4'	4'	4'
L-13	4'	3'	4'	4'	4'
L-14	4'	3'	4'	4'	4'
L-15	4'	3'	4'	4'	4'
L-16	4'	3'	4'	4'	4'
L-17	4'	3'	4'	4'	4'
L-18	3'	3'	4'	4'	4'
L-19	3'	3'	4'	4'	3'
L-20	3'	2'	4'	4'	3'

Chart data based on strut capacities provided by Airshore at 2:1 factor of safety and converted for spacings depending on Total L.
Reproduced from MUSAR Training Foundation. (2020). *Trench rescue shoring operations guide* (2nd ed.). Author. https://paratech.com/wp-content/uploads/2021/01/MUSAR-SOG-2020.pdf

TABLE 7-8 Wood Strut Chart

3' Maximum Vertical Spacing	
Strut Wood Type	Trench Width Up to 8'
Douglas Fir or Spruce/Pine/Fir	Maximum L-8

Notes:
Maximum L is based on 4"x 4" wood struts and Ellis Screw jacks.
Reproduced from MUSAR Training Foundation. (2020). *Trench rescue shoring operations guide* (2nd ed.). Author. https://paratech.com/wp-content/uploads/2021/01/MUSAR-SOG-2020.pdf

TABLE 7-9 Wale/Wood Strut Chart

8' Maximum Span/ 4' Maximum Vertical Wale Spacing						
Strut Wood Type	Trench Width					
	3'	4'	5'	6'	7'	8'
Douglas Fir	Max L-9	Max L-9	Max L-9	Max L-8	Max L-6	Max L-4
Spruce/Pine/Fir	Max L-7	Max L-7	Max L-7	Max L-7	Max L-6	Max -4

Notes:
Based on 4"x 4" wood struts, Ellis Screw jacks, and 7"x 7" LVL wales (3,100 psi bending strength)
Span is the distance between the struts supporting the wales.
Gaps between the wales and panels at the panel edges and both ends of wales must be filled with spacers and/or wedges.
Reproduced from MUSAR Training Foundation. (2020). *Trench rescue shoring operations guide* (2nd ed.). Author. https://paratech.com/wp-content/uploads/2021/01/MUSAR-SOG-2020.pdf

TABLE 7-10 Panel Chart

Composite Construction	Maximum Total L	Noncomposite Constuction	Maximum Total L
2 x 12 LVL/FinnForm	L-20	2 x 12 LVL/FinnForm	L-5
2 x 12 Wood/Finnform	L-11	2 x 12 Wood/Finnform	L-3
2 x 12 Wood/CDX	L-8	2 x 12 Wood/CDX	L-1

NOTES:
1. Composite construction requires the application of heavy-duty exterior construction grade adhesive that has an ultimate strength of at least 4,000 psi between the strongback and sheeting. Surface areas must be sanded prior to the application of adhesive.
2. Noncomposite construction includes attaching strongbacks to sheeting with nails, bolts, and screws. Simply positioning strongbacks against the sheeting is also considered noncomposite.
3. Capacities can be increased with the noncomposite application of additional strongbacks (a) 2 x 12 sawn lumber add 1L, (b) 2 x 12 LVL add 3L, (c) 6 x 6 sawn lumber add 6L, (d) 8 x 8 sawn lumber add 16L (e) 7 x 7 LVL add 20L.
4. Plywood (FinnForm and CDX) is 3/4".
5. "Wood " (strongback) values are based on #1 Douglas Fir.
6. LVL strongback values are based on fb-3,100 psi LVL.
Reproduced from MUSAR Training Foundation. (2020). *Trench rescue shoring operations guide* (2nd ed.). Author. https://paratech.com/wp-content/uploads/2021/01/MUSAR-SOG-2020.pdf

TABLE 7-11 Wale Chart

4' Maximum Vertical Spacing		
Wale Type	8' Span	12' Span
6 x 6	L-2	L-1
8 x 8	L-5	L-2
7 x 7 LVL	L-12	L-4
Paratech	L-7	L-2

NOTES: 2 Point Supported Wales
L values represent the maximum allowable Total L.
Span is the distance between the struts supporting the wale.

NOTES: All Wales
Gaps between the wales and panels at the edges and both ends of the wale must be filled with spacers and/or wedges.
Spacing wales @2' (vertical) will increase the maximum Total L capacity by 150% (Max L shown x 1.5).
6 x 6 and 8 x 8 timber capacities are based on #1 Douglas Fir.
7 x 7 LVL capacity based on a bending strength of 3,100 psi.
Reproduced from MUSAR Training Foundation. (2020). *Trench rescue shoring operations guide* (2nd ed.). Author. https://paratech.com/wp-content/uploads/2021/01/MUSAR-SOG-2020.pdf

TABLE 7-13 Sole Anchor Picket Chart

Total L	MINIMUM PICKETS
L-1	2
L-2	3
L-3	5
L-4	6
L-5	8
L-6	9
L-7	10
L-8	12

This chart includes picket requirements for sole anchor used for rakers, buttress and tension systems.
Pickets must be 1" diameter steel driven a minimum of 36" into weak cohesive soil or better.
Pickets spacing shall be a minimum of 6" apart unless they are used in a one-piece picket anchor system device.
Reproduced from MUSAR Training Foundation. (2020). *Trench rescue shoring operations guide* (2nd ed.). Author. https://paratech.com/wp-content/uploads/2021/01/MUSAR-SOG-2020.pdf

TABLE 7-12 Buttress Chart

Composite Panels	4' Deep Void	6' Deep Void
2 x 12 LVL/FinForm ¾"	Max Total L-6	Max Total L-4
2 x 12 Wood/FinForm ¾"	Max Total L-2	Max Total L-1

NOTES:
Capacities can be increased by placing an upright in front of the strongbacks and installing struts on the upright.

Added Upright	4' Deep Void	6' Deep Void
2 X 12 LVL	Added capacity L-2	Added capacity L-1
6 x 6 Wood	Added capacity L-4	Added capacity L-2
8 x 8 Wood	Added capacity L-10	Added capacity L-7
7 x 7 LVL	Added capacity L-20	Added capacity L-16

NOTES: Buttress Picket Details
The distance from void or sign of failure for a L≤5 failure shall have pickets can be installed no less than 4' from the void or failure point.
L≥6 failures shall have pickets can be installed no less than the 75% of the distance of the Simple L from the void or failure point.
Pickets must be 1" diameter steel driven a minimum of 36" into the soil.
Pickets spacing shall be a minimum of 6" apart unless they are used in a one-piece picket anchor system device.
See Sole Anchor Picket Chart.
Reproduced from MUSAR Training Foundation. (2020). *Trench rescue shoring operations guide* (2nd ed.). Author. https://paratech.com/wp-content/uploads/2021/01/MUSAR-SOG-2020.pdf

Prescriptive Shoring Designs

Many of the shoring designs being taught and/or used in the fire service have been neither engineered nor tested. There has been little involvement in the design of trench rescue shoring by professional engineers. As a result, members of the fire service alone have created a number of shoring system practices and designs. The designers of those systems lacked the training and engineering experience needed to determine the forces of moving masses of soil and to design a system capable of resisting those forces. Many of those practices and designs have found their way into trench rescue training curricula across the country. Because many of those systems have been repeatedly used with only minor failures, they have become part of an unofficial standard.

In the emergency medical services, trained emergency medical technicians and paramedics follow medical protocols designed by experienced physicians based on current medical evidence. This prescriptive approach combines experience, training, and science into protocols and procedures that can be seamlessly, safely, and successfully applied during an emergency.

Likewise, with **prescriptive shoring designs**, a professional engineer documents the designs of shoring systems for trench rescuers. The engineer addresses the common problems associated with shoring collapsed trenches and applies shoring designs that can be seamlessly, safely, and successfully applied during an emergency. The shoring designs in this text have been developed by engineers. They provide the prescriptive shoring designs that the authors recommend and that rescuers should follow.

Most of the shoring and the components in the prescriptive shoring designs are modular. The goal is to make the components easily transportable, easy to position and place, and flexible enough to accommodate unanticipated conditions. By applying the shoring designs, a trained operations level rescuer can assemble a trench rescue shoring system that quickly protects the victim and creates a working environment for the rescuers that will not collapse on them.

Criteria for Safe Zone in a Trench

The criteria for a **safe zone** in a trench is the same for operations and technician level incidents. A safe zone can be accomplished by the proper use of shoring, sloping, or shielding **FIGURE 7-13**. The subject of this chapter is shoring; therefore, the criteria for a safe zone provided in this section deals specifically with shoring. The equipment and techniques used by trench rescue response teams should be capable of shoring collapsed trench walls that are 20 feet (6.1 m) deep or less and are, for the most part, still standing in a nearly vertical position (70–90 degrees). By using this criterion, both operations level trenches (up to 8 feet [2.4 m] deep) and technician level trenches (Chapter 10, *Technician Level Trench Rescue Shoring*) are covered. The trenches can be either nonintersecting (operations) or intersecting (technician) and are commonly dug to install or repair underground utilities. Trenches dug next to buildings (for basement-wall waterproofing or repair) also can be safely shored with trench rescue shoring equipment and techniques.

Criteria for a zone made safe through the use of shoring are listed here. They apply to soil conditions where the water level is below the bottom of the trench, the soil is not oversaturated or not flowing, and the bottom of the excavation is not boiling. Shoring materials and equipment must be based on soil force calculations

FIGURE 7-13 A "safe zone" created with trench rescue panels and struts.
© Jones & Bartlett Learning. Photographed by Glen E. Ellman.

(T-L method) and tabulated data designed for use in T-L soil conditions. These criteria are as follows:

- Close sheeting (panel edges touching or near touching) must cover the height of the trench wall (vertical) and cover a horizontal area that is equal to or greater than the height of the walls.
- Struts must be placed within 10 degrees of level and 10 degrees of perpendicular (as viewed from above) to the trench walls.
- Struts must be installed on panels with between 1000 and 1250 pounds (454–567 kg) of activation force.
- Struts must be within 1 foot (0.3 m) minimum and 2 feet (0.6 m) maximum (below) the trench lip and within 1 foot (0.3 m) minimum and 2 feet (0.6 m) maximum (above) the bottom of the panel with a maximum 4 feet (1.2 m) spacing vertically between struts.
- Both ends of all struts must be secured to panels/strongbacks with two 16d nails.

Trench Rescue Shoring Plan

The fundamental trench rescue shoring plan must always start with providing immediate protection for the victim(s). The tactics used to support that plan must be easily expanded from protecting the victim to creating a safe working area for rescuers. The shoring system plan is then completed with a continuation of efforts that are necessary to maximize rescuer and victim safety during the extrication process. A planned approach can take what appears to be an overwhelming task and divide it into manageable pieces. By dividing the shoring into phases, a rescue team can create separate but linked goals for each phase. These goals must be recognizable and attainable even at complex trench rescue situations.

The trench rescue shoring plan presented here is structured around a fundamental strategy based on rapidly protecting a trapped victim from trench collapse. The plan is applicable to operations level trenches (nonintersecting trenches not more than 8 feet [2.4 m] deep) and technician level trenches (intersecting trenches, trenches deeper than 8 feet [2.4 m]). Tactics and procedures specific to each of the five general steps of the plan must be learned and practiced often.

Shoring Plan Briefing

It is essential that every member of the panel and shoring team involved in the shoring operation

FIGURE 7-14 The shoring plan briefing includes the operations officer, and officers from the entry team, shoring team, and panel team.
Courtesy of Chad Godfrey.

understands their assignment, duties, and responsibilities. The operations officer, rescue team manager, panel team officer, shoring team officer, and entry team officer will make the decisions and will coordinate, supervise, and implement the shoring plan **FIGURE 7-14**. Although the entry team seldom performs primary and secondary shoring duties, their input regarding victim care, extrication operations, and victim removal routes must be given the highest consideration in the shoring plan development. With a rescue situation, the focus must be on rapidly protecting the victim(s). The briefing must include mitigating and removing hazards, shoring assignments, tactics, and sequence for primary shoring. Following the installation of primary shoring, a briefing for secondary shoring tactics and sequence must occur. After secondary shoring has been installed, a briefing for complete shoring and ongoing shoring assessments needs to be provided.

Procedure for the Trench Rescue Shoring Plan

Step 1: Trench Size-Up

The purpose of the trench size-up is to gather the information needed to shore the trench safely. That

FIGURE 7-15 The shoring size-up assesses the lip conditions, the trench size and shape, the collapse potentials, the victim condition, and the equipment requirements (tabulated data).
Courtesy of Ron Zawlocki.

FIGURE 7-16 Determine the lip protection methods that will provide for the safe and efficient installation of shoring equipment.
Courtesy of Ron Zawlocki.

FIGURE 7-17 Determine the sections of trench wall that will likely collapse within the rescue area.
Courtesy of Cecil V. "Buddy" Martinette, Jr.

information includes the following assessments **FIGURE 7-15**:

- Assess the lip: Lip protection must be installed before shoring and panels team members can begin shoring operations. Given a collapse trench incident, the best practice method for lip protection is the use of lip bridges **FIGURE 7-16**.
- Assess the trench: Measure the depth of the trench floor to determine the level (operations or technician) of the incident. Measure the depth from the lip to the debris (collapse) pile to determine the number of panels needed. Measure the width of the trench to determine the length of struts needed. Measure the L for the trench and any surcharge loads to determine the total L. Estimate the wall angles.
- Assess the collapse potential: Identify signs of impending collapse. Look for fissures (cracks), cornices (undercut walls), bulging walls, surcharge loads within the L area, layered soil, and flowing water in the trench walls **FIGURE 7-17**.
- Assess the voids: Voids are created by sections of trench walls that have collapsed. The voids must be filled to both minimize the movement of the back wall of the void and to create an adequate point of resistance to any movement (load) being transferred from the opposing trench wall. To determine the best backfill technique for each void, rescuers must know the void size (height, width, and length), the angle of the void back wall, and whether it is an open or closed lip void (blown out or undercut) **FIGURE 7-18**.
- Assess the victim(s): The victim's condition will dictate the benefit and subsequent risk levels that will be acceptable. Calculated and

FIGURE 7-18 Determine the location, size, and shape of the existing voids.
Courtesy of Ron Zawlocki.

reasonable risks are appropriate for live victims. Minimal risks are appropriate for clearly deceased victims.

- Use tabulated data: The trench rescue shoring charts provided in this text have been engineered specifically for collapse soil conditions. The Michigan Urban Search and Rescue (MUSAR) rescue shoring charts have a minimum factor of safety of 2:1.

Step 2: Primary Shoring

The purpose of **primary shoring** is to rapidly provide protection to the victim(s) by stabilizing the area(s) of the trench adjacent to the core of the victim. The scope of primary shoring includes the use of strategically placed panels, struts, backfill, and occasionally, a **single point shore**. Strut pressures are sometimes temporarily set lower than the manufacturer's specifications and are used to temporarily hold the soil in place until backfill can be added and the initial shoring can be expanded into a full system. Primary shoring is usually concluded with the placement of two or three struts. A good primary shoring plan must allow for future installation of secondary shoring.

Step 3: Secondary Shoring

The purpose of **secondary shoring** is to provide a safe zone for rescuers working inside the trench. The scope of secondary shoring includes expanding and enhancing the area shored during primary shoring. A common goal of secondary shoring is the development of a safe zone that is at least 12 feet (3.7 m) wide (three-panel set). Appropriate vertical and horizontal strut spacing must be achieved within the shored area. During secondary shoring, backfill is put in place and struts are activated to the manufacturer's recommended pressures, collars are locked, and air pressure is released from the struts.

Step 4: Complete Shoring

The purpose of the fourth step—complete shoring—is to maximize the safety of the rescuers and the victim during extrication and victim removal operations. The scope includes the creation of a safe zone that is at least as wide as it is deep (e.g., a 15-foot (4.6-m) deep trench with three panels set and, with rescuers working inside, would be expanded to a four-panel set). During this step, all struts are pressurized to a professional engineer's or the manufacturer's specification, collars are locked, air pressure is released from the struts, all strut bases are nailed, and **supplemental shoring** (when needed) is in place.

Step 5: Shoring Performance Assessment

The purpose of the shoring performance assessment is to evaluate and adjust shoring to maintain safety. All shored areas, spoil piles, and trench walls adjacent to the shored area must be monitored throughout the rescue/recovery operation. The procedure is to physically inspect all shores twice during the first hour and then once per hour. If struts are loosening, use the strut activation mechanism (pneumatic pressure, screw jack, wedges) to adjust strut activation forces. Increased soil force means that the panels or wales are bending. Having wood in your shoring system provides warning signs (creaking, groaning, popping noises).

When signs of loading are present, add struts halfway between existing struts to reduce the unsupported surface area on the bending wood members. Visually check the walls (on both sides of the trench) that are next to the shored area. Look for signs of moving soil, such as widening fissures, bulging, leaning, sloughing, and raveling. Check under the lip protection for new or widening cracks. Finally, expand the shoring to support adjacent trench walls that are showing movement.

Emergency Procedures

In the event of soil movement or the recognition of any unmitigated hazard, an evacuation signal will sound and rescuers will exit the trench and the trench lip area and assemble at a designated emergency staging area for an incident action plan briefing.

Trench Rescue Equipment

The core trench rescue shoring equipment includes struts, panels, and wales.

Struts

Trench rescue shoring includes the use of struts. Struts are the horizontal braces (columns) that extend across the trench and transfer the forces from one trench wall, through the strongback/panels, and into the opposite wall. Struts are also known as shores or cross braces. The terms *struts*, *cross braces*, and *shores* are used interchangeably throughout the trench shoring industry. In this text, we will use the term *strut*.

Pneumatic, hydraulic, timber, and screw jacks are common types of struts used by trench rescue teams **FIGURE 7-19**. Each has advantages and disadvantages. The use of each of those types of struts is discussed in this chapter, but step-by-step instructions are provided for only pneumatic and timber struts. Rescuers using hydraulic and screw jack struts for rescue operations are encouraged to review the installation protocols established by manufacturers of these devices and by their rescue team's authority having jurisdiction.

The **strut activation force** creates pressure on the soil behind the strut and panels. When properly used (controlled, measured, and distributed), strut activation forces can help stabilize the soil. Additionally, correct strut pressure can actually increase the strength and performance of the strongbacks, panels, and wales.

A best practice for trench rescue shoring is to use struts that meet the following criteria:

- Can be installed with controllable and measurable strut pressure
- Can be installed and removed without entering the trench
- Have full strength with activation forces of between 1000 and 1500 pounds (454 and 680 kg) of force

The group of rescuers who install struts is called the shoring team. Teams are discussed in detail in Chapter 3, *Initial Actions*. The recommended minimum number of personnel on a shoring team is four. The duties of those members differ slightly based on the type and style of struts being used and will be addressed in the skill drills in this chapter.

> **TIP**
>
> Strut strength, regardless of the type, reflects the strut's length-to-diameter ratio. All other things being equal, as the length of a strut increases, its strength decreases.

Strut Force

Throughout this text, you will encounter the terms *strut pressure* and *activation force*, which are related but different measurements. *Strut pressure* refers to the amount of air pressure that the air system is sending to the strut. It is measured in psi (or kg/cm^2) and can be seen on the air system's gauge.

By comparison, the strut activation force is a measurement of the total force that the strut exerts on the panels. The total force is the pressure (psi or kg/cm^2) multiplied by the number of square inches of the strut piston surface area. A typical 2.5-inch (6.4 cm) diameter strut has a surface area of about 5 square inches (32.3 cm^2). If the air pressure used to shoot the strut is 200 psi (14.1 kg/cm^2), you would simply multiply the pressure by the surface area (in this case, 200 × 5 = 1000) to calculate the activation force. That is, the total force exerted by the strut is 1000 pounds (of force).

Strut testing, as conducted by MUSAR and validated by professional engineers indicated that a strut

FIGURE 7-19 Hydraulic struts. The construction-based shoring practices shown here are not safe for trench collapse conditions.

© Jones & Bartlett Learning. Photographed by Glen E. Ellman.

activation force of between 1000 and 1500 pounds (454 and 680 kg) of force enhances soil distribution pressure behind the panels, strongback and wale strength, and overall system performance. It also stabilizes the shoring system by tightening the panels against the walls. The primary purpose of struts, however, is to transfer energy, not create it. While some energy is created from the activation pressures of struts, the amount of pressure should be regulated and distributed (not concentrated) on the trench walls. Excessive strut pressure creates unwanted energy and can result in dangerous soil pressure behind the shoring system. Pushing too hard on soil that has already failed once could cause additional failure to occur. This is the reason that pneumatic struts are the preferred struts. They are only capable of a limited amount of activation force that does not reach "excessive force" levels.

TIP

The strut activation force creates pressure on the soil behind the strut and panels. When properly used (controlled, measured, and distributed), strut activation forces can help stabilize the soil. Additionally, correct strut pressure can actually increase the strength and performance of the strongbacks, panels, and wales.

A best practice for trench rescue shoring is to use struts that meet the following criteria:
- Can be installed with controllable and measurable strut pressure
- Can be installed and removed without entering the trench
- Have full strength with activation forces of between 1000 and 1500 pounds (454 and 680 kg) of force.

Pneumatic Struts

Pneumatic struts use air pressure to extend the piston inside the cylinders to create a force against the strongbacks, panels, and wales positioned on the trench walls **FIGURE 7-20**. With pneumatic struts, once the desired air pressure is achieved, the struts must be mechanically locked before entry. Depending on the strut being used, the mechanical lock can be accomplished by (1) twisting a collar that rotates on the cylinder and inserting pins through the piston into holes, (2) spinning a collar that moves on the piston, or (3) utilizing a built-in automatic (one directional) locking mechanism. Automatic locking struts do not require a rescuer to enter the trench to lock the collar. Collars that spin can be locked from outside the trench with a strut-locking tool but are sometimes locked by a rescuer inside the trench. Collars that are twisted and pinned require a rescuer to enter the trench to lock them. Whenever a rescuer is in a trench that has not been completely shored, the rescuer is exposed to potential collapse. The strut installation procedures used must minimize that exposure time.

Organizations that use struts that can be locked from outside the trench (non-entry) should use a shoring team consisting of a minimum of four members, including a shoring team officer, controller (shooter), and two strut handlers. The shoring team officer develops the shoring plan; assigns team positions; determines strut length, placement (verifies the strut is level within 10 degrees), sequence, and pressures; and manages and supervises the shoring team operations. The controller, or shooter, sets up the air system, shoots the struts at the pressure designated by the officer, and releases the strut pressure as ordered. Two strut handlers assemble the struts and ropes, carry the struts to the trench wall, position the struts for shooting, and verify that the strut is perpendicular (within 10 degrees) to the trench wall, and lock the collar from outside the trench.

Pneumatic Strut Pressure

Pneumatic systems allow rescuers to measure (gauge) and control (vary) strut pressure. When using such struts, the pressures can be varied to accomplish different tasks. For example, when a collapse has created a void in the trench wall, a panel set can be held in place with a strut shot at 75–100 pounds per square inch (psi) (5.3–7 kilograms per square centimeter [kg/cm^2]). If the strut were shot at 200–250 psi (14.1–17.6 kg/cm^2) on a panel that is directly over a void, it would bend the panel/strongback. By temporarily using a reduced strut pressure, the panel can be held in place while backfill is installed or completed behind the panel. After the backfill is complete, the strut pressure can be increased to the pressure (psi) recommended by a professional engineer or the strut manufacturer. Knowing how and when to adjust the air pressure, based on the situation presented, is an important part of strut installation procedures.

FIGURE 7-20 Pneumatic strut.
© Jones & Bartlett Learning. Photographed by Glen E. Ellman.

> **SAFETY TIP**
>
> Reduced strut pressure should be only temporary. Ultimately, every strut in the shoring system needs to be brought up to the pressure recommendations of a professional engineer or those found in the manufacturer's tabulated data. If after backfill is completed, the pneumatic pressure of a strut causes a shoring system to slide, bend, or move (after backfill is in place), that system is inherently weak and would have no chance of resisting the forces of an active trench wall. The potential soil force could be many times larger than the activation force.

Pneumatic Strut Placement

The placement and sequence of strut installation is important to ensure safe and efficient rescue shoring. The proper strut placement and installation sequence at a rescue incident depends on several factors, including how deep the trench is, what is likely to collapse next, and whether a rescuer must enter the trench to lock the collar manually.

Trench rescuers have tried to comply with OSHA shoring standards such as shore from the top down or other prescribed strut sequence recommendations such as middle-bottom-top. Both of these prescribed sequences can work well on trench walls that have not collapsed. However, at a trench rescue incident, strut placement and sequencing must be based on protecting the trapped victim. Shoring rules like shore from the top down or middle-bottom-top are not applicable to collapsed walls because of the variety of collapse conditions and the variety of resulting collapse patterns that are possible. Rescuers need to sequence the placement of struts so that the initial strut positions and holds the panels in place while providing a level of protection from the next potential collapse. The second strut needs to resist the backfill forces and subsequent struts, if needed, and must be placed to comply with vertical spacing requirements. In most rescue situations, protecting the victim from additional collapse requires the placement of panels, struts, and backfill in the void area. This tactic is called primary shoring.

> **TIP**
>
> Rescue organizations are advised to obtain written approval from strut manufacturers or registered professional engineers for installation practices (spacing and sequence) that are not specified in the tabulated data.

Pneumatic Strut Spacing

The need for close sheeting at cave-in incidents and the design of trench rescue panels provide an apparent answer to most strut spacing questions. The 4-foot (1.2-m) rule—no space greater than 4 feet (1.2 m) between any horizontal or vertical shoring point—is a great rule of thumb and the cornerstone of most trench rescue shoring systems. If the top strut can be installed 18 to 24 inches (46–60 cm) from the trench lip, with the bottom strut installed within 24 inches (60 cm) of the bottom of the panel/strongback, and the resulting distance between the top and bottom strut is not more than 4 feet (1.2 m), then only two struts would be necessary for most 8-foot (2.4-m) deep trenches. If the distance exceeds 4 feet (1.2 m), simply install another strut midway between the struts already in place.

By placing the panels right next to each other and shoring to the center of the strongbacks, you automatically maintain the 4-foot (1.2-m) horizontal spacing. While the 4-foot rule is a practical rule for trench rescue shoring, it is not considered in the tabulated data developed by most strut manufacturers.

Hydraulic Struts

Hydraulic struts have been successfully used to shore construction trenches for many years. The rescue service borrowed this technique from construction workers. The activation pressure is created by pumping hydraulic fluid from a reservoir through hoses and into the strut cylinders, which house the moving pistons. Hydraulic struts do not have collars that (mechanically) prevent the movement of the pistons and cylinders. Instead, the (compressive) strength of the strut remains dependent on the hydraulic pressure throughout the entire duration of their use in the trench. As a result, the pressures used for hydraulic struts are much higher than the pressures needed for pneumatic, timber, and screw jack struts. These higher pressures are effective on trench walls that are intact and have stable soil conditions. Hydraulic struts are not a good practice for trench rescue shoring, however, and therefore skill drills are not presented for their use. Users of these struts should request tabulated data and shoring procedures from the manufacturer that is specific to cave-in soil conditions.

Screw Jack Struts

Screw jacks are used with timbers **FIGURE 7-21**. The screw allows the strut to be tightened (compression) against the trench walls. The activation pressure created by turning the screw, although much more gentle than pounding in an oversized strut, is not a

FIGURE 7-21 The capacities of pipe-style screw jacks do not cover a wide enough range of collapse condition soil forces to be considered a good choice for rescue teams.
© Jones & Bartlett Learning. Photographed by Glen E. Ellman.

measurable or predictable force. Attempting to create a reliable activation force is difficult and is based on the installer's touch rather than gauges. Because the installer must spend time in a trench before the struts have been tightened, it is critical to minimize exposure time to the shoring team during installation. Many of the screw jack struts used by rescue teams have a significantly lower breaking (buckling) strength than pneumatic and hydraulic struts. This is due in part to the smaller diameter of the screw jack struts (length–diameter ratio). Any rescuer needs to consider that the strength of a timber strut using a screw jack is the smaller of the timber capacity or the screw jack capacity. Screw jack struts are not a recommended best practice for trench rescue shoring, and therefore skill drills are not presented showing their use. Users of these struts should request tabulated data and shoring procedures from the manufacturer that is specific to cave-in soil conditions.

Timber Struts

Timber struts have a long history in both construction and rescue shoring. In either application, they are time consuming and create additional risks because they expose installers to the trench before it is safe. The minimum recommended size of timber struts is 4 × 4 inches and larger. OSHA provides a comprehensive guide for timber shoring. Rescue teams that deviate from that standard should have tabulated data for rescue timber shoring approved by a registered professional engineer.

In some communities, the fire department's rescue and truck companies carry enough timbers to start initial shoring at a trench rescue incident. Using timber shores as a stopgap method to protect a trapped victim is a legitimate technique that all rescuers should have in their tool box. However, timber shores have limited application at technician level trench rescues and at rescues where time is of the essence.

Installing timber shores begins after panels are in the trench by measuring the distance between panels and cutting the timber to fit. This requires the panels to be tight against the walls to get accurate measurements. Timber shores are then set from the top down. Working from the top down on a ladder allows the rescuer to install the shores without being more than waist deep in the unprotected part of the trench. It is hoped that if a secondary collapse does occur, the rescuer will most likely be buried only up to the waist, which normally is a survivable incident.

Prenailed bottom **scabs** can help the rescuer properly hold and position the timber strut for installation. In some cases, the strut needs to be forced into place using a maul or hammer. Another technique to make installation and adjustment of timber shores easier is to rail the panels with 2 × 4-inch stringers. Individual struts can then be moved up and down inside the vertical rails. The rescuer has to put only two scabs in place to finish the installation.

The top shore is installed between 12 and 24 inches (30 and 60 cm) below the trench lip. The middle and bottom shores would follow, in that order. The number of struts necessary to protect the trench is based on the L value of the trench. You can never go wrong by placing too many struts, however. If you are not comfortable with the environment or the conditions, add more struts than you would otherwise use.

Timber Strut Placement

The shoring sequence for the first set of panels is top, middle, and then bottom. This sequence is the safest application for timber shore use; however, it often does not provide for the most efficient method of protecting the victim from impending collapse. Because timber shoring requires the installer to descend the ladder to install struts, the installer is continually exposed to the risk of collapse. That risk of can be somewhat reduced if the installer's head and chest remain above a properly installed strut. However, this also limits the vertical distance between struts. A vertical distance of between 24 and 30 inches (60 and 76 cm) is usually achievable using this method.

Timber Strut Activation Forces

All struts must have a mechanism that provides a compressive force which pushes the panels/wales tightly to the trench walls and holds the struts in place. With timber struts that compressive force is commonly developed by using wedges, screw jacks, or oversized struts that are driven into place with hammers. Timber struts that are activated by using wedges or screw jacks can be tightened by further driving the wedges or by use of the screw jacks if the struts loosen as other struts are installed. This helps keep the struts tight, but results in an inconsistent and unpredictable activation force. The oversized strut technique is called cut-to-fit. The distance between strongbacks is measured and an additional ½ to 1 inches (13–25 mm) is added to the strut length. The longer-than-needed strut is then placed between the strongbacks on an angle and is beat into place with a hammer. The cut-to-fit technique often bends the strut and loads it eccentrically rather than axillary. It also results in unequal strut pressures with some struts being loose while others are tight. Typically, the first strut installed by this method loosens as subsequent struts are installed. This results in more activation force on lower struts and less activation force on upper struts.

The authors have tested this method and have recorded excessive strut activation forces that could cause rather than prevent collapse. For these reasons, the authors do not condone the use of cut-to-fit techniques.

> **TIP**
>
> While conducting a test to determine the forces that a timber strut and an Ellis screw jack could create, the authors randomly selected 10 firefighters and asked them to tighten the screw jack as much as they could using a 22 oz framing hammer handle as a lever to tighten the screw. Tightening a screw jack to the max is not recommended procedure because there is no way (no gauges) to determine when an acceptable amount of force has been applied. As a result, stronger firefighters usually apply the most force. The goal of the test was to determine if firefighters could inadvertently apply excessive strut activation forces (i.e., forces that could cause a collapse). In this test, all 10 firefighters (of various sizes and strengths) applied excessive strut activation forces.

Trench Rescue Panels

Panels are required to both collect the loads from the unstable wall and distribute them to the opposite wall. They are very effective when unstable soil and collapsed wall conditions are encountered.

The recommended best practice for construction of a **trench rescue panel** by the authors is ¾- or 1-inch (19- or 25-mm) thick, 4 × 8-foot (1.2 × 2.4 m) sheets of FinnForm, Shorform, or Chudoform with a 2 × 12-inch LVL strongback attached (composite).

Strongbacks should be permanently connected (screwed and glued, not bolted) to the panels, as this structure offers a greater system strength and, therefore, greater system safety. This panel construction, although expensive based on the cost of the premium materials, will result in the strongest panel with the ability to handle nearly any incident. Rescuers need to remember that they are dealing with dynamic forces and shoring time requirements that are different from what is found in shoring guidance provided for construction shoring. Most tabulated data provided by shoring manufacturers is not designed for shoring rescue soil conditions.

Most trench rescue teams rely on default shoring methods that include the use of 4-foot (1.2-m) wide by 8-foot (2.4-m) tall panels with strongbacks positioned in the center. Placing panels right next to each other and installing struts on the strongbacks provides horizontal spacing at 4 feet (1.2 m) on center. Horizontal and vertical spacing of 4 feet (1.2 m) or less will safely shore the majority of soil conditions found at trench emergency incidents if composite panels are used. Application of this method need not wait for the soil to be tested and does not complicate the issue by asking a rescuer, who may be unfamiliar with soil testing, to make a critical decision based on soil analysis.

> **TIP**
>
> Trench rescuers must recognize that trenches dug for training usually result in much straighter and more vertical walls than will be found at real-world trench rescue incidents. Instructors should work with the excavator operator to produce ugly trenches rather than pristine and squared-up walls. Trenches that have collapsed often require cutting, shaping, and overlapping panels to make the equipment work. At that point, you should not worry about making the shores pretty. Concentrate instead on covering the unstable walls with panels and preventing additional collapse with strong trench rescue panels and struts.
>
> The only consideration with adding more panels or cutting down panels is the increase of struts in the trench. More columns of struts can hinder access and extrication, so make sure the plan to remove the victim is considered.

Although most shoring manufacturer's tabulated data state that panels should not be given any structural value, recent tests conducted by the MUSAR Training Foundation provided the following results:

- The ¾-inch (19-mm) FinnForm is more ductile than 2 × 12-inch sawn lumber strongbacks

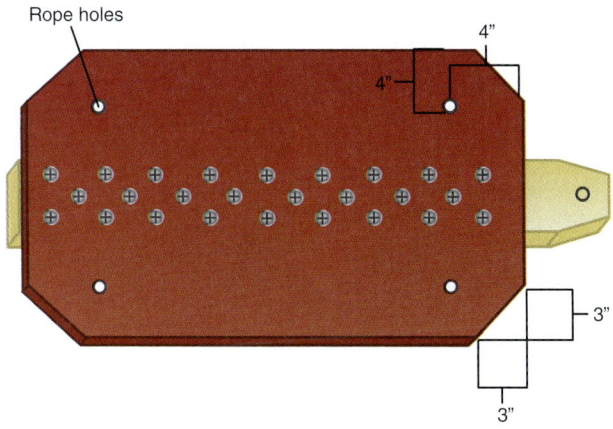

FIGURE 7-22 Corners are cut at a 45-degree angle 3 inches (8 cm) from each end. Rope holes are drilled 4 inches (10 cm) from each end and side.

(both treated and untreated white pine 2 × 12s were tested). This provides more warning before panels fail.

- The ¾-inch (19-mm) FinnForm is significantly stronger than 2 × 12-inch sawn lumber strongbacks (both treated and untreated white pine).
- When glued and screwed together, ¾-inch (19-mm) FinnForm and 2 × 12-inch LVL lumber strongbacks are significantly stronger than FinnForm and strongbacks that are bolted together. Screwing and gluing help the two components react to force together instead of separately (see Chapter 4, *Personal Protective Equipment and Equipment Basics*, for more information).

The bottom line is that panels should always be used to shore collapsed walls. Shoring that includes panels provides safety, strength, and redundancy. The effects of **strut pressure** on soil pressure (stability) using ¾-inch (19-mm) FinnForm panels with strongbacks greatly improves the distribution of forces. In turn, strut pressures that are well distributed are less likely to cause collapse and are more likely to stabilize soil walls.

Setting panels in a trench is accomplished using several different methods. The method used depends on the condition of the trench and the location of the victim **FIGURE 7-22**.

Wales

Wales are essentially beams used in trenches to span large areas of trench walls (without intermediate struts). The large area needing to be spanned may be a result of the voids created by the cave-in, the

FIGURE 7-23 Inside wales used to span a trench panel.
Courtesy of Ron Zawlocki.

trench shape, or an obstruction at the wall. Wales are also used to create room for extrication work and victim removal. Wales essentially act as headers do in building construction. The wale takes the place of struts, transferring soil load on a panel into the opposing wall by way of struts on opposing wales and panels. See Chapter 4, *Personal Protective Equipment and Equipment Basics*, for an additional discussion on wales.

For years, trench rescue training courses have taught two different wale techniques: inside wales and outside wales. Inside wales are installed after panels have been positioned on the trench walls **FIGURE 7-23**. The panels provide a large surface area to collect or distribute the load that is transferred through the struts to the wale. Outside wales are installed directly on the trench wall. Panels are then installed over the wales and struts are installed on the panel/wale interface. The surface area of a wale is relatively small and that creates a concentrated load on the soil. That concentrated load can cause the soil to fail (collapse). The use of outside wales limits the direct collection of soil

FIGURE 7-24 Outside wales concentrate the load from opposing walls which can allow or even cause collapse.

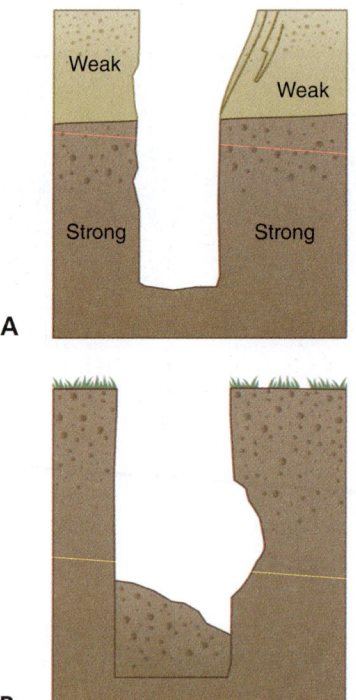

FIGURE 7-25 A. Open void. **B.** Closed void.

load to that which contacts the wale. The rest of the distribution requires the soil to collect the load internally. The 6 to 8-inch (15–20 cm) gap that is created by placing a wale between the trench wall and the panels (see **FIGURE 7-24**) provides the soil with an area to move toward (point of least resistance) and gives the moving soil a chance to gain momentum, which will increase the force on the shoring system. Because of these potential problems, the authors do not recommend the use of outside wales.

Shoring Voids

Collapsed trench walls have voids that are the result of a variety of different types of soil failures. (This is discussed in detail in Chapter 2, *Soil and Collapse Mechanics*.) Depending on the type of soil and the conditions present during the time the trench is open, voids will vary in type, the angle of the remaining void back wall, and the size of the void. Voids must be filled behind panels and struts to support soil forces from trench walls that are across the trench. Rescuers must be able to make rapid and accurate assessments of voids to select appropriate backfill options.

Void Assessments

Void assessments are necessary to determine appropriate backfill options. We must know the void type (open lip or closed lip), the angle of the void back wall, and the size of the void to determine proper backfill options.

- Void type: Open lip voids start at the lip, whereas closed lip voids begin below the lip area, commonly at the midpoint in the wall but sometimes at the toe **FIGURE 7-25**.
- Wall angle: The angle of the back wall of the collapse must be determined to select the correct backfill options. The angle can be determined by using an angle finder application on a smartphone or with a carpenter style angle finding tool. High angles are those between 70 and 90 degrees with 90 degrees being vertical. Low angles are those less than 70 degrees. The installation of struts, panels, wales, or airbags on low-angle walls will create sliding forces that result in shoring systems that are not capable of resisting the loads associated with unstable and dynamic soil conditions.
- Size: A tape measure is used to measure the height, length (simple L), and width of the

void. A cubic yard is 3 feet (height) × 3 feet (length) × 3 feet (width) or 27 cubic feet (0.9 m × 0.9 m × 0.9 m or 0.76 m^3). Using this measurement as a reference, a small void is less than ½ cubic yard (<13.5 cubic feet [<0.38 m^3]), a medium void is ½ to 1 cubic yard (between 13.5 and 27 cubic feet [0.38 and 0.76 m^3]), and a large void is greater than 1 cubic yard (>27 cubic feet [>0.76 m^3]). In trenches that are 8 feet (2.4 m) or less in depth you will likely encounter small- and medium-size voids following an initial collapse.

Another default method for protecting a trench from the collapse of soil is the use of shields. Utility and construction companies, and now the fire service, occasionally use trench boxes while working in a trench; however, in a trench rescue incident, they are typically found on the trailer instead of in use. If a shield system is onsite but not in the trench, rescuers might consider using the trench shield to create a safe working zone within the trench.

Modular shields, also referred to as **trench boxes**, consist of steel or aluminum side walls and spreaders that are assembled at the scene and are rated for human safety **FIGURE 7-26**. Modular shoring is available in a variety of lengths, heights, widths, side wall thickness, and weights. In addition to coming in fixed sizes, some of these units are available with adjustable struts **FIGURE 7-27**.

The modular systems can be quickly assembled and lowered into the trench for use as rescue isolation devices or safe zones for rescuers. Even a small modular aluminum shield can weigh several hundred pounds, but with a rope system and a carefully coordinated rescue squad, a small shield can be lowered into a trench. Each brand and individual models have specific assembly and installation protocols and procedures that must be followed. Rescuers who use shields as part of their response must receive shield-specific training and follow the manufacturer's tabulated data and practices. More importantly, rescuers should train regularly with the system so that installation becomes second nature.

It is possible to arrive at a trench rescue incident to find a shield in place. Rescuers need to evaluate the shield and the way it is installed to determine if it is safe for use. The knowledge needed to make such an evaluation is found later in this section.

Shielding Systems

Shield systems must be designed by a registered professional engineer and will be designated for use only up to certain depths and widths in certain classifications of soil. Those restrictions may be referenced as the tabulated data for the shield. The tabulated data for each box or system is required to be on the job site and must be referenced only for a particular shield or system. Shields may consist of large double-wall steel panels with specially designed pipes that span the width of the trench to hold the panels apart, or may consist of aluminum corner posts, spreaders, and panel sections that can be easily customized and assembled on site. The modular aluminum shields can also be constructed as a four-sided box, protecting the people in the trench from collapses

FIGURE 7-26 Modular shield.
Courtesy of Ron Zawlocki.

FIGURE 7-27 The trench shield is a stable piece of commercial equipment that is used to provide worker protection.

while they are working in small repair-hole-type excavations.

The selection and use of a trench shield is based on soil type and trench width and depth. The criteria for use are based on the manufacturer's data and the analysis of a competent person. A trench box is a safe and effective method for protecting construction workers in a trench. Using one at a rescue scene, however, requires a well-developed plan. If the use of a trench box is deemed necessary, the following issues should be considered:

- The use and placement shall be in accordance with all specifications, recommendations, and limitations issued or made by the manufacturer.
- The top of the box must extend 18 inches (46 cm) above an otherwise unprotected area of the trench. This means the box must extend 18 inches (46 cm) above the lip, or if the top of the box is below the lip, that any soils above the box must be properly sloped or benched back and must intersect the box 18 inches (46 cm) below the top of the box. This 18-inch (46-cm) extension not only keeps soil out, but also provides protection from construction materials such as pipe that may be at grade level from rolling into the working area of the trench.
- The bottom of the trench box may be a maximum of 2 feet (0.6 m) from the bottom of the trench, and only if a competent person determines the soil will stand with no possible loss from behind or below the shield and if the shield is rated for the full depth of the trench.
- The trench box height may not be extended with steel plates unless approved sheet piling techniques are used.
- The shield should be inspected to ensure it is in good condition, relatively free from excessive damage, and that all pieces of the assembly are properly secured. For a two-sided shield, there should be a minimum of two braces or spreader pipes at each end of the protective shield, each secured with pins and keepers. The spreaders should be free from significant dents or bends.
- The authors recommend that the gap between the trench wall and trench box sides should be 6 inches (15 cm) or less. This limits the impact force of soil that may break free from a trench wall as it hits the trench box. This minimizes the potential damage to the shield and reduces the potential movement of the box within the trench. Backfill (e.g., air bags, backshores, timber/soil) may be required to resolve larger gaps.

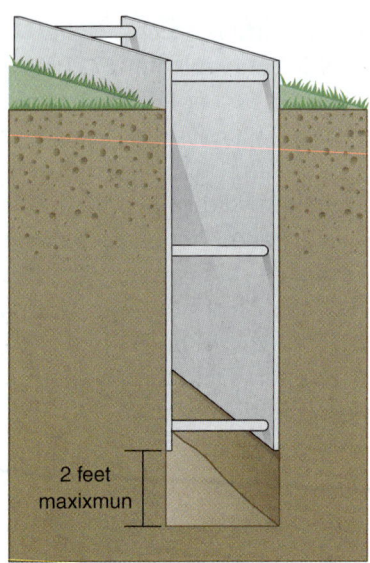

FIGURE 7-28 Diagram of a trench box.

- Unless the shield has four complete sides, the authors recommend that the box extend 4 feet (1.2 m) beyond the rescue work area in both directions of the trench length.
- Rescuers may never work outside the protection of the trench box or other protected area of the trench **FIGURE 7-28**.

SAFETY TIP

Installing a trench shield that requires the use heavy equipment and operators is a technician level skill. Operations level rescuers should not attempt to install heavy shields because lowering or dragging a trench box with heavy equipment, which can weigh several tons, into the area of a trench containing a victim may cause additional victim trauma. The rescue team must have strong indications that the trench box insertion will stay clear of the victim's position and location.

Installation of a trench box may also cause additional collapse. The issues of surcharged (backhoe/crane) loads, vibrations (heavy equipment), and collision of the trench box with the trench wall must be considered and resolved.

TRENCH RESCUE SHORING PRACTICES

Practices of Trench Rescue Shoring

The overwhelming majority of trench rescues involve soil failure and the collapse of one or more trench walls. Shoring trenches that have collapsed and contain trapped victims is significantly different and much more dangerous than shoring trenches with walls that are intact and do not contain victims (construction shoring). Trench rescue training programs that teach practices and principles that are used by the construction industry and neglect to teach in the conditions found at trench rescue incidents give rescuers a false sense of security. Unfortunately, that false sense of security is not exposed until the rescuers are faced with a real trench rescue event. Training for trench rescue shoring must include a trapped victim, cave-in debris on the trench floor, and a variety of collapsed wall conditions that must be addressed. A concrete trench expert who is suddenly responding to a trench with an overhanging lip will not have the strategies to deal with the common complications found at trench collapse incidents. Training with these complications builds a toolbox of strategies that rescuers can use when they encounter difficult situations.

Lip Protection

The majority of trench rescue shoring activities should be completed from outside of the trench. In addition, these activities must take place on trench **lip protection**. Ground pads are placed to distribute rescuer and equipment weight around the trench lip, and lip bridges are used to transfer and distribute the weight even farther away from the lip. These actions are discussed in detail in Chapter 3, *Initial Actions*. Ideally, a ground pad will be centered at the victim's location, so that when you are working above the victim you are standing at the center of a ground pad. This will provide maximum load distribution at the most critical area of the trench. Proper installation of the lip protection is discussed next.

At each trench, rescuers will need to decide which kind of lip protection is best suited for the situation. For example, at a trench with stable walls, rescuers might decide to use 2 × 12-inch boards (or half sheets of ¾-inch [19-mm] plywood) on the spoil pile side of the trench and a 4 × 8-foot (1.2 × 2.4 m) piece of ¾-inch (19-mm) plywood on the other side (ground pads). Conversely, if a trench wall has collapsed or looks unstable, rescuers should decide to use lip bridges instead of ground pads. To install ground pads, follow the steps in **SKILL DRILL 7-1**.

Be aware that ground pads help distribute the weight of rescuers on the lip but do not eliminate the load above weakened walls **FIGURE 7-29**. In the event of a secondary collapse, the ground pad and everything on it (including rescuers) can fall into the trench. Ground pads and lip bridges are covered in detail in Chapter 3, *Initial Actions*.

SKILL DRILL 7-1
Installing Ground Pads (NFPA 1006: 12.2.2, 12.2.4)

1 When a spoil pile interferes with the placement of lip protection, the best solution is to begin by removing the the spoil while standing on a 2 foot (0.6 m) wide ground pad (either two 2 x 12 [0.6 x 3.6 m] or 2 x 8 [0.6 x 2.4 m] boards, or a 2 x 4 [0.6 x 1.2 m] section of plywood). After determining that the lip is stable enough for ground pad use, clear an area about 30 inches (0.7 m) back from the lip and as far forward as your shovel will reach. Rescuers should not walk or stand closer than 4 feet (1.2 m) from the trench wall on a section of unprotected trench lip.

(continues)

SKILL DRILL 7-1
Installing Ground Pads (NFPA 1006: 12.2.2, 12.2.4)

2 After clearing the spoil pile, step back and move the ground pad forward. Continue shoveling and advancing the ground pad forward. Be sure not to step off the ground pad onto an unprotected lip and avoid standing and walking on the spoil pile.

3 Repeat Step 2 until enough of the spoil pile has been removed to allow the ground pad to protect the lip in the rescue area. The "rescue area" is generally considered to be 6 to 8 feet (1.8 and 2.4 m) on both sides of the trapped victim. With a spoil pile obstruction, a larger walkway may need to be cleared and protected with ground pads in order to safely access the rescue area.

Courtesy of Ron Zawlocki.

FIGURE 7-29 This four-sided lip bridge was needed to install shoring over a repair hole where three of the walls collapsed on a city worker.
Courtesy of Ron Zawlocki.

SAFETY TIP
Plywood sheets (4 × 8 feet [1.2 × 2.4 m]) will bend, bow, and cup. To minimize tripping hazards, make sure that the edges (rather than the center) of the plywood contact the ground.

A lip bridge is built with girders (timbers), beams (ladders or timbers), decking (prebuilt wood platforms, manufactured aluminum stages, or ladders/lumber), and supports (sections of timbers placed away from the compromised wall and used to elevate the bridge above the lip) **FIGURE 7-30**. To install lip bridges, follow the steps in **SKILL DRILL 7-2**.

CHAPTER 7 Operations Level Trench Rescue Shoring   159

FIGURE 7-30 Fire service ladders with 2 × 12-inch lumber can be used as lip bridge beams and platforms.
Courtesy of Ron Zawlocki.

TIP

Platforms can be prebuilt or can be fire service ladders (beams) with 12-foot (3.7-m) lengths of 2 × 12-inch lumber (decking) placed on the rungs. Using 2 × 12-inch lumber alone (without beams) does not provide the strength to safely support the load of rescuers and equipment. Sixteen-foot (4.9-m) aluminum stages (or prefabricated LVL lumber) are the preferred lip bridge platforms **FIGURE 7-31**.

SKILL DRILL 7-2
Installing Lip Bridges (NFPA 1006: 12.2.4)

1 Place two ground pads on each side of the trench an equal distance (6–8 feet [(1.8–2.4 m)]) from the victim's head/chest. When using lip bridges, it is acceptable for adjacent ground pads to meet at the victim's location since all loads will be carried by the lip bridges.

2 Place lip bridge supports (6 × 6 inches) on pad at least 4 feet (1.2 m) from the trench wall.

Courtesy of Ron Zawlocki.

(continues)

SKILL DRILL 7-2
Installing Lip Bridges (NFPA 1006: 12.2.4)

3 Place girders across the trench on the supports.

4 Place platforms on the girders and slide the platform to the edge of the trench.

Courtesy of Ron Zawlocki.

FIGURE 7-31 Aluminum stages with three-person (750 lb (340 kg) safe working loads are a strong and durable option for lip bridge platforms (decking).
Courtesy of Ron Zawlocki.

Trench Rescue Shoring

Shoring trench walls that have collapsed and the resulting voids are skills that are critical to the fundamental concept of trench rescue shoring. Victims of a trench collapse are usually under or near the void in the wall that resulted from the soil failure (collapse). Protecting the victim from additional collapse requires shoring (panels and struts) the walls above and adjacent to the victim's position in the trench **FIGURE 7-32**. Most shoring activities should take place outside of the trench. Those activities can and should take place while standing on lip protection.

Until a minimum number of panels and struts are in place, rescuers working in the trench or on an unprotected trench lip are at high risk to be injured or killed by a collapse. Struts that require rescuers to

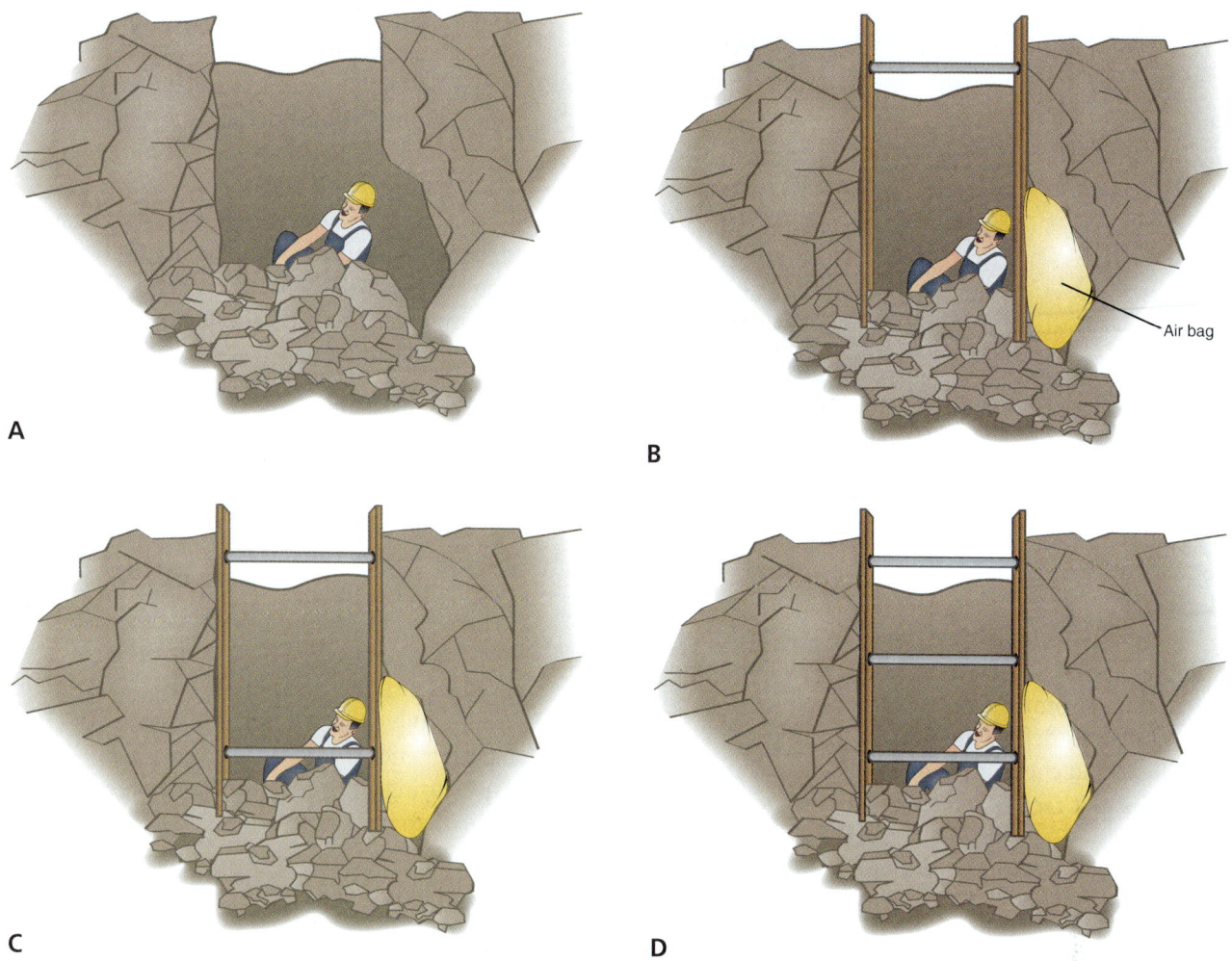

FIGURE 7-32 Shoring voids with struts that do not require entry. **A.** Position of a victim in a cave-in. **B.** Position of the first strut. **C.** Position of the second strut. **D.** Position of the third strut.

enter the trench to install them expose rescuers to that high risk. Struts that allow rescuers to install and remove them from properly installed lip protection minimize that risk. In this section, we have divided strut use into two categories: (1) entry shoring and (2) non-entry shoring.

Entry Shoring Overview

Entry shoring includes timber struts, pipe screw jacks, and pneumatic struts that cannot be mechanically locked without entering the trench. Entry shoring is inherently dangerous, because the rescuers are entering an unshored trench to install struts, but it becomes less dangerous as more struts are installed. Equipment exists that allows for installation without entry, which significantly reduces the danger and risk. The authors have included this section to help keep rescuers as safe as possible when they are limited to using this type of equipment.

To minimize hazards to rescuers, working with struts that require rescuers to enter the trench during installation necessitates the areas next to the collapse be shored before the area above the victim can be shored. This makes the operation much more dangerous for the victim, because they are exposed to a potential secondary collapse for longer periods, and it is possible that the adjacent shoring work could induce a secondary collapse if driving struts or wedges via pounding. Struts that require rescuers to enter the trench for installation are installed from the top down. This reduces the rescuer's exposure to collapsing walls located above them. However, while the rescuer is in the process of installing the top strut, there is a collapse exposure from both walls. Once the top strut is properly installed, the rescuers climb down the ladder only deep enough to install the next strut, while keeping their head and chest above the previously installed strut. The shored area at and above the strut level offers some degree of protection but it does not prevent the collapse of the trench walls below the strut. Such a collapse could knock rescuers off the ladder and bury them in the trench. A typical shoring sequence

FIGURE 7-33 Entry shoring practices expose rescuers to collapse hazards.
© Jones & Bartlett Learning. Photographed by Glen E. Ellman.

for shoring a small collapse/void with entry struts is described next.

Note: Entry shoring practices expose rescuers to collapse hazards **FIGURE 7-33**. The technique shown here reduces some of that exposure to the extent possible, but also delays the goal of protecting the victim's head and chest (airway). For these reasons, entry shoring is not recommended as a best practice by the authors.

Take the following steps to apply entry shoring at an open lip collapse:

1. Set up lip protection.
2. Lower panels into the trench on both sides of the void, and lower panels into the trench directly opposite the first two panels. If at all possible, the gap between the panels at the void location should be wide enough to fit a 4-foot (1.2-m) panel between them.
3. Measure the distance from strongback to strongback at the top of trench. Select, build, or cut the strut so that it will fit securely/snugly between the strongbacks.
4. Position a ladder in the trench for a shoring team member (installer) to climb in so that their head and chest stay above the trench lip.
5. While standing on lip protection, the shoring team members (strut handlers) lower the strut with ropes between the strongback to a depth of 18 to 24 inches (46–60 cm) below the lip.
6. From the end of the trench, the safety trench officer will determine the depth and will visually sight the strut to direct level placement (within 10 degrees), while the strut handlers verify that placement is within 10 degrees of perpendicular to the trench wall. Alignment of the struts can be verified with a smartphone angle finder app or with a carpenter's angle finder found in most lumberyards or home improvement stores.
7. The installer will descend the ladder to activate and lock the strut (wedges or screw jack) and will secure both ends with a minimum of two (16d) nails per end.
8. The installer will exit the trench until the next strut is prepared and ready to be lowered.
9. The installer will then descend the ladder, keeping their head and chest above the previously installed strut.
10. The shoring team will repeat steps 6 to 9 until three struts are installed on each **panel set**.
11. The panel team will install a panel at the void (between two previously placed panels) and one directly across the trench from it.
12. The panel team will lower wales to rest on the top struts on both sides of the trench.
13. While working within the shored panels, the installer(s) will activate and lock the strut (wedges or screw jack) to the wales and will secure both ends with a minimum of two (16d) nails per end.
14. The shoring and panel teams will repeat steps 11 to 13 to install bottom wales and struts on both sides of the trench.
15. Install appropriate backfill to the void area.

Non-entry Shoring Overview

Non-entry shoring can be assembled using rescue struts that can be completely installed, mechanically locked, and removed without entering the trench. Paratech and ResQtech struts are two currently available pneumatic systems that can be installed and removed without entering the trench. Non-entry struts allow the area directly above the trapped victim to be shored first. Rapidly protecting the victim(s) from additional collapse should be a strategic goal for all trench rescue operations.

With small/localized collapse, shoring the area directly above the victim is best accomplished by using the following:

- Panels: Lowering panels into opposite sides of the trench centered at the victim's head/chest
- **Positioning strut**: Installing a strut near the vertical center of panels to position and hold the panels on the trench walls (and provide some protection)
- **Backfill strut**: Installing a strut (variable activation force) near the center of the void area to contain and resist the pressure from the backfill

- **Compliance strut**: Adding a strut (when necessary) to comply with strut spacing requirements

To apply non-entry shoring, follow these steps:

1. First, set up lip protection. Position a panel set (**primary panels**) centered at the victim's head and chest. Install a strut near the vertical center of the panels to position and hold the panels on the trench walls (positioning strut).
2. Install a strut at the void area (backfill strut) at 500 pounds (227 kg) of strut activation force, and secure both ends with a minimum of two (16d) nails per end.
3. Place appropriate backfill behind the panel.
4. Install a strut to comply with vertical spacing requirements (compliance strut) as needed.
5. Using a wye, simultaneously bring all struts up to recommended operating pressures and secure both ends with a minimum of two (16d) nails per end when it is safe to enter the trench. Position panel sets (secondary panels) on both sides of the primary panel set. Repeat steps 3 to 5.

Panel Installation

When placing panels for a rescue, the first two panels are set centered on the victim (primary shoring), and then two more are set on each side (secondary shoring). To accomplish that, rescuers need to be competent with several panel installation options. We call these options the one-side panel set and the two-side panel set.

A one-side panel set is most often used when an obstruction (spoil pile, construction equipment, etc.) exists on one of the trench lips and impedes the transportation and installation of panel team equipment such as 4 × 8-foot (1.2 × 2.4 m) ground pads, wales, and panels. The side of the trench with the obstruction is referred to as the **weak side** and the side without obstructions is called the **strong side**. A one-side panel set involves installing both panels from the strong side of the trench. Most of the work is done on the strong side.

A two-side panel set is used when there is enough room on both trench lips to allow panel team operations to take place. A two-side set is accomplished by lowering panels into the trench from both sides of the trench. The panels are lowered into place from the same side of the trench on which they will be positioned.

For a one-side set, the first panel is installed on the wall that is across (opposite) from the side used to lower it. This technique works best when rails (usually 4 × 4-inch timbers) are positioned from the accessible lip and extend down to the toe on the opposite side of the trench; these are used to slide the panels across the trench and position it on the opposite wall.

When it is not possible to set the panel from the same side of the trench, you can use a one-side technique to set the panels using two 4 × 4-inch timber rails. To perform a one-side panel installation, follow the steps in **SKILL DRILL 7-3**.

SKILL DRILL 7-3
One-Side Panel Installation (NFPA 1006: 12.2.4)

1 Place two 4 × 4-inch timber rails (about 18 inches [46 cm] apart) in the trench so the bottoms of the rails create a designated destination for the strongback on the panel. The rails will go from the accessible trench lip to the toe on the opposite wall.

(continues)

SKILL DRILL 7-3
One-Side Panel Installation (NFPA 1006: 12.2.4)

2 Place a roped panel with the strongback facing down on the lip protection and slide it out using the lifting ropes for control; lower the panel onto and then down the rails.

3 The panel can then be pushed or pulled to the opposite side using ropes or a pike pole where a member of the panel team will stand it up on the trench wall and hold it until the opposing panel and a strut can be put in place.

SKILL DRILL 7-3
One-Side Panel Installation (NFPA 1006: 12.2.4)

© Jones and Bartlett Learning. Courtesy of MIEMSS.

4 After removing the rails, the panel directly across the trench from the first panel will be lowered and positioned from the strong side. These "primary panels" should have struts installed as soon as possible to protect the victim from subsequent collapse. Repeat steps 1 - 4 to install the "secondary panels." In a straight (non-intersecting) trench up to 8 feet (2.4m) deep the minimum recommended panels sets are three (six panels).

To prepare to install a two-side panel set, tie lifting ropes on the panel and move the panel into place on the lip protection. Align the panel so the strongback will be in the proper orientation for the position used in the trench (strongback facing the interior of the trench). Then perform the steps in **SKILL DRILL 7-4**.

TIP
The one-side panel set is also a good technique when a victim is present in the trench and additional care must be taken with panel placement. The rails can protect the victim by diverting the panel toward the wall and away from the victim, protecting them against panel drops.

TIP
For easy panel installation, use a chainsaw to cut off the rail just above the lip protection.

SAFETY TIP
As a rule, strive to create a *minimum safe* area (secondary shoring) of 12 linear feet (3.7 m) in the trench.

> **SAFETY TIP**
>
> Secondary collapse is a real possibility, especially after soil removal. Make sure that the victim is protected in such an event by setting panels and shores over the victim first.

For nonintersecting trenches up to 8 feet (2.4 m) deep, the recommended number of panels is six (three sets). Of course, this setup is not possible if the trench is less than 12 feet (3.7 m) long. In the latter case, after installing the primary shores, be sure to have panels set at both ends of the trench that will facilitate the

SKILL DRILL 7-4
Two-Side Panel Installation (NFPA 1006: 12.2.4)

1 Orient the panel so the strongback will face the interior of the trench. Place the panel on the trench lip protection and tip the bottom of the panel into the trench.

2 The two rescuers at the sides of the panel will use the ropes to lower the panel into the trench. The rescue in the center will push out on the top (this action brings the bottom of the panel to the trench wall) so that the panel is set tightly against the trench wall in the desired location.

3 Repeat this procedure from the lip protection on the opposite wall and position the second panel directly across from the first. Repeat steps 1-4 to install the "secondary panels." In a straight (non-intersecting) trench up to 8 feet (2.4m) deep the minimum recommended panels sets are three (six panels).

Courtesy of Ron Zawlocki.

installation of struts that are within 2 feet (0.6 m) of the end wall. This may require panels to overlap or to be cut to fit. If the end walls are not sloped or benched, shoring will have to be added to support the end walls.

Installing Pneumatic Shores–Non-entry

After the panels are installed, the shores (or struts) are installed. As a first step, the shoring team should conduct a shoring size-up, assign team positions, place an escape ladder, work on lip protection, and have the needed shoring equipment staged and assembled prior to beginning shoring operations.

To install pneumatic shores without entering the trench, follow the steps in **SKILL DRILL 7-5**.

Installing Pneumatic Shores–Entry

Organizations that use pneumatic struts that must be installed from inside the trench should use a shoring team consisting of a minimum of five members, consisting of a shoring team officer, a controller, strut handlers, and an installer. The controller sets up the air system, shoots the struts at the pressure designated by the shoring team officer, and releases the strut pressure as ordered. There should be two strut handlers who assemble the struts and attach ropes, carry the struts to the trench wall, and assist the installer with positioning, verifying the strut is placed perpendicular (within 10 degrees) to the trench walls, and supporting the struts during installation. The installer climbs the ladder into the trench, aligns the struts, calls for the struts to be shot, locks the collar, and nails the bases.

To install pneumatic shores that require rescuer entry, follow the steps in **SKILL DRILL 7-6**. This skill assumes that struts can be installed from top to bottom with the installer working from a ladder while maintaining a position that keeps his/her head and chest above the trench lip or above a previously installed strut. To prepare, the shoring team will conduct a shoring size-up, assign team positions, place an escape ladder, work on lip protection, and have the needed shoring equipment staged and assembled prior to beginning shoring operations.

TIP
When installing more than two struts on a strongback, use a wye adapter to pressurize all struts at the same time.

SAFETY TIP
When using struts that require a person to enter the trench to install them, the only person who should call for the shore to be shot is the installer. This will help keep the person from being injured if the shore is accidentally activated.

SKILL DRILL 7-5
Installing Pneumatic Struts Without Entering the Trench (NFPA 1006: 12.2.4)

1. The officer takes a position at the end of the trench. The controller (shooter) works on lip protection near the primary panel set. Two strut handlers work on lip protection from both sides of the trench. The controller connects the air hose to the strut and adjusts the regulator to the proper pressure to achieve the proper activation force.

(continues)

SKILL DRILL 7-5
Installing Pneumatic Struts Without Entering the Trench (NFPA 1006: 12.2.4)

2 Each strut handler attaches a rope to the strut. (It is advisable to use different colored ropes on each end of the strut to ease commands to raise and lower.) The strut handlers position themselves across the trench from each other at the primary panel set and use the ropes to lower the strut to the proper position and verify the strut is level and perpendicular (within 10 degrees) to the trench walls.

3 The officer acts as a spotter at the end of the trench to determine whether the shore is level (within 10 degrees) and in the correct location (depth). The handlers determine when the strut is aligned across the trench. The officer asks the handlers if they have aligned the strut: "Handlers ready?" After telling the controller the desired pressure of the strut, the officer gives the command, "Shoot and hold."

4 After the strut has been shot and is in proper position, the strut collars are locked. If not using automatic locking struts, the strut handler on the collar side of the trench locks the collar with the collar-locking tool. When the collar has been locked, the strut handler tells the officer, "Collar locked." With the collar locked, the strut is at its strongest point. Air pressure is no longer needed and the officer tells the controller "Down on air" and all air is released from the strut.

CHAPTER 7 Operations Level Trench Rescue Shoring   **169**

SKILL DRILL 7-5
Installing Pneumatic Struts Without Entering the Trench (NFPA 1006: 12.2.4)

5 The process is repeated to install all needed struts on the primary panel set. Using struts with collars that can be locked without entering the trench can allow an entire rescue shoring system to be installed without exposing rescuers to the risk associated with collapse. Using the same procedures, additional shoring (panels and struts) can be added to expand the safe area in the trench.

6 A rescuer, equipped with a hammer and nails, enters the trench on a ladder placed within the shored panel area and installs two nails to all strut bases. Strut bases are nailed to the strongbacks (from top to bottom) by a rescuer climbing down a ladder.

SKILL DRILL 7-6
Installing Pneumatic Shores with Entry Operations (NFPA 1006: 12.2.4)

1 Connect the air system to the strut, and set the regulator to achieve the proper activation force.

(continues)

SKILL DRILL 7-6
Installing Pneumatic Shores with Entry Operations (NFPA 1006: 12.2.4)

2 Two strut handlers work on lip protection on opposite sides of the trench behind the primary panels. They lower the shore into place using ropes attached to either end (it is advised to use different colored ropes on each end the ease commands to raise and lower). Verify that the strut is perpendicular to the trench walls.

3 When the strut is lined up in the proper place (usually determined best by the safety trench officer positioned at the end of the trench) the installer gives the signal to shoot and hold the shore.

4 The shooter activates the air at the pressure recommended by a professional engineer or the manufacturer's tabulated data that is specific to soil conditions, including collapsed trench walls. This extends the strut and creates pressure on the trench wall.

5 An installer enters the trench on a ladder and manually locks the strut. With the collar locked, the strut is at its strongest point.

SKILL DRILL 7-6
Installing Pneumatic Shores with Entry Operations (NFPA 1006: 12.2.4)

6. After securing the strut manually, the installer nails the strut bases to the strongbacks and then says, "Release" (to release the pressure). The air hose is then removed and placed on another strut as needed. Following the installation of lip protection and with panels placed tightly against the wall, the shoring team will install timber struts.

Shoring comes down to these simple facts: if the victim needs immediate life-saving treatment, one strut is risky but may have to suffice for victim care to begin; if the victim appears stable, two struts with less than 4 feet (1.2 m) of spacing between them may be enough. If the situation is a recovery, it may be best to install three struts per panel and complete shoring before entry is made to ensure rescue personnel are as safe as possible. In all cases, rescuers should use the trench entry protocols developed by their local rescue organization.

> **TIP**
> Remember to always nail the bases of all struts to the strongbacks or wales. With this technique, if subsequent shores cause the system to become loose, previously installed shores will not fall into the trench, potentially striking the victim or rescuer.

Installing Timber Shores

Organizations that use timber struts that must be installed by rescuers entering the trench should have a shoring team consisting of a minimum of five members. Those positions include the shoring team officer, the strut handlers, the installer, and the cutter. There should be two strut handlers, who build the struts and attach the ropes, carry the struts and compression device (wedges/scabs or screw jack) to the trench wall, and assist the installer with positioning and supporting the struts during installation. The installer climbs the ladder into the trench, aligns the struts, tightens (compresses) the struts by driving opposing wedges, and nails the struts in place. The installer also measures the distance between strongbacks for each additional strut. The cutter works at the cut station and cuts the timber struts and scabs at the direction of the shoring team officer.

Because accurately cutting the timbers for shoring is difficult at best, scabs are used to hold the struts in place. It is usually easier to nail the bottom part of both scabs to the strongbacks and to use this as a shelf on which to rest the strut while installing the top and side scabs. Timber struts are usually tightened (put into compression) by using wedges or timber screw jacks or by oversizing the length of the strut and driving it in with hammers. In unstable soil conditions with irregular wall shapes, the oversizing method is the least desirable. Rails made from 2 × 4 inch lumber can be installed on the strongback prior to setting the panels. This technique, which is referred to as railing the strongback, will make shore adjustment easier for the installer **FIGURE 7-34**.

As a first step, the shoring team will conduct a shoring size-up, assign team positions, place an escape ladder, develop a cutting station, work on lip protection, and have the needed shoring equipment staged and assembled prior to beginning shoring operations.

Timber shores require a somewhat different procedure for their installation than do pneumatic shores.

Following the installation of lip protection and with panels placed tightly against the wall, the shoring team uses the following steps to install timber struts.

1. After the lip protection is in place and with panels placed in the trench, strut handlers, the safety trench officer, and the installer work together to position the first (top) strut.
2. The installer activates strut compressive forces (wedges or screw jack) and secures both ends of the strut with nails.
3. The process is repeated with subsequent struts as they are installed deeper into the trench. After each strut is installed with activation forces applied, it must be secured to the strongback with nails. This usually involves the use of scabs (short 2 × 4s, boxing the end of the strut.

FIGURE 7-34 Railing (stringers) the strongback can make the installation and adjustment of timber struts easier for the shore installer.
© Jones and Bartlett Learning. Courtesy of MIEMSS.

SKILL DRILL 7-7 is designed to provide the basic skills needed to install timber struts. While this is not a recommended best practice, we have included it here because many trench rescue resources use timber struts. Timber struts require rescuers to be in the trench during the installation process. This skill drill assumes that there is a near vertical trench wall with no significant voids.

SAFETY TIP

Minimize the time the installer spends on the ladder. Have the installer exit the trench between the installation of each strut if the strut is not readily available.

Installing Inside Wales

At the operations level, wales are most commonly used to open up the shored area, in essence skipping a column of struts, to make room for patient care, extrication, and patient removal techniques. A 12-foot (3.7-m) long wale should have minimum crew of three members (panel team) for installation. A 16-foot (4.9-m) wale will need four or five members.

The shoring and panel teams will conduct a shoring size-up, assign team positions, install lip protection, place an escape ladder, and have the needed shoring equipment staged and assembled prior to beginning shoring operations.

SKILL DRILL 7-7
Installing Timber Shores (NFPA 1006: 12.2.4)

1 After the timber shore has been measured and cut, install the top timber.

CHAPTER 7 Operations Level Trench Rescue Shoring 173

SKILL DRILL 7-7
Installing Timber Shores (NFPA 1006: 12.2.4)

2 Minimize the time the installer spends on the ladder. Strut handlers can assist the installer by lowering the timber with ropes to the designated spot. After the strut is installed, it must be secured to the strongback with nails. Use scabs to secure all timber struts.

3 Install the middle timber shore.

4 Install the bottom timber shore.

© Jones and Bartlett Learning. Courtesy of MIEMSS.

To install inside wales with non-entry struts, follow these steps:

1. Lower two sets of wales (check the tabulated data to assure the wales have the capacity required for the specific trench L value) to the trench floor before installing struts can be advantageous; however, if a secondary collapse is imminent, it is important to immediately shore the area above the trapped victim. Wales can be maneuvered below and above struts later in the operation.
2. Install struts on the primary panels, giving yourself enough room to place wales above the top struts and below the bottom struts.
3. After the primary shoring has protected the victim, position panels on both sides of the primary panels (secondary).
4. Based on the current collapse or signs of failure, determine if the next likely collapse will occur near the top or bottom of the trench wall and install the first set of wales (one on either side of the trench) at that level.
5. Position struts at the intersections of the strongbacks and wales and activate strut pressure to both the struts on the first set of wales installed, then locking the strut collars. If necessary due to the trench wall variations, install wood shims between the wale and the strongbacks. Activation of the struts should not bend the wales. Repeat strut installation on the second set of wales.
6. At this point, the struts on the primary panels can be removed. The wales will act as headers to carry the load behind the primary panels.

TIP

All areas of the trench need to be backfilled when there is an open space between the protective system and the trench wall. If you are concerned with the stability of your protective system, you can always add more strongbacks, struts, or wales, shorten the shoring zone, or call an engineer for advice.

Shoring Voids

In this section, you will learn to shore voids in operation level trenches using non-entry shoring techniques. Voids in trench walls result from collapse. Voids need to be filled both to minimize soil movement behind the panels and to provide adequate resistance to soil forces created in the trench wall that is directly across the trench from the void. Backfill is a generic term for material and equipment used in the voids to help collect the load at the previously failed surface, distribute the load from the opposite wall to the failed surface, and minimize soil movement in void areas.

Air Bag Backfill

One type of backfill is an air bag. Air bags are best for voids between 18 and 48 inches (46 and 122 cm) behind the original trench face. Air pressure should be set at 0.5 psi for each total L for the specific trench conditions. Open lip voids must have back wall angles between 90 and 70 degrees for air bag use as backfill.

To utilize an air bag to fill a closed lip void, use the following steps. First, the shoring team and panel team will conduct a shoring size-up, assign team positions, install lip protection, place an escape ladder, and have the needed shoring and backfill equipment staged and assembled prior to beginning backfill procedures.

1. Lower a partially inflated air bag into the void (behind the panel) to make contact (80% of void area) with the panel and back wall of the void. Tie the bag off to a picket or other anchor point to suspend the bag in the center of the void.
2. Working from lip bridges, position a panel in line with the victim's head and chest if visible or at the center of the void and place a second panel directly across from it.
3. Use rails and the one-side panel installation procedure to position the primary panel set.
4. Install a positioning strut (up to 500 pounds [227 kg] of strut activation force) on the panel at the center of the void. In this case, the positioning strut will also serve as the backfill strut.
5. Install the first compliance strut (up to 500 pounds [227 kg] of strut activation force) between 12 and 24 inches (30 and 61 cm) below the lip.
6. Install the second compliance strut (up to 500 pounds [227 kg] of strut activation force) 12 to 24 inches (30–61 cm) above the bottom of the panel.
7. Slowly increase the pressure on the air bag (stopping to increase strut pressures whenever the panel bends out more than 1 inch [2.5 cm]) until the air bag pressure reaches 0.5 psi for each L for the specific conditions of the trench.
8. Using a wye, increase all strut activation forces to 1000 pounds [454 kg] of force, then lock the strut collars and release the air pressure.

To utilize an air bag to fill an open lip void, use the following steps. First, the shoring team and panel team will conduct a shoring size-up, assign team positions, install lip protection, place an escape ladder, and have the needed shoring and backfill equipment staged and assembled prior to beginning backfill procedures.

1. Working from lip protection, position a panel in line with the victim's head and chest if visible or at the center of void and place a second panel directly across from it.
2. Install a positioning strut just below the bottom of the failure to hold the panels in place.
3. Install a backfill strut (500 pounds [227 kg] of strut activation force) on the panel at the center of the void.
4. Lower a partially inflated air bag into the void to make contact (80% of void area) with the panel and back wall of the void.
5. Install compliance struts as needed to have a strut between 12 to 24 inches (30–61 cm) below the lip and 12 to 24 inches (30–61 cm) above the bottom of the panel, and vertical spacing of struts at a maximum of 48 inches (122 cm).
6. Fill the areas on both sides of the air bag with soil backfill and tamp it down. *Soil should be placed into the voids on both sides of the air bag to stabilize it.*
7. Slowly increase the pressure on the air bag (stopping to increase strut pressures whenever the panel bends out more than 1 inch [2.5 cm]) until the air bag pressure reaches 0.5 psi for each L.
8. Using a wye, increase all strut activation forces to 1000 pounds [454 kg] of force, then lock the strut collars and release air pressure from the struts.

SAFETY TIP

Like all components of a trench rescue shoring system, pickets must be monitored twice within the first hour of installation and then at least once each hour after that. Monitor the front side of the pin for heaving soil and the back side of the pin for voids, which are an indication of failure and reinforcement is necessary.

Wood Backfill

Wood backfill includes timber, lumber, and shims. It is used to fill small voids accessible from the lip between 2 to 12 inches (5–30 cm) beyond the original trench wall (L-1 or less. Wood backfill is only practical for use with open lip voids. Wood should not be placed into the void until at least one strut has pressurized the panel set. It is very important to create a minimum 24 × 24-inch (61 × 61-cm) contact area between the panel and the existing wall directly behind the strut.

To utilize wood backfill, follow these steps. The shoring team and panel team will conduct a shoring size-up, assign team positions, install lip protection, place an escape ladder, and have the needed shoring and backfill equipment staged and assembled prior to beginning backfill procedures.

1. Working from lip protection, position the primary panel set to protect the victim.
2. Install a positioning strut just below the bottom of the failure to hold the panels in place.
3. Working from a lip bridge, place cribbing (horizontally) across the back of the panel. No more than two pieces of wood may be stacked in the same direction.
4. Install a backfill strut (500 pounds [227 kg] of strut activation force) on the strongbacks in the center of the void.
5. Install and tighten shims between the cribbing and back wall.
6. Install compliance struts as needed to have a strut between 12 to 24 inches (30–61 cm) below the lip and 12 to 24 inches (30–61 cm) above the bottom of the panel and vertical spacing of struts at a maximum of 48 inches (122 cm).
7. Fill in open areas between shims with soil backfill.
8. Using a wye, increase all strut activation forces to 1000 pounds (454 kg) of force, lock all strut collars, and release strut air pressure.

Buttress

A buttress is best for large lip shear voids accessible from the lip that have left low angled walls. A buttress can be constructed using timber or aluminum (pneumatic) struts. A buttress supports the back of a panel at the level of the lip. It can be used to resist the movement of the opposite wall but should only be counted on to resist the force transferred from the strut positioned 2 feet (0.6 m) below the lip. A buttress does nothing to prevent soil movement from the wall on which it is built.

To utilize a buttress, follow these steps. The shoring and panel teams will conduct a shoring size-up, assign team positions, install lip protection, place an escape ladder, and have the needed shoring and backfill equipment staged and assembled prior to beginning backfill procedures.

1. Working from lip protection, position the primary panel set to protect the victim.
2. Install a positioning strut just below the level of failure to hold the panels in place.
3. Working from a lip bridge, hang a wale (cut to fit) on the backside of the panel.
4. Install the buttress anchor (pickets and timber) at least twice the distance of the farthest failure point (L) from the original trench wall. (Minimum of four pickets and minimum of one picket per total L.
5. Install a backfill strut (500 pounds [227 kg] of strut activation force) on the strongbacks between 12 to 24 inches (30–61 cm) below the trench lip.
6. Install two back shore struts from the wale to the anchor (picket) system.
7. Install compliance struts as needed to have a strut between 12 to 24 inches (30–61 cm) below the lip and 12 to 24 inches (30–61 cm) above the bottom of the panel, and vertical spacing of struts at a maximum of 48 inches (122 cm).
8. Using a wye, increase all strut activation forces to 1000 pounds (454 kg) of force, lock strut collars, and release air pressure from the struts.

Soil Backfill

Soil backfill is best utilized when voids are accessible from the lip. Soil from the spoil pile may be shoveled into voids from the lip after the panels and at least one strut is in place. Stone can be used and will not require as much compaction effort if it is readily available at the incident site. Soil must be compacted behind the panels after the struts are in place. Use of soil for backfill is only practical when a void opening exists at the lip **FIGURE 7-35**.

To utilize soil to backfill voids, follow these steps. The shoring team and panel team will conduct a shoring size-up, assign team positions, install lip protection, place an escape ladder, and have the needed shoring and backfill equipment staged and assembled prior to beginning backfill procedures.

1. Working from lip protection, position the primary panel set to protect the victim.
2. Install a positioning strut just below the bottom of the failure to hold the panels in place.
3. If the void is wider than the panel (48 inches [122 cm]), use shims, wedges, or adjacent panels to temporarily contain the soil behind the panel. After additional (secondary shoring) panels are positioned and shored, complete the soil backfill behind them.

FIGURE 7-35 Soil backfill.
Courtesy of Ron Zawlocki.

4. Install compliance struts as needed to have a strut between 12 and 24 inches (30 and 61 cm) below the lip and 12 to 24 inches (30–61 cm) above the bottom of the panel and vertical spacing of struts at a maximum of 48 inches (122 cm).
5. Using a wye, increase all strut activation forces to 1000 (454 kg) pounds of force, lock strut collars, and release air pressure from struts.

Open Lip Void Shoring Procedure with Entry Techniques

The shoring team and panel team will conduct a shoring size-up, assign team positions, install lip protection, place an escape ladder, and have the needed shoring and backfill equipment (including a fully equipped cut table if using timber struts) staged and assembled prior to beginning void shoring operations. To perform this skill, follow the steps in **SKILL DRILL 7-8**.

Shoring System Disassembly and Removal

Part of the termination phase of a trench rescue incident includes the disassembly and removal of shoring equipment. The removal of shoring equipment can be a hazardous endeavor. The release of compressive forces (strut activation forces) and removal of panels can cause the soil to move and, if not done correctly, can result in a collapse involving rescuers on the lip and/or in the trench.

Equipment removal needs to be carefully planned and executed to avoid unnecessary mistakes and injuries. When possible, it is best to use fresh crews to

SKILL DRILL 7-8
Shoring Voids with Struts that Require Entry (NFPA 1006: 12.2.4)

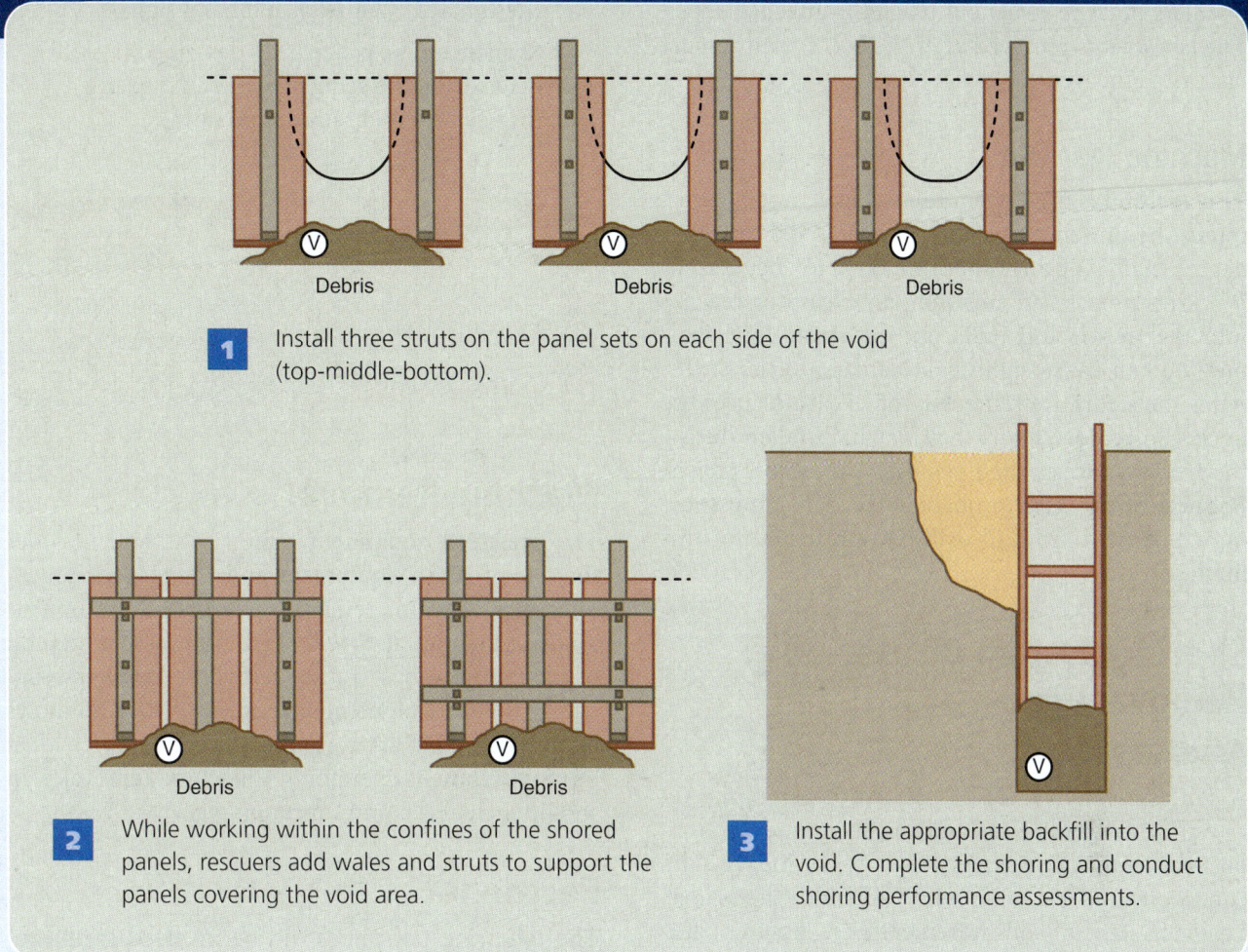

1. Install three struts on the panel sets on each side of the void (top-middle-bottom).

2. While working within the confines of the shored panels, rescuers add wales and struts to support the panels covering the void area.

3. Install the appropriate backfill into the void. Complete the shoring and conduct shoring performance assessments.

perform equipment removal. The safety trench officer, who oversees the equipment removal operation, must create a plan that includes an assessment, the removal method selected, specific assignments, a safety plan, communications plan, and a clear sequence of events. A detailed briefing must be held before the removal operation begins.

Removal Methods
Machine Removal
Machine removal is the safest method and the best practice. This is a combination of limited manual and heavy equipment operations. It is only possible with struts that have collars that can be remotely operated (outside of the trench). Rescuers utilize the ropes connected to the panels to rig and attach to a lifting beam that is attached to a certified lifting point on the excavator. Rescuers then enter the safe zones of the trench and secure the strut bases with four nails per base. With no one in the trench, and while working on lip protection (lip bridges recommended), rescuers systematically (last in-first out) loosen and back off the strut collars 4 to 8 inches (10–20 cm). If the soil pushes the panel/strut, the process is stopped and the removal method is changed to machine tear out.

Manual Removal
This method puts personnel at the most risk. An entry team rescuer descends into a shored panel set on a ladder and attaches ropes to the struts for the strut handlers to control, removes nails from the strut

(reverse order of installation) bases, and removes strut activation forces (wedges, screw jacks, or loosens locking collars on pneumatic struts). The entry team rescuer then moves into a safe zone before the strut handlers pull the struts out using the attached ropes. The last strut is removed after the entry team rescuer has exited the trench.

Machine Tear Out

This method is used when the soil assessment determines the presence of active (moving) soil to such a degree that using any other process is too dangerous. Working from a safe position, an excavator reaches into the trench and pulls out the struts first. This method can cause significant damage to the struts. After the struts are removed, the excavator uses the bucket to remove struts that may have fallen during the initial process, then removes wales and panels. Because of the high probability of damaging shoring equipment, no skill drills are presented for this method.

Shoring System Disassembly/Removal Plan

Assessment

A soil assessment must be performed before the removal method is selected and must continue throughout the removal process. First, examine the lip conditions under each ground pad and/or lip bridge. Look for signs of soil movement (e.g., fissures, sliding, or shifting). Next, evaluate the condition of the walls around the shoring (e.g., for fissures, bulging, or sloughing). If cracks, fissures, sloughing, and other indicators of active soil are present, do not use manual removal operations.

Briefing

A briefing should be conducted by the safety trench officer with all personnel who will be engaged in the disassembly and removal process. The briefing will include defining the removal method to be used, clearly defining the assignments, safety plan, communications plan, and sequence of events.

- Assignments: Each type of disassembly and removal process has specific assignments that are based on the types of struts that were used for shoring and the resources available.
- Safety plan: This plan details personal protective equipment (PPE) requirements, identifies current and potential hazards, and identifies mitigation procedures.
- Communications plan: This plan describes the type(s) of communications that will be used. Common types of communications used for this process include: verbal (face-to-face), radio (designated channels), and hand signals.
- Sequence: The procedures described in skill drills detail the sequence of each type of disassembly and removal process.

> **TIP**
> If a pneumatic strut collar cannot be unlocked, additional struts can be placed above or below the locked-up strut and activated with full pressure to loosen the pressure on the locked strut.

Machine Removal

The preferred equipment removal method includes the use of heavy equipment and trained equipment operators. Machine removals are possible with pneumatic struts. Rescuers will need to develop a method of communications (radio, verbal, and/or hand signals) with the equipment operator. **FIGURE 7-36** shows hand signals that are commonly used by professional heavy equipment operators. A machine removal plan should include a communication plan.

Personnel

The following personnel will be involved in the removal process. The shoring team officer plans, manages, and supervises the removal operation. The technical rescue safety officer develops the removal safety plan and monitors and oversees safe operations during the removal process. The panel team ties off the panel ropes to the lifting beam, looks for soil movement during the removal process, and notifies the safety trench officer of any soil movement. The entry team rescuer enters the trench to connect air hoses and secures all strut bases with 4 (16d) nails bent over. The controller manages the air pressure to struts and air bags (if used). The signal person communicates the orders of the shoring team officer directly to the heavy equipment operator. A trained and experienced heavy equipment operator contributes to the shoring equipment removal plan and follows the directions of the shoring team officer. The heavy equipment operator directly communicates with the signal person. The disassembly crew is responsible for disconnecting panels and shoring from the heavy equipment, disassembling struts from panels/wales, and reservicing equipment.

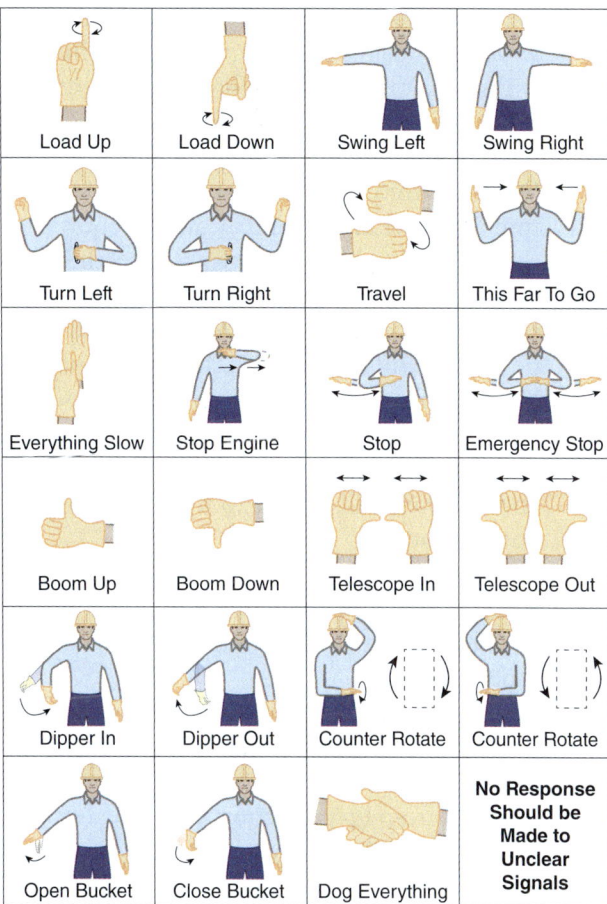

FIGURE 7-36 Excavator hand signals.

To remove trench shoring using machines, follow these steps:

1. The panel team (one on each side of the trench) brings the panel ropes up and ties a knot 2 to 3 feet (0.6–0.9 m) above the top of the panel. This helps lift the load higher (gain). The team attaches the panel ropes to a lifting beam that is connected to the lifting eye on the excavator or excavator bucket, and raises the bucket to take out the slack on the ropes.
2. The entry rescuer enters the trench within the confines of a properly shored set of panels and connects air hoses to the struts, and secures strut bases with four (16d) nails within that panel set. The entry rescuer and all other personnel must be out of the trench to continue on to the next steps.
3. At the direction of the shoring team officer, the controller, working from the lip protection (lip bridges when possible), decreases the pressure on air bags or backshores that are opposing the struts that will be loosened.
4. Beginning with the last struts installed (reverse order of installation), the safety trench officer will direct the controller to pressurize the struts with air until the collar loosens. Riggers remotely (on lip protection) spin the collars back (6–12 inches [15–30 cm]). The controller then slowly decreases the air pressure in the struts as all personnel watch for soil movement. If any significant soil movement occurs (more than minor raveling), remove all personnel from the trench lip and then change to the machine tear out method.
5. If the soil does not move significantly after all struts in the panel set have been decompressed, the safety trench officer, signal person, and heavy equipment officer will coordinate the lifting out of the panel set and struts and then land it in a designated (safe) area away from the trench. Repeat these steps for the remaining panels and struts.
6. Assign disassembly crew personnel to the landing zone to disconnect the lifting system from the heavy equipment, remove struts from the panels, and restore equipment. After all shoring is out of the trench, the panel team removes the lip protection.

> **TIP**
>
> This procedure describes machine removal techniques for the sequential (one at a time) removal of panel sets (two opposing panels) and the struts connected to them. With practice and additional training, multiple (simultaneous) panel sets can be removed using machine removal techniques.

If heavy equipment and operators are unavailable or their use is not feasible, manual removal techniques can be used. The shoring system disassembly/removal plan must be used with the assignments and sequence detailed here. The safety trench officer, technical rescue safety officer, entry team rescuer, and controller all perform the same tasks as the previous skill.

The strut handlers use ropes attached to struts to lift the struts out of the trench while the panel team removes panels, wales and, lastly, lip protection.

To manually remove shoring, follow these steps:

1. The entry rescuer, working within the confines of a secure panel set, connects hoses and ropes to all the struts within the panel set.
2. At the direction of the safety trench officer, the controller, working from the lip protection (lip bridges when possible), decreases the pressure on air bags or backshores that are opposing the struts that will be loosened.
3. The entry rescuer removes the nails from the strut bases and then coordinates and gives direction to the controller to pressurize the struts to loosen the locking collars and spin strut collars back 6 to 12 inches (15–30 cm).

4. The entry rescuer then moves to a safe zone, other personnel move off the lip, and then the controller, at the direction of the safety trench officer, decreases the air pressure in the struts. All personnel will watch for significant soil movement. If any significant soil movement occurs (more than minor raveling), the controller will reshoot the strut. Then, remove all personnel from the trench lip, and change to the machine tear out method.
5. If the soil does not move significantly (more than minor raveling) after the strut has been decompressed, the strut handlers will use the connected ropes to remove the struts.
6. Repeat these steps for all struts in that panel set, moving in the reverse order of installation.
7. Working from lip protection, the panel team will use ropes to remove panels and wales.
8. Repeat steps 1 to 5 (this is probably steps 1 to 8 for equipment removal) in subsequent panel sets. After all shoring is out of the trench, the panel team removes the lip protection.

To manually remove timber, the personnel and assignments remain the same as the previous removal skills. Follow these steps:

1. The entry rescuer, working within the confines of a panel set, connects ropes to all of the struts within the panel set.
2. Whenever an area that has been backfilled with an air bag is encountered, the controller will work at the direction of the safety trench officer to decrease the pressure on the air bags.
3. Working from a safe area, starting at the bottom and working up, and at the direction of the safety trench officer, the entry rescuer will slowly decrease the compression (e.g., opposing wedges, screw jacks) on the strut. All personnel will look for significant soil movement (more than minor raveling) as the shore is loosened, and if significant soil movement occurs, the entry rescuer will exit the trench, and the team will change to the machine tear out method.
4. If no significant soil movement occurs, the entry rescuer will move to a safe area and strut handlers will remove the struts with ropes.
5. Repeat these steps for all struts in that panel set, moving in the reverse order of installation (bottom to top).
6. Working from lip protection, the panel team will use ropes to remove panels and wales.
7. Repeat steps 1–6 for equipment removal in subsequent panel sets. After all shoring is out of the trench, the panel team removes the lip protection.

Voices of Experience

At approximately 10:00 AM on a sunny morning in May of 2002, a report came in for a trench collapse in Harlem. In addition to the first-alarm assignment of two engines, two trucks, and a battalion chief, other resources—Rescue 3, Squad 41, and the special operations command chief—responded. When the radio report confirmed that a victim was indeed pinned in the trench, the dispatcher also assigned Rescue 1 to the call.

When the first companies arrived, they discovered one man trapped and partially buried in a trench. The trench was approximately 25 feet (7.6 m) long and 8 feet (2.4 m) wide and was dug from the street directly to a building. The trench varied in depth from 6 feet (1.8 m) to 8 feet (2.4 m).

The north side of the trench wall had sloughed in the entire length of the trench. Above the trench on all sides were concrete sidewalk panels that were approximately 6 inches (15 cm) thick. The north side panels were still in place and unsupported for several feet, which created a cantilever of concrete over the open trench.

The construction workers had 2 × 12-inch shoring in place, although it was not sufficient to prevent a collapse. (Author's note: This is an example of a trench collapse with shoring in place: 2 × 12-inch skip shoring is not safe on trench walls that are not compact and intact or that show any signs of movement [e.g., fissures, sloughing, raveling, collapsing].) When the north wall failed and the sheeting let go, it pinned the victim to the bottom of the trench under yards of soil and several pieces of lumber. He was buried up to his waist in soil, and his legs were pinned by two pieces of 2 × 12-inch shoring. His back was pinned up against the south wall of the trench.

When Rescue 3 arrived, we immediately started our assessment while the officer went directly to the chief to gather what information he had already obtained. At the time, there were two firefighters and a police officer in the trench trying to ascertain the victim's condition and determine how badly he was pinned. Rescue 3 immediately started to shore the remaining trench walls. The protective system consisted of plywood sheeting and was placed on both the north and south walls of the trench and secured with wales and shores. This secured the trench and created a fairly open space in which to work.

As rescue personnel started digging around the victim, it quickly became evident that this would be a difficult rescue. The effects of the collapse had pinned the worker's legs under the lumber originally used for sheeting and shoring. The lumber entrapping his legs was approximately 12 feet (3.7 m) long and completely buried; thus, the decision was made to dig down along the worker's legs and under the buried lumber. This would allow minimum movement of soil while also accomplishing the objective of freeing his legs.

The area in the trench was tight, so entrenching shovels and buckets were used to remove the dirt. The digging operation continued until we had removed approximately 3 feet (0.9 m) of soil from around the worker's entrapped legs—although the deeper we dug, the less room there was to work in.

To enhance the digging operation, we called for Con Edison to respond with a vacuum truck. It is standard practice for two vacuum trucks to be on duty 24 hours per day in the city. One of the trucks arrived in less than 30 minutes and was immediately put to work. The vacuum truck operation made things tremendously easier. Even though the ground was hard clay with some rocks mixed in, firefighters used tools to pry and loosen the dirt until it was of a size that could be vacuumed up through the hose and into the truck's debris tank.

Voices of Experience

When both of the worker's legs were finally freed, he was stabilized in the trench, splinted, spine stabilized, and lifted out of the hole in a Petzl harness/seat. He was then lowered onto a Stokes basket and transported to the hospital. During the extrication, operation personnel started an IV (intravenous line), and the victim was covered in blankets to help protect him and keep him warm. Oxygen was provided. He suffered only chest contusions and a broken leg and ultimately recovered fully.

In this situation, rescue personnel made a good assessment and determined the worker had a strong chance of survival. For this reason, the primary objective was to establish a protective system suitable to allow the extrication to start and to prevent any type of secondary collapse. The use of the vacuum truck also strongly enhanced our ability to gain access to the worker's legs. It is always quicker to use these types of systems, although the dirt needs to be reduced to the proper size to be vacuumed. Finally, in addition to protecting the worker, we were able to provide advanced life support care, that included precautions for crush syndrome, traumatic fracture injuries to the victim's legs, possible cervical spine injuries, and shock.

Lessons Learned

- Shoring in the underground construction world (large construction projects and utility installation and repair) differs from shoring of a trench for rescue in purpose, soil conditions, time, and planning.
- Trench rescue shoring systems must be capable of collecting the load, transferring the load, distributing the load, and resisting the load.
- Operations level rescuers should not attempt to install heavy shields because lowering or dragging a trench box may cause additional victim trauma or additional collapse.
- The fundamental trench rescue shoring plan must always start with providing immediate protection for the victim(s).
- It is important that every member involved in the shoring operation (panel team and shoring team) understands their assignment, duties, and responsibilities. The trench rescue shoring plan includes the following steps: a trench size-up, primary shoring, secondary shoring, complete shoring, and shoring performance assessment.
- The purpose of the trench size-up is to gather the information needed to shore the trench, which includes assessments of the lip, trench, collapse potential, voids, victim(s), and tabulated data. Pneumatic struts use air pressure to extend the piston inside the cylinders to create a force against the strongbacks, panels, and wales positioned on the trench walls.
- Rescuers must determine the void type (open lip or closed lip), the angle of the void back wall, and the size of the void to determine proper backfill options.
- The removal of shoring equipment can be a hazardous endeavor and needs to be carefully planned. Removal methods include machine removal, manual removal, and machine tear out.

John O'Connell
Firefighter (Retired)
1st Grade Rescue Company No. 3
New York, New York

After-Action

IN SUMMARY

- The shoring systems used at most trench rescue incidents across the United States consist of panels, struts, wales, and support equipment.
- The sheeting and shoring techniques used by rescuers have evolved from the underground construction industry.
- A best practice for trench rescue shoring is to use struts that meet the following criteria:
  - Can be installed with controllable and measurable strut pressure
  - Can be installed and removed without entering the trench
  - Have full strength with activation forces of between 1000 and 1500 pounds (454 and 680 kg) of force
- Hydraulic struts and screw jack struts can create excessive activation forces that can cause rather than prevent trench wall collapse.
- Timber struts have a long history in both construction and rescue shoring. In either application, they are time consuming and create additional risks because they expose installers to the trench before it is safe.
- Wales are horizontal members used to span openings along the trench walls. They should only be used inside the panels (except for back shoring or buttress), and can be made of timber, laminated beams, or metal. Ladders should not be used as wales.
- At each trench, you will need to decide which kind of lip protection is best suited for the situation.
- Equipment removal needs to be carefully planned and executed to avoid mistakes and injuries.

KEY TERMS

Backfill strut Strut designed to (1) oppose the force of the backfill and (2) transfer the force of the opposing wall into the backfilled area of the shoring system.

Compliance strut Additional struts placed in order to meet the strut spacing requirements as listed in the tabulated data.

Lip protection Ground pads or bridges placed on or over the trench lip to distribute the load of rescuers over a greater area, or further away from the trench lip.

Panel set A panel set is any two panels directly across the trench from each other that have or are intended to have struts installed between them.

Positioning strut The first strut installed between the primary panels must be placed in a position that will hold the primary panels in place so that backfilling can begin. The positioning strut must be placed in an area where sections of stable and accessible trench walls exist.

Prescriptive shoring designs Modular rescue shoring designs prepared by a rescue engineer that when assembled per the guidance will safely carry the loads to the indicated level.

Primary panels The first sets of panels that are positioned to protect the victim (centered around the victim's head and chest) from trench collapse.

Primary shoring The initial panels and struts installed to protect the victim from trench collapse.

Safe zone Area within the trench that is fully shielded by adequate trench rescue shoring.

Scabs Small pieces of lumber, usually 2 × 4 inches, that are used to secure timber struts to panels by boxing-in the strut end and other connective devices in place.

Secondary shoring Panels, struts, and wales (as needed) installed to enhance the area being protected by the primary shores.

Simple L (SL) The distance (length) measured in feet (meters) from the original trench wall perpendicular to the farthest point of soil failure or cracks/fissures.

Surcharge loads (ScL) Spoil piles and equipment that are within the area between the original trench walls and the farthest point of soil failure. Measured in feet (meters) perpendicular to the trench wall.

Single point shore A single strut used with a partial panel (minimum 2 × 2 foot [0.6 × 0.6 m] or larger) to shore around obstructions or as supplemental shoring placed as victim access excavation progresses.

Strong side The side of the trench without the obstruction.

Strut activation force A measurement of the total force that the strut exerts. The total force is calculated by multiplying the strut surface area (square inches) by the pressure (psi).

Strut pressure The amount of pressure that the air or hydraulic system sends to the strut. It is measured in psi and can be seen on the system's gauge.

Supplemental shoring Shoring consisting of cut-down sections of panels and struts used in conjunction with full trench rescue shoring panel and strut systems.

T-L method A method designed for use by trench rescuers to obtain rapid and accurate determinations of actual earth pressures and lateral forces based on the visible soil failure conditions.

T-L soil conditions A classification of soil that assumes the worst-case conditions are present.

Total L (L) The simple L (SL) plus the surcharge loads (ScL [spoil pile and equipment]), if present.

Trench boxes Boxes that consist of steel or aluminum side walls and spreaders that are assembled at the scene used to provide worker protection; also called modular shields.

Trench rescue panel A combination of plywood sheeting and lumber strongback.

Trench rescue shoring The use and application of prescriptive shoring techniques designed to resolve the collapse conditions at the majority of trench rescue situations.

Weak side The side of the trench with the obstruction.

REFERENCE

S. M. LaBaw, Earth pressure determination in trench rescue shoring systems, Ph.D. thesis, University of Maryland, College Park, MD, USA, 2009.

On Scene

1. What equipment do you think you should carry on your trench rescue trailer in order to respond safely and effectively to most operation level trench rescue incidents?

2. What are the benefits of non-entry shoring?

3. What are the risks that need to be addressed with entry shoring?

4. What are the benefits of pneumatic struts with automatic locking collars?

5. Why would you want to use ground pads and/or lip bridges?

6. Why would you choose to undertake machine removal rather than manual removal?

CHAPTER 8

Operations Level

Victim Care and Extrication

KNOWLEDGE OBJECTIVES

After studying this chapter, you should be able to:

- Identify the mechanisms of injury associated with trench emergencies. (**NFPA 1006: 12.2.5**, pp. 186–187)
- Review non-entry methods of victim removal.
- Describe the process of removing soil to release a victim from entrapment. (**NFPA 1006: 12.2.5**, pp. 190–194)
- Describe the use of vacuum equipment to release victims from entrapment. (**NFPA 1006: 12.2.5**, pp. 191–194)
- Describe considerations for disentanglement of a victim from objects in a trench. (**NFPA 1006: 12.2.5**, pp. 189–190)
- Describe considerations when assessing an entrapped victim. (**NFPA 1006: 12.2.5**, pp. 194, 197–198)
- Explain actions that can be taken to stabilize and protect an entrapped victim. (**NFPA 1006: 12.2.6**, p. 198)
- Explain victim packaging considerations in the trench rescue environment. (**NFPA 1006: 12.2.6**, pp. 201–202)
- Conduct a risk versus gain analysis to inform victim removal decisions. **NFPA 1006: 12.2.5**, p. 202)
- Identify victim packaging equipment for trench rescue incidents. (**NFPA 1006: 12.2.2, 12.2.6**, pp. 201–202)

SKILLS OBJECTIVES

After studying this chapter, you should be able to:

- Release a victim from soil entrapment. (**NFPA 1006: 12.2.5**, pp. 193–194)
- Disentangle a victim from objects in a trench. (**NFPA 1006: 12.2.5**, pp. 189–190)
- Package a victim for removal from a trench. (**NFPA 1006: 12.2.6**, pp. 201–202)
- Remove a victim from a trench. (**NFPA 1006: 12.2.6**, pp. 201–202)

You Are the Rescuer

Upon arrival, you find a worker's foot trapped under a pipe in a trench where the contractor's protective system was in place and functional at the time of the incident. When responders arrived at the scene, the victim was conscious, but not alert; this was a true emergency. In fact, the victim appeared to be in some sort of trauma-related shock.

Following the initial victim assessment from outside of the trench and after determining that a proper protective system was in place, a ladder was placed in the trench and a paramedic entered to conduct a victim survey. The paramedic immediately became confused and disoriented; he slumped over at the bottom of the ladder. Realizing that something was wrong, the incident commander ordered a firefighter to enter the trench with a self-contained breathing apparatus (SCBA). This firefighter, wearing full protective clothing and SCBA, climbed down the ladder and retrieved the paramedic. He was also able to place a supplied-air breathing apparatus (SABA) on the victim.

Fortunately, the worker trapped by the pipe was not seriously hurt because the pipe had trapped only his foot in the soft sand of the trench floor. The victim's confusion and disorientation were mistaken as signs of shock, but he was simply overcome by the toxic vapors from the solvent that he was using to clean the pipe. The incident commander had assumed that the trench was safe because the contractor's protective system was in place.

1. Before rescuers and paramedics enter the trench, what actions must be completed to ensure their safety?
2. How could the hazards from the solvent fumes have been detected?
3. What life-saving measure could the firefighters have implemented without entering the trench?

Access Navigate for more practice activities.

Introduction

At the heart of trench rescue is the victim. To be successful in a trench rescue, rescuers need to mitigate the incident in a manner that brings no further harm to the victim and absolutely no harm to the rescue personnel. To accomplish this, rescuers must use all the information at their disposal to decide about the most appropriate method of gaining access to the victim, providing the appropriate level of care, and removing the victim.

As we will discuss later in this chapter, we have categorized trench incidents into two types: those that involve a cave-in (collapse) and those that do not involve a cave-in. We have done this to simplify the size-up process and accelerate the brief incident report process. In addition, trench emergencies tend to produce predictable injuries, which allows for the swift assessment of a victim. In this chapter, we discuss mechanisms of injury and victim care, packaging, extrication, and removal options.

Mechanism of Injury

The mechanism of injury at the majority of trench and excavation emergencies is related to the collapse of soil walls. About three of every four (75%) fatalities at trench/excavation sites result from cave-ins. Soil, mostly composed of clay, often breaks off in large chunks and compacts when it hits the victim and the trench floor. Large falling chunks of soil may hit victims in the trench with thousands of pounds of force, often resulting in impact-related injuries to the head, neck, spine, pelvis, and extremities, as well as blunt organ injuries. Granular soils can also hit with high impact, but they are also likely to enter and block the airway. The weight and engulfing capability of soil frequently cause traumatic asphyxia and crush-related injuries.

Other common injuries are caused by moving or falling loads (such as equipment and materials) and can include impact injuries, crush injuries, cuts, bruises, and burns. Electrocution injuries can cause cardiac arrest; muscle, nerve, and tissue damage; thermal burns; and trauma from a fall after contact with electricity.

Cave-In Incidents

Incidents with a cave-in can be organized into two different categories: those involving a partially buried victim and those in which the victim is completely buried. Both types will be challenging because each may involve a substantial amount of work, depending on the amount and type of the entrapping mechanism—soil—but they also may include other materials resulting from the collapse. If a victim is partially buried, rescuers can look for signs of life (chest movement, lack of cyanotic skin color, absence of traumatic head or chest injuries, and lack of excessive bleeding). A partially buried victim may be conscious and talking. A risk versus gain analysis is simpler when it is apparent if the victim is alive or dead.

When a victim has been completely buried for several minutes, the chance of survival is extremely low. However, completely buried victims have occasionally been successfully rescued. Completely buried victims have survived when some type of object (such as pipe, plywood, boards, concrete block, etc.) is in the trench. Those objects can both absorb some of the impact of the falling soil and can provide a sheltered area where breathing is possible via an air pocket. However, being unable to make a visual assessment of the victim's condition should result in lower-risk tactics.

Incidents Without a Cave-In

About 75 percent of trench/excavation incidents result from cave-ins. Of the remaining 25 percent, most of the injuries and fatalities result from moving or falling construction equipment or electricity. At incidents without cave-ins, victims in a trench may have experienced a medical emergency such as a heart attack, seizure, or blood glucose emergency. Trench accidents that involve a live victim will almost always be more difficult to contend with than accidents in which the victim is clearly deceased. This difference arises because it is much more stressful to deal with a situation when someone's life is hanging in the balance. At incidents that do not involve a collapse, rescuers can usually see and often can communicate with the victim. When the victim is alive or is unconscious but appears to have a good chance of survival, the stakes go up and the need to perform a rescue in a timely manner increases.

At trench incidents without a cave-in, rescuers need to determine why the trench has not collapsed. Has it been properly protected with shoring, sloping, benching, a trench box, or by other engineered methods? Of course, for a rescuer to determine proper protection, the rescuer must understand the requirements. For instance, the requirements that are listed by the Occupational Safety and Health Administration (OSHA) for sloping and benching soil that has not collapsed can be used. Likewise, if a trench box is properly installed (see OSHA requirements), the tabulated data from the manufacturer of the box can be referenced for depths and soil type. OSHA soil analysis is required for all sloping, benching, and trench box selection. Because rescuers are neither experienced nor equipped to analyze soil types, an experienced competent person must confirm that the trench is safe for entry by fire and rescue personnel.

If, however, the trench has not collapsed but has not been properly protected, rescuers will need to develop and implement a shoring plan before entry operations begin.

To say that rescuers' skills will be tested in these situations would be an understatement. Whether the situation involves a pipe that has broken a rigging strap and fallen, an equipment failure, a toxic environment, victim injury, or medical condition, success will be determined by following a safe and logical incident action plan (see Chapter 3, *Initial Actions*).

Non-entry Rescue and Victim Self-Rescue

Non-entry rescue and victim self-rescue are the preferred methods because they reduce the risk to rescuers. They should be considered in every incident. This approach may be as simple as a worker who is only partially buried being able to dig themselves out (self-rescue) or a non-entry rescue where rescuers extricate the victim without ever going into the trench (via a ladder).

The best case scenario is the one in which the victim just needs a ladder to get out of the trench. A fire service ladder should be used, rather than a ladder that the contractor has onsite. Keep in mind that even individuals with broken bones will climb a ladder if they think a potential for collapse exists. If they are hurt and only partially able to help themselves, pass down a modified packaging device such as the chest portion of a **class 3 harness** (rope rescue harness) or an LSP Cinch Rescue Ring **FIGURE 8-1**. If victims are able to follow instructions, rescuers can instruct them on how to secure the harness and then pull them up the ladder to safety using the retrieval line.

In other cases, a non-entry rescue could take the form of a hauling device attached to an elevated anchor (high directional) that removes the victim vertically. Access to the trench and proximity to the lip for the safe operation of an aerial ladder or a piece of heavy equipment (such as an excavator or backhoe with a rated lifting eye) will need to be considered as a part of the operational plan in this situation.

It is unlikely that the victim has been trained in how to don a class 3 harness or other webbing configuration, so a simple harness should be provided. Rescuers should resist the temptation to send someone down into the trench to secure the packaging system **FIGURE 8-2**. In all cases, never attempt to pull a partially buried victim from the trench.

FIGURE 8-1 A modified packaging device like the LSP Cinch Rescue Ring or a "screamer suit" can be passed down to a victim in a trench.
Courtesy of CMC Rescue, Inc.

FIGURE 8-3 A rescued worker being pulled out vertically.
Courtesy of Captain David Jackson, Saginaw Township Fire Department.

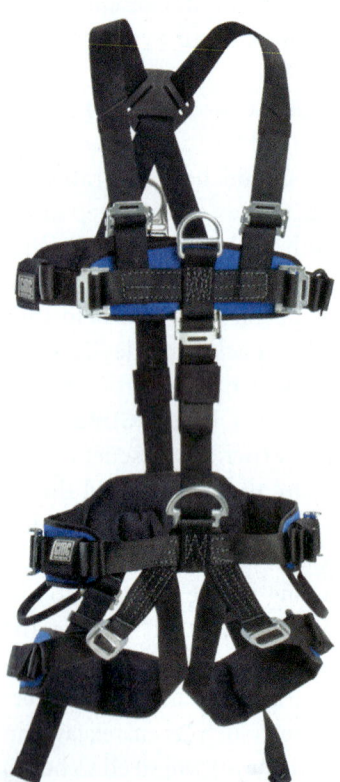

FIGURE 8-2 Class 3 harness.
Courtesy of CMC Rescue, Inc.

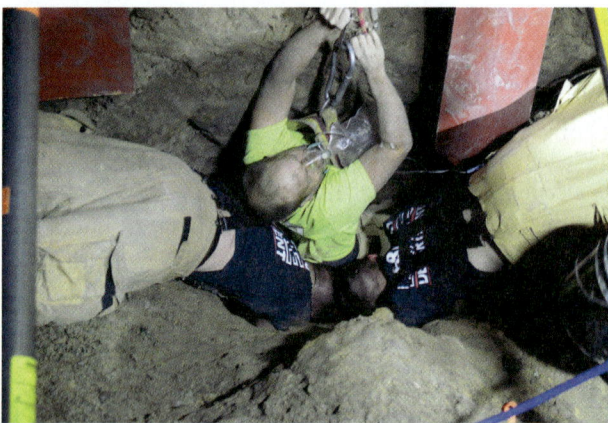

FIGURE 8-4 Injured but conscious and alert workers can be difficult for rescue personnel to deal with from a risk–benefit perspective.
Courtesy of Dennis Walus.

Another simple but slightly more risky method of performing non-entry rescue is to have the victim place a pair of **wristlets** on their wrists. The best way to remove a victim with wristlets is a rope-based mechanical advantage system and a high directional anchor. A faster and even more risky way would be to pass a rope through a carabiner connected to the wristlets with the rope extending to both sides of the trench where rescuers on both sides simply pull the victim up and out. It all comes down to the risk versus gain analysis **FIGURE 8-3**.

Whether the victim can self-rescue or even partially assist in the removal process, mitigating the incident and avoiding endangering rescue personnel is always the preferred method **FIGURE 8-4**.

Entry Operations

If victim self-rescue or non-entry rescue have been eliminated as options, a comprehensive incident action plan is needed, especially if the trench is unprotected and requires shoring to be installed (review Chapter 3, *Initial Actions*) and an entry team assembled, per local standard operating guidelines. Entry operations include locating victims, victim access, victim stabilization, and victim transfer (extrication, packaging, and removal).

Pre-entry Briefing

Before any rescuers enter the trench, all personnel involved must be briefed on the incident action plan. Conducting a pre-entry briefing allows the incident commander, operations officer, or rescue team leader to think out loud one more time. It also enables personnel to question the assumptions on which the incident commander devised the plan. Items to include in the pre-entry briefing are: the overall goal of the operation, position assignments, protective system design, hazards and mitigation efforts, safety requirements, accountability system, and emergency procedures.

Other than the initial action activities, no action in the immediate rescue area should take place until the pre-entry briefing is conducted. The pre-entry briefing is important because rescuers need to understand the desired outcome and the strategic and tactical steps that are part of the overall rescue plan.

Personal Protective Equipment for Entry Team Operations

Because rescuers cannot dress to protect themselves from the effects of soil collapse, the trench walls must be made safe prior to entry. Rescuers working inside of the trench must stay within the confines of the safe zone to minimize the risks associated with secondary collapse. However, certain personal protective equipment (PPE) can protect rescuers from other hazards associated with trench and excavation emergencies. Although structural firefighter PPE can be used, it is often too bulky, hot, and cumbersome for the types of activities needed for entry operations. As a reminder, the following PPE is recommended for entry team personnel: helmet, protective clothing, eye protection, gloves, footwear, respiratory protection, and knee and elbow pads. Chapter 4, *Personal Protective Equipment and Equipment Basics*, covers PPE in detail.

Entry Team Duties

After concluding pre-entry operations, the entry team will focus on locating, accessing, stabilizing, and transferring (or removing) victims from trenches and excavations.

1. **Locate:** Working within the protected areas of the trench, the entry team will probe and dig to locate the victim.
2. **Access:** Entry teams will access the protected areas of the trench by using ladders or properly sloped walls. On rare occasions, entry personnel may be lowered into the trench on suspended lines (rope system). The rapid intervention team (RIT) will have a secondary means of egress immediately available for entry team personnel.
3. **Stabilize:** The amount and level of victim care to be administered inside of the trench is based on victim needs, availability of space, and the hazards present. Once the victim has been located, the entry team will:
   a. Work to quickly uncover the head and chest area.
   b. Determine and report the victim's condition (rescue or recovery).
   c. Provide appropriate levels of emergency medical care.
4. **Transfer:** Transfer operations include extrication from entrapment and removal from the trench. Entry team members will:
   a. Apply an appropriate victim packaging (transfer) device to the victim.
   b. Certify rope anchors and rigging.
   c. Work from safe areas.
   d. Communicate and coordinate.

> **TIP**
>
> The cardinal rule for gaining access to the victim is to remove any entrapment mechanism and uncover the victim's head and chest as quickly as possible.

Extrication

At the operations level, rescuers must be capable of releasing a victim from soil entrapment (or collapse). That typically begins with shovels and buckets. Although a typical long handled shovel might work well at the top of a trench (for spoil pile removal), it has little value at the bottom of the trench. The entrenching tool is a small, collapsible version of the larger shovel that is designed to be used in situations where space is limited because of trench walls and shoring. Due to its smaller size, it provides rescuers with a better "feel," to know if they are hitting a differing material while they are digging in or around a victim. **Probes** and a special shoveling technique must be used to prevent injuries to buried victims.

Additionally, trench rescue teams should have pre-arranged agreements with local operators of vacuum trucks or vacuum trailers. Vacuum trucks or trailers can dramatically reduce the time spent releasing a victim from soil entrapment, but only if they are immediately dispatched and the rescuers have conducted adequate training in their use.

At the technician level, rescuers learn how to release trapped victims from other components, including heavy objects (pipes, slabs, equipment etc.). The equipment, techniques, and skills needed for those types of operations can be found in Chapter 9, *Lifting and Load Stabilization*.

Soil Entrapment

If the in-place protective system has been evaluated or a trench rescue shoring system installed and it is discovered that soil is the entrapping mechanism, then it is time get out the entrenching tools and start the extrication process. Depending on the type of soil (compactness and weight) and the amount of soil, this might take minutes or several hours. The magnitude of this type of situation is never more evident than the first time a rescuer has to use a rope to raise a 5-gallon (19-L) bucket half filled with dirt. In many cases, tons of dirt may have to be removed from the trench by hand **FIGURE 8-5**.

Regardless of the type of digging equipment used to remove a buried victim, there are a few rules to follow:

- Never use a mechanical device or backhoe to dig up or pull out a partially buried victim.
- Never attempt to pull out a partially buried victim. Remove dirt completely. The victim needs to be uncovered and rescuers need to see both of the victim's feet before pulling.
- Be extremely careful when hand digging when near or around the victim **FIGURE 8-6**. Rescuers do not want to cause further injury to the victim or cause physical damage to a body. Do not step on the shovel or push downward. Instead use scraping motions to systematically de-layer the soil.
- Resist the urge to pull or otherwise try to remove the victim before completely freeing the victim from the entrapment mechanism. The author has witnessed 15 minutes of digging that is then followed by 10 minutes of pulling. The cycle then repeats itself for several hours while someone stands over the trench and says, "You might as well keep digging; he is not going to come out."

TIP

Resist the urge to try to pull out a partially buried victim until you can see the victim's feet.

FIGURE 8-5 Dirt bag being used to excavate spoil from a trench.
Courtesy of Cecil V. "Buddy" Martinette, Jr.

FIGURE 8-6 Be careful when digging near the victim.
Courtesy of Ron Zawlocki.

TIP

"I have actually stood over the trench and said to a group of firefighters, 'I know you are not going to be able to resist pulling on the victim, but he is not going to come out until the last piece of dirt is removed from around his ankle.' This was followed by repeated cycles of the "pull and dig" game until the last piece of the victim's foot was uncovered. It gave me only limited pleasure for the firefighters to exit the trench and say, 'You know what? You were right about getting him completely free before we could get him out.'"

Courtesy of Chief "Buddy" Martinette, Jr.

Vacuum Systems

Removing soil that has collapsed to release a trapped victim from soil entrapment is a time-consuming endeavor. A mobile **vacuum system** can significantly reduce the time needed to dig a trapped victim out of a cave-in. Some estimates state an 80 percent reduction in time spent removing the soil entrapment. In a more heavily populated area, there is a good possibility that a local utility company has a vacuum truck. Municipal vacuum trucks have hose extensions that enable them to access and clear storm drains or clean up loose road debris such as leaves and dirt. Additional equipment that enhances the use of vacuum equipment at rescue incidents include an inline safety release that should be positioned on lip protection near the victim area and a tip that allows the system to be used without entering the trench **FIGURE 8-7**. Depending on local SOGs, it would be ideal to have a vacuum truck and related equipment be dispatched as soon as the report of a trench rescue incident is received.

Vacuum Trucks

A **centrifugal vacuum truck** uses a large fan to create suction in the intake line. Truck-mounted centrifugal vacuums average 5000 cubic feet (141.6 m^3) per minute airflow, but have a relatively low vacuum pressure. In trench rescue applications, the low vacuum pressure means that such equipment is not well suited for removing large pieces of soil. Instead, for this system to work effectively, the soil needs to be granular or loose in consistency. If the soil is of a consistency that makes it easy to vacuum, the airflow picks up the debris and transports it through a flexible hose and deposits it in a debris tank. When this tank becomes full, it can be emptied through a dump hatch located at the rear of the vehicle **FIGURE 8-8**.

Because of the combination of low vacuum pressure and limited length of the intake hose, centrifugal vacuum trucks must be positioned close to the actual trench to be effective. Such vacuum trucks usually weigh between 50,000 and 60,000 pounds (23,000 and 27,000 kg), so positioning them close enough to the trench to be effective when there is limited access or an unstable trench environment can be a real risk versus gain challenge. In any case, a vacuum truck should not be positioned closer to the trench wall than two times the trench depth (2 × depth) or three times the distance of the farthest point of failure (3 × simple L).

Positive-displacement vacuum trucks may also be available. This type of vacuum truck, although not as popular or numerous as the centrifugal type, has a much greater vacuum pressure than centrifugal units and is generally used in instances that require a high lifting capacity. Because of the high vacuum pressure at the hose tip, using this type of vacuum around a victim can be dangerous; rescue personnel should use extreme caution to avoid injury.

FIGURE 8-7 Municipal vacuum trucks can expedite the removal of soil.
Courtesy of Larry Collins.

FIGURE 8-8 Centrifugal vacuum truck.
Courtesy of RescueVac.

FIGURE 8-9 Hydro vacuum trucks reduce the soil to a product that can be picked up easily by the vacuum.
Courtesy of Badger Daylighting.

FIGURE 8-10 Technology like the Rescue Vac (brand name) vacuum system combines older municipal vacuum technology with a product specifically designed for rescue personnel and dirt excavation.
Courtesy of RescueVac.

FIGURE 8-11 Centrifugal vacuum truck.
Courtesy of RescueVac.

If a 60,000-pound (27,000-kg) truck is not enough to remove all of the soil from the trench, a hydro vacuum truck may be used. With this technique, water is used as a reduction method to make the soil run so that it can be picked up and then vacuumed to the debris tank **FIGURE 8-9**. As mentioned earlier, water can be a destabilizing element if not handled correctly. When using this type of evacuating equipment, apply only enough water to reduce the soil so that it can be successfully vacuumed from the trench.

Vacuum trucks use hoses with couplings that allow specially designed nozzles to be quickly attached **FIGURE 8-10**. These nozzles can create enough vacuum pressure that the vacuum truck itself can be placed far enough away from the trench to minimize vibration hazards, surcharge loads, and the loud noises associated with operating vacuum trucks. Rescue-specific attachments that can be added to existing vacuum trucks owned by municipalities are available and are highly recommended.

The vacuum system can also be operated from outside the trench using a combination of handles and guide ropes. In certain conditions, soil can be removed without rescue personnel entering the trench, and the process can start before the protective system is in place **FIGURE 8-11**. However, removing soil can create an unsafe situation if it is done before a protective system is established.

Rescue-specific attachments for vacuum trucks add to the safety and efficiency of victim release operations. If any type of vacuum system is utilized, it will most likely have to be applied in tandem with some manner of soil reduction. Water, as mentioned earlier, may be used as the reduction method; however, far more prevalent is the use of air in the form of an **air knife**. In this situation, air is injected into the soil at approximately 100 pounds per square inch (psi; 7 kg/cm^2) to break the soil into smaller particles. The smaller particles can then effectively be picked up and moved through the vacuum system **FIGURE 8-12**.

All of these methods remove soil to accelerate the process of gaining access to the victim, and all of them are much better than using 5-gallon (19-L) buckets and a rope. This can be great news for rescuers who would have had to haul buckets, as well as for the victim who needs more room to expand their chest to

CHAPTER 8 Victim Care and Extrication

FIGURE 8-12 The air knife uses compressed air to reduce the size and change the consistency of soil so that it can be vacuumed more easily.
Courtesy of RescueVac.

FIGURE 8-13 A trench portable on demand storage, or POD.
Courtesy of Cecil V. "Buddy" Martinette, Jr.

breathe. However, this newfound speed in soil removal could add risks to the situation, as the speed can cause rescuers to lose track of the limited area of the protective system and sometimes finds them outside of the safe zone. Careful consideration should be given to ensuring all elements of the protective system are in place and to maintaining a proper balance among speed, efficiency, and the safety of everyone involved in the incident.

TIP

Remove all dirt from the trench when you are digging to access a victim. Resist the temptation to just shove dirt to one side; otherwise, you will most likely be moving it again.

Vacuum Truck Response

As we can see, a vacuum truck, along with additional equipment and several lengths of hose extensions, can help a municipal fire department's response to trench rescue. To ensure a quick and well-equipped response, consideration for deploying the vacuum truck with additional equipment should be planned in advance. To prepare for an emergency response, all supplemental equipment should be loaded onto an easily deployed vehicle or trailer. Portable on demand storage (PODS) units have become popular with fire departments due to their ability to store multiple discipline equipment (trench, confined space, hazardous materials, etc.) in relatively inexpensive storage units while only needing one vehicle, which can quickly load and unload the storage units and transport them to the scene **FIGURE 8-13**.

A prearranged emergency response plan that includes the local department of public works (DPW) workers and trench response team is important for a successful collaboration. An emergency response plan should include predetermined communications (including portable radios and frequencies) and compatible vacuum truck and rescue equipment. Consideration should be given to providing cross-training for city workers who may respond with trench response teams on the incident management system (IMS) and trench rescue concepts, not to mention training together on trench rescue exercises.

When on scene at a trench rescue, the vacuum system operation is vital. First, the vacuum truck needs to be close enough to reach the trench with hose extensions, but there are several other considerations. As discussed earlier, once filled, the vacuum truck dump tank needs to offload the soil. Securing a path to shuttle back and forth is important to ensure continuous operation. A constant water supply is also necessary. Although most vacuum trucks have their own hydro systems, the authors have found that it is best to integrate a fire engine for additional lines to help clear out debris stuck in the hose extensions and even to use in the trench to break up large soil or heavy clay.

Operations

When utilizing a vacuum system, a group needs to be set up, including the following:

- Group leader
- Operator

- Safety brake
- Additional support: assembly, handlines, and debris removal

Communications should include an initial briefing on the assignment and establishing the emergency stop signal and type of communication (ideally radio with backup hand signals) to be used.

This group should first lay out hose extensions to find the best path to the trench before connecting each extension (be careful: the hose extensions are directional). Once the hose extensions have been laid out, clamps secure each extension. Using a reducer allows for a more maneuverable hose at the end; however, maintaining a larger diameter will also allow more soil removal and, more important, fewer problems with clogging. At the end of the hose, finish with a nozzle and grab bar attachment. This operation can be physically demanding, especially when using larger diameter hose. Coordinating with on-scene personnel to have an overhead anchor with a mechanical advantage can help take the weight off the rescuer holding the nozzle during the operation or allows a defensive position to start removing soil from a distance with tag lines **FIGURE 8-14**.

As discussed earlier, the vibrations and movement in soil from the vacuum system can create a hazard. Primary shoring and a safe work area must be in place before vacuum system operation. Removing soil by itself can cause clogging of the hose. By introducing water to create a soil/water mixture, rescuers will keep the operation going without having to stop for clogs.

Rescuers can create a sump by reading the trench floor for a low spot. Using a handline in the trench to break up the soil and working the vacuum nozzle from the sump to the victim area can remove a lot of soil quickly. This process can injure the victim, so it is ideal to remove soil from around the victim by plunging down from different sides with the vacuum system and then hand digging across toward the victim. This allows the fastest removal of soil without injuring the victim.

In rescue mode, all rescuers are working at a fast pace. Vacuum system operations can remove significant amounts of soil quickly, but there are inherent safety concerns that must be addressed. When removing soil, monitor the pressure on the shoring system for any changes. The current shoring was based on the floor level of the trench that collapsed. Plan ahead for supplemental shoring to minimize the delay in operations and maintain safe shoring practices.

The continuous use of a vacuum system during a trench rescue requires constant coordination of the rescue team manager, entry team, safety officer, and vacuum truck operator. Considerations for long-term operations are as follows:

- Water supply
- Vacuum truck tank
- Fuel
- Related safety issues

Being able to incorporate vacuum systems into the trench rescue response is a valuable asset. This system has already shown its value in trench rescue scenes by helping to remove large amounts of soil or soil that is difficult to remove with shovels, like clay. Planning in advance to ensure an efficient response and accounting for this tactic early in operations can be the difference in a rescue versus recovery.

Victim Care Considerations

People who have been involved in rescue work for any length of time understand that in each situation, personnel can be called on to perform a variety of tasks. With specialized rescue situations such as trench collapses, the interaction between functions becomes more critical as the situation becomes more technical. In those situations, it is imperative that each rescuer, regardless of assignment, helps those performing other functions to improve the **victim survivability profile**—a determination based on a thorough risk versus gain analysis and other incident factors that addresses the potential for a victim to survive or die with or without rescue intervention. Under the IMS, each person has a primary assignment to perform and a team officer to report to, even though they may be cross-trained. Assisting other members with tasks that do not fall under the primary assignment is permissible but must be cleared by the team officer.

FIGURE 8-14 Vacuum truck hose with high directional anchor and rope mechanical advantage system.
Courtesy of Ron Zawlocki.

Voices of Experience

On Friday, October 19, 2012, the Spring Arbor Fire Department (Michigan) was called to the campus of Spring Arbor University for a man trapped in a trench cave-in. Upon arrival, members of Spring Arbor Rescue #1 assessed the scene as one person buried to his chest in a trench that was 12 feet (3.7 m) deep. The victim was conscious and trying to dig himself out with the help of his co-workers. The Spring Arbor Fire Department requested a trench rescue team from the nearby Summit Fire Department, which is part of Michigan's (Region 1) USAR Response System.

Hearing the call from home was Summit Fire Department's Lieutenant Scott Stoker, who resides in Spring Arbor. The off-duty lieutenant recognized that the incident was only a few blocks away and responded to offer assistance. Upon arrival, he noted numerous workers and firefighters in the trench attempting to dig out the victim. The trench was not shored, and soil conditions were extremely unstable. The entire length of the trench was marked by walls that had sloughed, raveled, and caved in. Other obvious hazards included exposed utilities and unsupported concrete slabs hanging over the rescue area. Lieutenant Stoker's immediate concern was to remove the workers and firefighters from the trench. After some convincing, he was able to accomplish that life-saving goal. Site control measures included placing a ladder into the trench and eliminating vibrations by having most heavy equipment shut down. Heavy equipment still operating at the site included two excavators running at the trench lip.

When the Summit Fire Department trench rescue team arrived, they placed ground pads on the lip and took measurements for primary shoring operations. Shortly after that, the victim freed himself from the collapsed soil. The injuries that he suffered to both legs prevented him from being able to climb the ladder to safety. The focus of the tactical operations quickly changed from shoring to non-entry rescue. Rescuers repositioned the ladder (already in the trench) so the victim could support his body and hold onto a rung with both hands. From a safe area on the trench lip (ground pads), rescuers were able to utilize a moving ladder slide technique to extricate the victim from the trench. Within minutes of extricating the victim from the trench, a second cave-in occurred.

The victim was treated on the scene and transported to Allegiance Hospital in Jackson, Michigan. He was treated for a fractured lower leg and was expected to return to work following a rehabilitation period.

Lessons Learned

- The Spring Arbor trench rescue was a good example of the best of all rescue opportunities (non-entry rescue). The fact that the stage was set for a successful non-entry rescue was a tribute to good training and command presence. Gaining control of the scene eliminated the potential for additional victims. Site control (ground pads and stopping vibrations) improved the window of opportunity for rescue by postponing the secondary collapse.

- Quick and decisive initial actions were key to the success at this rescue. They saved the lives of first-arriving firefighters and construction workers. The rapid change in tactical objectives allowed the rescuers to remove the victim from the trench just minutes before the impending cave-in occurred. Both training and experience can be credited with the lieutenant's ability to make those decisions and to implement the appropriate actions.

Voices of Experience

- Recognizing the soil dynamics and the impending secondary collapse caused the rescuers to change tactics from a shoring-based entry rescue to a quicker and safer non-entry rescue operation. Shutting down the excavators may have delayed the secondary collapse those few extra minutes that allowed for the rescue to be completed.
- Active soil creates very unsafe lip conditions, and ground pads are not always capable of preventing rescuers working on the lip from riding a secondary collapse into the trench. Lip bridges would have been a much safer method of lip protection at this scene. As a result of this incident, the Summit Fire Department now carries and trains with lip bridges.
- Having equipment stored on your trench rescue unit in the order it needs to be applied to the rescue is paramount to efficient operations. We recommend that at least one panel set (two panels) be completely assembled and rigged for quick placement of primary shores.

Aaron Osburn
Lieutenant, Summit Fire Department
Rescue Squad Officer, MI-TF1

The information included here is not meant to take the place of any specific local medical protocols. Rather, it is provided so that trench rescuers, regardless of their specific function on the trench scene, will have an idea of the factors that they should consider when dealing with a viable victim.

Providing Victim Care

It is vital to consider who, exactly, will provide the victim care. The removal of a partially buried victim will likely take some time, and victim care typically starts before victim removal and continues during packaging. Consequently, the right people with the right medical skills should be part of any trench response **FIGURE 8-15**.

After the rescue team establishes a safe zone around the victim, the entry team officer should assign a paramedic in the trench to perform a primary assessment and begin victim care. Keep in mind that this person needs to be nimble and fit because there is usually limited room in which to provide victim care after rescue shoring is in place. It is important that the person assigned to provide victim care be comfortable with the trench environment.

> **TIP**
>
> The authors highly recommend that in addition to trench rescue, confined space rescue, and rope rescue, members of a trench rescue team complete the Urban Search and Rescue (USAR) Medic course (Federal Emergency Management Agency [FEMA] curriculum or equivalent) and that those members become the core of the entry team.

> **TIP**
>
> It is recommended that all the team doctors and paramedics be trained in trench, confined space, and rope rescue. Such training provides victim care personnel with a sufficient understanding of the environment and the tools and, most importantly, helps create trust that rescue personnel have the skills and talent necessary to maintain their safety while they are providing victim care.

Victim Assessment and Initial Care

All providers of victim care should be aware of and practice local authority having jurisdiction (AHJ) requirements regarding **standard precautions**. Such precautions cover protection of both the provider and the victim from exposure to blood and other body fluids, and/or airborne products that could pass from the rescuer to the victim or from the victim to the rescuer. Examples of these protections include handwashing, gloves, gowns, masks, eye protection, and respiratory protection.

Assuming rescuers are properly protected, the trench has been made safe, and rescuers are ready to enter and begin the assessment and treatment of the victim, remember the first rule of medicine: Do no harm. Always protect the victim from further injury and proceed with caution. Most trench rescue scenes will be cramped, and there will be limited room in which to work, making assessment and treatment a challenge. Often the trench will be muddy and contain water, creating a slippery, uncomfortable, and intimidating situation. Rescuers should block out these distractions and concentrate on victim care.

During scene size-up, determine whether the victim has suffered some type of injury or experienced a medical problem. If an injury has occurred or the victim has fallen, employ cervical spine stabilization precautions, if possible. Start the assessment with a primary survey, checking CAB (circulation, airway, and breathing). First, check for extreme bleeding. If

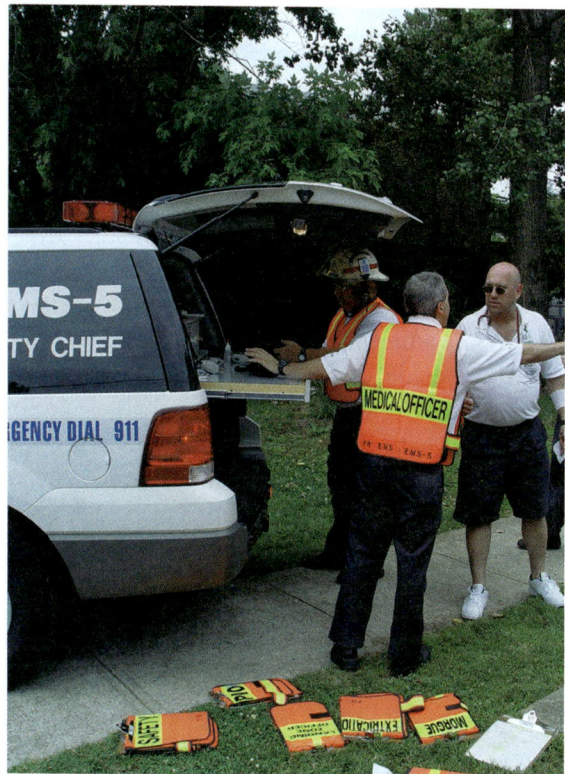

FIGURE 8-15 Emergency medical care providers should understand the trench environment and be comfortable operating in the special conditions they present.
Photo by Martin C. Grube.

major extremity hemorrhage is present and cannot be controlled with a pressure dressing or direct manual pressure, use a tourniquet and a blood-clotting product if allowed by local protocols. A blood pressure cuff applied over a dressing can be inflated to apply direct pressure to an extremity when manual pressure cannot be achieved during the removal process.

Next, the emergency medical services (EMS) provider checks for an open airway and takes steps to secure the airway as necessary. If the rescuer is trained in advanced airway management, an oral airway may be placed if the victim is unconscious and will tolerate it, but the rescuer should be prepared to intubate if required.

After securing the airway, the victim's breathing is assessed. If the victim is not breathing adequately, supplemental oxygen is administered and the rescuer assists with ventilations, if required. One trick when administering oxygen in a cramped space is to keep the oxygen bottle outside of the trench and add extension tubing to the delivery device on the victim via an inexpensive, double-male adapter. If the victim is having breathing difficulty for no apparent reason, reevaluate for the possibility of a toxic or oxygen-deficient atmosphere—the atmosphere should have been checked at least once before the victim is assessed. Confirm that the trench is well ventilated and constantly monitored, while ensuring that airflow on or near the victim does not cause hypothermia.

Finally, the victim's circulation is assessed. This can be accomplished by checking capillary refill and taking radial, brachial, or carotid pulses or a blood pressure by palpation. The trench environment may be noisy, making it difficult to auscultate blood pressures.

After the primary survey has been completed, proceed with the secondary survey, and check for any additional life-threatening injuries **FIGURE 8-16**. If none is found, prepare to package the victim for removal. It will be much easier to continue victim care after the victim is out of the trench and in the ambulance.

FIGURE 8-16 When deciding on victim packaging devices, consider the limiting factor of the shoring system.
Courtesy of Brad Ferguson.

Victim Stabilization

If additional life-threatening problems are found during the secondary survey, they should be treated as quickly as possible. Moderate bleeding can be stabilized with a bulky pressure dressing. There will be additional time to control minor and moderate bleeding after the victim has been removed from the trench.

Fractures that are not life threatening should be stabilized by securing the victim to a long backboard. Do not take the time to splint these injuries in the trench unless the victim's condition is completely stable, and the scene is safe. After the victim has been removed from the trench environment, continue crush syndrome protocols and evaluations by an EMS provider while en route to the trauma center.

During a prolonged extrication, the trench environment will be cooler than the surrounding area, and hypothermia could be a concern, even in summer. The earth below grade remains at a constant temperature, usually in the low 50s (degrees Fahrenheit; 10 degrees Celsius), year round. Prolonged contact with cool soil can lower the victim's core temperature, and because most trench environments are damp or contain water, the victim may be surrounded by water or mud, increasing the body's rate of heat loss. Inclement weather can also lower a victim's body temperature. Given these factors, rescuers should consider the use of heated forced-air blowers or the administration of heated intravenous (IV) fluids as a method to prevent hypothermia when dealing with prolonged trench extrications in colder climates.

Try to keep the victim as dry as possible, and limit their contact with the ground if possible. Place some form of insulation, such as an isothermal blanket, under and around the victim to prevent heat loss. The use of a foam pad, extra turnout gear, or blankets will also help. Heated, humidified oxygen may be administered to warm the core body from within. Hot packs under the armpits or around the neck/head region will warm the victim. Rescuers can also direct portable quartz lights toward the area to add heat if the environment is safe.

Victim Care Involving a Collapse

If the victim is completely covered, rescuers should try to determine where the victim's head is and uncover the head and chest first. The victim's mouth and airway may be full of dirt and foreign matter, so the airway should be cleared as quickly as possible by any means available. Rescuers may use their fingers and suction from at least one portable suction device to remove the obstructing debris, and then attempt to ventilate the victim. If this fails to open the airway, the rescuer should look deeper in the victim's airway for additional debris.

After the airway is clear, the rescuer checks the victim for adequate breathing. If the victim's chest and abdomen are covered by soil, breathing may be restricted and compromised. If the dirt in the trench is dry or sandy, it will easily move and flow around the victim and cause additional chest restriction. Each time the victim exhales and the chest deflates (traumatic asphyxia), dirt will flow in and fill the void, causing more restriction; therefore, it is essential to clear the dirt from around the victim's chest to allow proper lung expansion as soon as possible.

After the head and neck are clear, a cervical collar should be placed to stabilize any possible cervical spine injury. If the victim can communicate, the rescuer should ask whether they are aware of any injuries in the areas still covered by dirt.

In loose granular soil conditions, the victim might slide downward during the digging process, causing further restriction of the chest and diaphragm. A victim harness connected to a rated sling or rescue-quality rope can keep a victim from sliding down deeper into a hole during digging. The rescuers should make sure that the device does not restrict the victim's breathing or complicate the victim's condition. The best practice is to develop an elevated attachment point **FIGURE 8-17**. The primary reason for this practice is the need for a method that can both support an unresponsive victim during the digging process and remove the victim once the victim is free. Generally, as the entrapping mechanism is removed from around the victim, they become flaccid and difficult to manage. Supporting the victim from an elevated attachment point keeps the victim in a mostly vertical position throughout the extrication process. Additionally, removing the victim vertically allows removal from the trench while staying entirely between the safe areas of the panels.

The method used to stabilize the victim during digging operations may be as simple as wristlets attached to an elevated anchor point **FIGURE 8-18**. This system may be all that is required until soil removal allows placement of a hasty harness around the victim's hips and legs.

Uncovering a victim's buried limbs can be slow work, and rescuers will need to provide emotional support to the victim during this process. In addition,

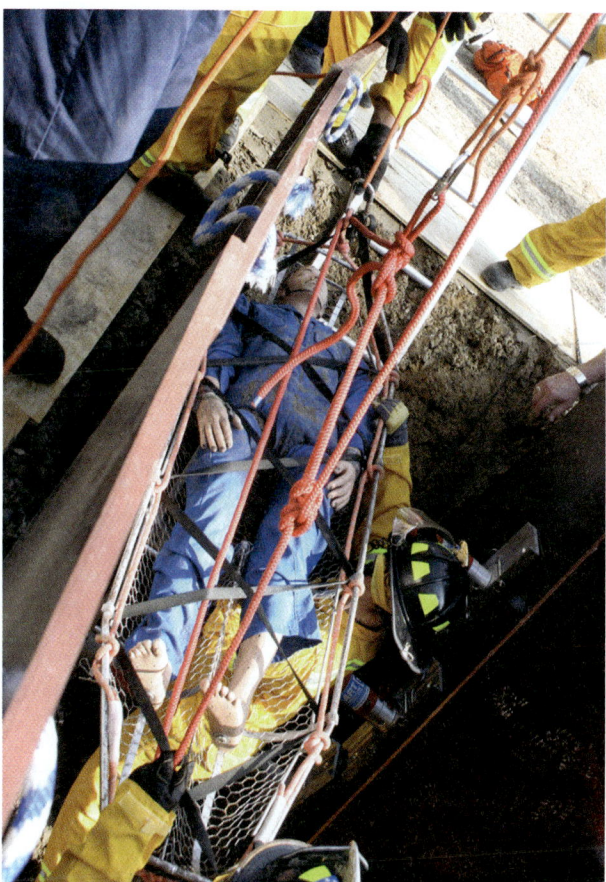

FIGURE 8-17 The vertical lift is usually the preferred method for removing victims from a trench.
Courtesy of Lance Cpl. Angel J. Velasquez/U.S. Marines.

rescuers can also use this time to plan how to manage additional injuries. For example, if the victim says that they have a broken upper leg and it feels wet, rescuers should be prepared to handle an open femur fracture and have the necessary equipment ready for when the injury is uncovered.

As soon as an extremity is uncovered and accessible, EMS providers may establish an IV line of normal saline with a 16- or 18-gauge catheter. This line can be used for fluid replacement or medication administration, if necessary. The IV should be well secured because there will be a lot of victim movement and many people working in the area.

If the victim's chest is uncovered and circumstances allow, EMS providers may place a cardiac monitor on the victim and check for abnormal heart rhythms. As the extrication proceeds, EMS providers monitor the victim's condition, while constantly providing emotional support. Such support is important and will have a positive impact on the victim's outcome.

Special Considerations

A condition known as **crush syndrome** occurs in prolonged entrapments where the victim's body tissue is

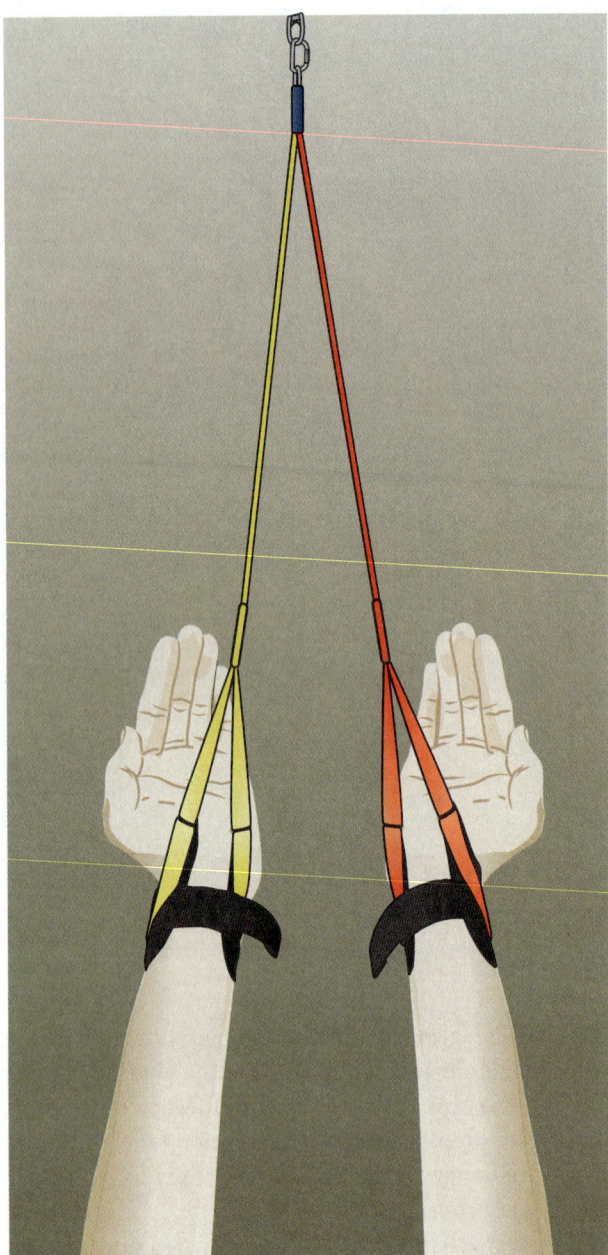

FIGURE 8-18 Wristlets are lowered to the victim to provide stabilization during digging operations.

TIP

When administering medication via IV, consider drawing up the medications from the glass vials into plastic syringes before passing them down to the medical personnel in the trench. This will eliminate the possibility of the glass vial breaking and will save the medical rescuer some time.

crushed and circulation to the tissue is restricted. Because the blood flow is reduced or absent, the affected tissue becomes acidotic, and lactic acid builds up. When the crushed tissue is relieved and circulation restored, blood rushing into the central circulatory system may cause problems such as cardiac arrhythmias and electrolyte imbalances. Although an exhaustive review of crush syndrome is beyond the scope of this chapter, operations level rescuers should be able to recognize the potential for this condition and have paramedics in place to start treatment and monitor the electrocardiogram for changes before and during the removal of the load on the victim.

Victim assessments must always include the mechanism of injury. Entrapment related to heavy objects (e.g., soil, pipes, heavy equipment, etc.) is a common mechanism of injury at trench and excavation emergencies. When large portions of the body are trapped under heavy objects for prolonged periods, crush syndrome should be expected and appropriate medical protocols implemented. Recognizing the potential for crush syndrome based on symptoms can be difficult. Victims are typically conscious, have palpable distal pulses, and numbness in the extremity which may be masking actual levels of pain. Therefore, early recognition of the mechanism of injury is essential to the successful treatment of crush syndrome.

Compartment syndrome describes increased pressure within a muscle compartment of the arm or leg. This pressure increase causes nerve damage due to decreased blood supply. It is most commonly due to physical trauma such as a bone fracture (or a broken bone; up to 75% of cases) or crush injury. Common signs and symptoms of compartment syndrome include pain, faint or absent distal pulses, numbness, pale skin tone, and weakness.

Surgery is the only treatment for acute compartment syndrome. The muscle compartment is cut open to allow muscle tissue to swell, decrease pressure, and restore blood flow. If not treated within 6 hours, permanent muscle or nerve damage can result. Complications may include muscle loss, amputation, infection, nerve damage, and prolonged perfusion, acidosis, and kidney failure–related death.

When a victim is entrapped for an extended period of time, they may deteriorate to a point where defibrillation or cardiopulmonary resuscitation (CPR) is required. Rescuers must be prepared for this situation and its potential hazards. Remember that a victim trapped in a trench environment may be wet. If defibrillation must be attempted, make sure that the victim is dried off as much as possible, that defibrillation pads are securely in place, and that the area is as clear as possible.

If the victim's condition continues to deteriorate and they progress to asystole (absence of electrical activity in the heart), carefully consider all factors before starting resuscitation efforts. Effective CPR will be difficult, if not impossible, when a victim is trapped in a trench environment. If removal of the victim is expected to be a lengthy process, rescuers should reconsider the entire situation and the local termination of care protocol before starting resuscitation efforts. If there is a question regarding whether CPR should begin, medical control should be contacted.

Removing a Victim from a Trench

Trench rescue **victim packaging** and removal techniques are not all that different from the other technical rescue (rope and confined space rescue) victim packaging techniques that you have learned up to this point. Nevertheless, a few extra considerations warrant your attention.

While working to remove a victim from a trench, extreme care must be taken not to dislodge any of the shoring material. This can, and frequently does, happen during extended digging operations when many personnel are working at the same time. Someone on the scene should continuously monitor the integrity of the protective system **FIGURE 8-19**.

Essential to the successful removal of a buried victim is careful and comprehensive preplanning of activities leading up to removal. It is not uncommon for the shoring team to place struts so close to the victim's location that it is impossible to remove the victim. Take the time to forecast movement patterns and consider how the type of packaging device will affect the victim's removal. If planning reveals that adjustment or moving of the struts is necessary, start the process before victim packaging begins.

Removing a properly packaged victim from a trench is best done using rope rescue techniques. Either vertical lifts or diagonal slides are commonly used routes for removal. The path for removal of a trapped victim will need to be closely evaluated. When lifting a victim out of the trench (vertically), the use of a high directional anchor point and a rope-based mechanical advantage system is a best practice **FIGURE 8-20**.

A

B

FIGURE 8-19 Constantly monitor the protective system, and never let your guard down or rush to remove the victim during the extrication process.
Courtesy of Captain David Jackson, Saginaw Township Fire Department.

FIGURE 8-20 Commonly used high directional anchor points for vertical victim removal operations. **A.** Leaning ladder. **B.** Aerial ladder.
Courtesy of Dennis Walus.

See Chapter 9, *Lifting and Load Stabilization*, for more information.

With vertical lift outs, the path can stay within the confines of the protective system that is in place. When diagonal slides (ladder slides) are used, the victim's route may include passing through an unprotected area of the trench FIGURE 8-21. That adds some risk to the rescue operation and should only be attempted when the time it would take to add additional shoring would not jeopardize the victim's chance of survival. If a path through an unprotected area of the trench is determined to be necessary, make sure that the path is clear and unobstructed. Also make another assessment of both trench walls along the removal path. Make sure that no signs of soil failure are present.

Following a team briefing of the removal plan that includes specific assignments, have the victim packaged and secured with all equipment and personnel in place before the removal begins. A well-executed ladder slide may remove a victim with less than 20 seconds of exposure to the unprotected area of the trench. Always keep in mind that the victim's condition and the risk versus gain considerations ultimately will determine the speed at which the victim needs to be removed and the mechanism employed to effect the removal.

Victim Packaging Equipment

Although it is likely that the local fire department and EMS provider will have the victim care equipment needed to treat most trench-related injuries, it is less likely that they will have the specialty equipment needed to package and remove an injured person from a trench. Packaging an injured victim for removal from a trench is a rope rescue skill (a prerequisite for trench rescue). The equipment needed for this skill must include attachments for rope rescue rigging and must include or be compatible with cervical spine motion restrictions. Two basic victim packaging options meet that criteria. They are a victim harness with integrated cervical spine motion restrictors and litter baskets that are compatible with backboards and cervical collars.

Victim Harness

The Yates Spec Pak and the LSP Half Back Extrication/Lift Harness are rated for lifting and lowering with ropes and both have cervical spine motion restriction capabilities FIGURE 8-22.

Litter Basket

Be sure to utilize a litter basket that is rated for vertical lifting operations. Cervical spine immobilizers can be added when necessary. Litter baskets can be used both for vertical lifts and for diagonal slides (ladder slides) as victim removal options.

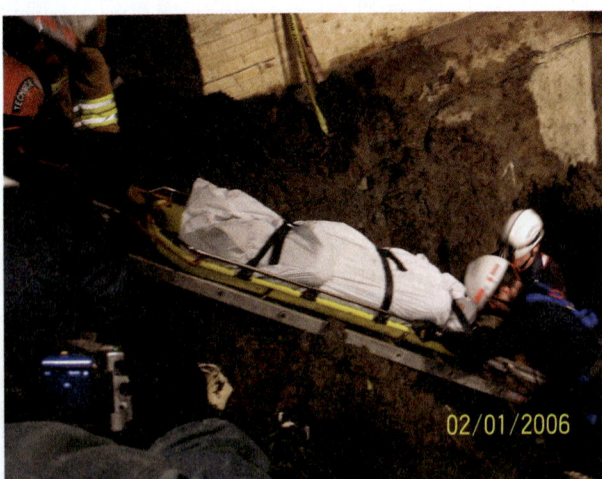

FIGURE 8-21 A ladder slide and litter basket being used to recover a trench fatality.
Courtesy of Ron Zawlocki.

FIGURE 8-22 A rated lifting harness with cervical spine protection is a necessity for trench rescue teams.
Courtesy of Yates Gear, Inc.

After-Action

IN SUMMARY

- To be successful in a trench rescue, rescuers need to mitigate the incident in a manner that brings no further harm to the victim and absolutely no harm to the rescue personnel.
- Non-entry and victim self-rescue are always the preferred rescue methods because they reduce the risk to rescue personnel and they should be considered in every incident.
- Conducting a pre-entry briefing allows the incident commander, operations chief, or rescue team leader to think out loud one more time; it also enables personnel to question the assumptions on which the incident commander devised the plan.
- Excavation techniques include working with existing systems and vacuum systems.
- After the rescue team establishes a safe zone around the victim, the entry team officer should assign a paramedic in the trench to perform a primary assessment and begin victim care.
- Always protect the victim from further injury and proceed with caution.
- Clear the victim's airway as quickly as possible by any means available.
- During a prolonged extrication, the trench environment will be cooler than the surrounding area, and hypothermia could be a concern—even in summer.
- Trench rescue victim packaging and removal techniques are not all that different from the other technical rescue (rope and confined-space rescue) victim packaging techniques.

KEY TERMS

Air knife Tool that injects air into the soil at approximately 100 pounds per square inch (psi; 7 kg/cm^2) for the purpose of breaking the soil into smaller particles.

Centrifugal vacuum truck Apparatus that uses a large fan to create suction in the intake line.

Class 3 harness A harness designed with lower body and upper body attachments.

Compartment syndrome A condition in which increased pressure within a muscle compartment of the arm or leg causes nerve damage due to decreased blood supply.

Crush syndrome A condition that is the result of a prolonged entrapment where the victim's body tissue is crushed and circulation to the tissue is restricted.

Probes A non-metallic rod that is pushed into the cave in spoil pile to determine the buried victim's location.

Standard precautions Precautions (e.g., handwashing, gloves, gowns, masks) that cover protection of both the provider and the victim from exposure to blood and other body fluids, and/or airborne products that could pass from the rescuer to the victim or from the victim to the rescuer.

Vacuum system Equipment that attaches to most municipal vacuum trucks that uses a standard hose with couplings and specially designed nozzles to create vacuum pressure to expedite the removal of soil.

Victim packaging The application of devices (e.g., wristlets, cinch rings, harnesses, and litters) on the victim for removal from the trench.

Victim survivability profile Determination based on a thorough risk–benefit analysis and other incident factors that address the potential for a victim to survive or die with or without rescue intervention.

Wristlets Webbing or other material designed to wrap around the wrist for the purpose of providing a lifting or pulling attachment point.

REFERENCES

Labors' Health & Safety Fund of North America. https://www.lhsfna.org/index.cfm/occupational-safety-and-health/trenches-and-excavations/#:~:text=Trench%20fatalities%20are%20a%20serious,to%20struck%2Dbys%20or%20electrocutions. Accessed April 26, 2021.

Labors' Health & Safety Fund of North America. https://www.lhsfna.org/index.cfm/occupational-safety-and-health/trenches-and-excavations/#:~:text=Trench%20fatalities%20are%20a%20serious,to%20struck%2Dbys%20or%20electrocutions. Accessed April 26, 2021.

On Scene

1. Which steps would be most appropriate to assist a lightly buried worker in a self-rescue?

2. When is it safe to send an EMS provider into a trench to begin assessment on the trapped worker?

3. What injuries or conditions are most likely to be a concern for your medical team for a worker trapped for two hours?

SECTION 3

Technician Level

CHAPTER **9** **Lifting and Load Stabilization**

CHAPTER **10** **Technician Level Trench Rescue Shoring**

CHAPTER 9

Technician Level

Lifting and Load Stabilization

KNOWLEDGE OBJECTIVES

After studying this chapter, you should be able to:

- Describe the principles of lifting mechanics. (**NFPA 1006: 12.3.9**, pp. 207–210)
- Explain the use of levers to gain mechanical advantage. (**NFPA 1006: 12.3.9**, p. 209)
- Identify factors that must be analyzed during a load assessment. (**NFPA 1006: 12.3.8, 12.3.9**, pp. 210–212)
- Explain considerations when creating a stabilization plan. (**NFPA 1006: 12.3.8**, p. 214)
- Describe compression-based stabilization. (**NFPA 1006: 12.3.8**, pp. 214–216)
- Identify tools and equipment used for compression-based stabilization. (**NFPA 1006: 12.3.8**, pp. 214–216)
- Describe construction of box crib systems. (**NFPA 1006: 12.3.8**, pp. 214–216)
- Describe tension-based stabilization systems. (**NFPA 1006: 12.3.8**, pp. 217–218)
- Describe inspection procedures for stabilization tools and equipment. (**NFPA 1006: 12.3.8**, pp. 218–219)
- Explain considerations when creating a lifting plan. (**NFPA 1006: 12.3.9**, p. 219)
- Describe the purpose of a lifting plan. (**NFPA 1006: 12.3.9**, p. 219)
- Identify the steps covered in a lifting plan. (**NFPA 1006: 12.3.9**, p. 219)
- Identify topics to be covered in a lifting plan briefing. (**NFPA 1006: 12.3.9**, p. 219)
- Explain surfaced-based lifting techniques. (**NFPA 1006: 12.3.9**, pp. 220–222)
- Explain overhead lifting techniques. (**NFPA 1006: 12.3.9**, pp. 222–224)
- Describe inspection procedures for lifting tools and equipment. (**NFPA 1006: 12.3.9**, p. 222)
- Define terms used in relation to heavy equipment operation. (**NFPA 1006: 12.3.10**, pp. 226–228)
- Identify considerations for use of heavy equipment at a trench rescue incident. (**NFPA 1006: 12.3.10**, p. 228)
- Explain methods of communication for heavy equipment operations. (**NFPA 1006: 12.3.10**, pp. 227–228)

SKILLS OBJECTIVES

After studying this chapter, you should be able to:

- Design a load stabilization plan. (**NFPA 1006: 12.3.8**, p. 211))
- Construct a compression-based load stabilization system. (**NFPA 1006: 12.3.8**, pp. 214–216)
- Design a lifting plan. (**NFPA 1006: 12.3.9**, p. 219)
- Lift a load with surface-based methods. (**NFPA 1006: 12.3.9**, pp. 220–222)
- Lift a load with overhead methods. (**NFPA 1006: 12.3.9**, pp. 222–225)
- Coordinate lifting a load with heavy equipment. (**NFPA 1006: 12.3.10**, pp. 226–229)

You Are the Rescuer

You are the technician level rescuer on the first-due engine responding to a call at an excavation site about one-half mile (0.8 km) from your firehouse. Upon arrival, you find that a quick-connect excavator bucket has pinned a worker in the trench. The bucket dislodged from the quick-connect couplings as it was passing over the worker in the trench. As it fell, it hit one of the trench box spreaders before it landed on the worker's legs. The bucket now is about half full of dirt and is resting on the worker's lower extremities inside of the trench box.

After hearing the brief initial report, the officer of the rescue company radios that they are about 5 minutes out. He asks you to begin gathering information for a lifting plan.

1. How will you determine the weight of the bucket and the dirt inside of it?
2. How will you estimate the center of gravity of the bucket and the dirt within the bucket?
3. What rescue stabilizing and lifting techniques will be most appropriate for this load (bucket/dirt)?
4. Should the on-scene excavator and the operator be considered for lifting the bucket off of the trapped worker?

JONES & BARTLETT LEARNING NAVIGATE — Access Navigate for more practice activities.

Introduction

Heavy objects and equipment are common at every construction site. Accidents involving an **excavator** such as a front end loader, involving pipes, slabs, or steel plates rolling or falling into trenches have been documented. Additionally, construction sites that are near roadways have been encroached by out-of-control vehicles, causing injuries to drivers, passengers, and construction workers alike. Rescuers responding to construction site accidents must be prepared to stabilize and move heavy objects to make the rescue work zone safe.

Most of the lifts that technician level rescuers will perform at a trench incident will meet the definition of a **critical lift**. A critical lift is any lift with a human being below the object and/or any lift where the object exceeds 75 percent of the lifting or stabilizing tool's rated capacity. Critical lifts require a lifting plan that includes a load assessment, equipment inspection, ongoing stabilization, lifting procedures, and coordination between stabilization and lifting team members. Although many ways to stabilize and lift heavy objects are described here, in this chapter, we will explore methods that use equipment that is commonly found at trench rescue incidents.

Tools used for lifting and stabilizing must be inspected before each use. At a rescue scene, quick actions are often the difference between a victim rescue and a body recovery. Therefore, thorough inspections must be completed immediately after lifting and stabilizing equipment is used. Then, a quick visual inspection can be accomplished when the equipment is being put into action at the rescue scene.

TIP

SAFETY TIP

This chapter discusses options for load stabilization and lifting heavy objects that are commonly found at trench and excavation sites. The techniques and practices associated with these tactics must be learned from instructors who are subject matter experts in stabilizing and lifting.

Lifting Mechanics

The theory of lifting **mechanics** is subdivided into two areas: **energy** and **work**. The effective use of rescue tools is often determined by a thorough understanding of these concepts and their application. Lifting mechanics deal with energy and forces in relation to bodies. Something that creates a positive output in a given situation is called **mechanical advantage**. The efficiency, or advantage, is a measurement of the distance traveled compared with the force used to effect the movement.

Energy is the capacity for doing work and overcoming resistance. In other words, it deals with how hard it is to push or pull something over a distance. Energy is measured in foot/pounds, as either kinetic (in motion) or potential (at rest) energy; therefore, energy is what it takes to accomplish work.

Work is distance times force, or force as it is applied to set a body in motion. When climbing a mountain, the climbing is work, and the rate of climbing is power. The total work performed is equal to the amount of energy expended.

Simple machines can help us understand how mechanical advantages work. Simple machines that are often used at trench rescue incidents include the **inclined plane**, **lever**, and pulley. Each simple machine provides a mechanical advantage that can place the energy and workloads required to perform a task within the range of human strength and endurance.

A mechanical advantage is the ratio of the output force that a machine exerts compared with the input force furnished to that machine to do work **FIGURE 9-1**. A simple machine is measured in terms of its efficiency and effectiveness when doing work. For example, if you get five units of work out of a machine by putting in one unit of work, the machine has a five-to-one (5:1) ratio of efficiency.

As rescuers, the challenge is to take these theoretical concepts and put them to use. How does this apply to rescue? Suppose you are operating at a trench rescue with a 200-pound (91-kg) man injured in the bottom of a 12-foot (3.7-m) deep trench. Carrying a this man up the ladder would be a difficult task for a firefighter and the trench width often does not permit the placement of side by side ladders to effect a two-person (two-ladder) team lift out.

By using the mechanical advantage of an inclined plane (a simple machine) you can theoretically affect this rescue with a less effort **FIGURE 9-2**. As illustrated in **FIGURE 9-3**, the ladder used as an inclined plane becomes the mechanical advantage. By placing a 24-foot (7.3-m) extension ladder on the end wall of the trench (12-feet [3.7-m] deep), a rescuer can develop a 2 to 1 mechanical advantage and lift the 200-pound (91-kg) man out of the trench with 100 pounds (45 kg) of force.

When we place a ladder in a trench and then use the angle of the ladder to facilitate victim removal, we use the inclined plane concept. Using the angle of the ladder to decrease the slope divides the energy over a period of time. If you want to figure the advantage gained by using the ladder, divide the height of the trench into the ladder's height at the point it contacts the ground. In this case, if the base of the ladder is 4 feet (1.2 m) from the wall and the trench is 8 feet (2.4 m) deep, you would divide 8 by 4 (2.4/1.2), giving a theoretical 2:1 mechanical advantage Figure 9-3.

The advantage of the trench ladder is figured as distance of the ladder base from the trench wall, divided by the height of the trench where the wall meets the ladder.

Levers can be used to move, haul, pull, or raise a load. Every lever has a fulcrum, which serves as a pivot point for the lever; a force, which provides the power; and a load. The position of the fulcrum in relationship

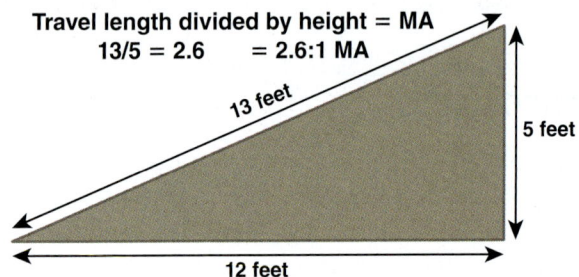

FIGURE 9-2 Advantage is calculated as travel distance divided by height.

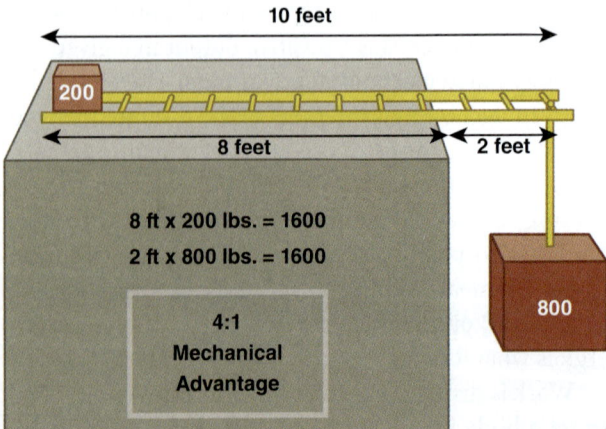

FIGURE 9-1 In this case, the efficiency of the lever is based on the distance that the load is from the fulcrum or turning point.

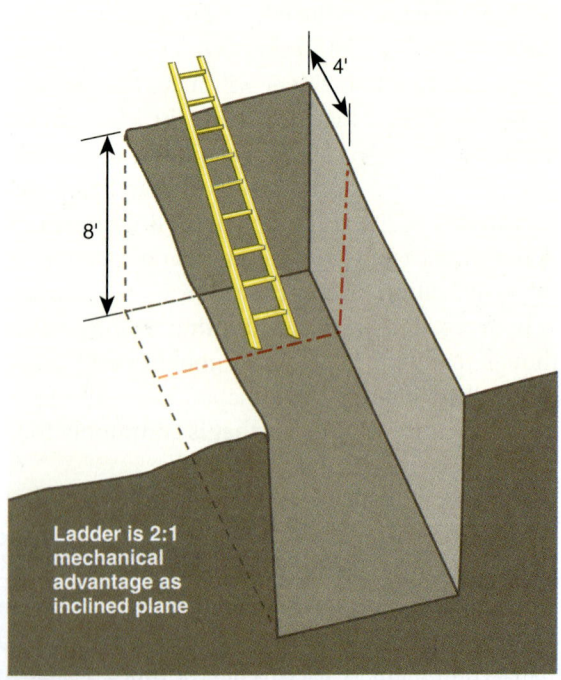

FIGURE 9-3 The mechanical advantage of this fixed ladder slide is 2:1.

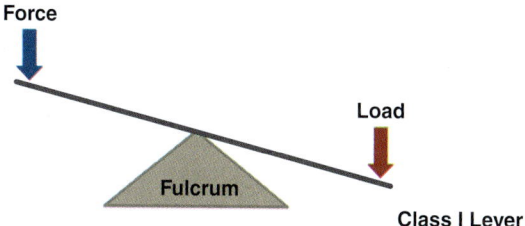

FIGURE 9-4 The fulcrum is located between the force and the load.

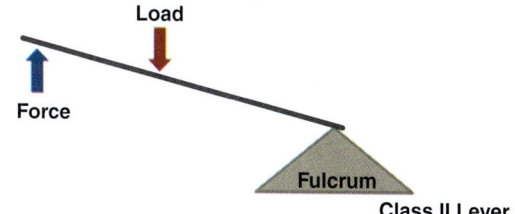

FIGURE 9-5 The load is located between the force and the fulcrum.

to the load determines the classification of the lever. Levers are categorized as class I, class II, or class III.

Levers

To understand the application of a lever, let us look first at the **class I lever**. The best example of this type of lever is a pry bar. If you have to lift a concrete slab to perform a rescue, the pry bar can provide a purchase point under the edge of the concrete. In this example, the concrete represents the load. If you place a block of wood under the pry bar, the wood becomes a fulcrum. The force, applied by the rescuer in a downward motion, will effectively manage the lift. The fulcrum is located between the force and the load **FIGURE 9-4**.

You can easily calculate the advantage that the class I lever has provided in this situation. The distance from the fulcrum to the force, divided by the distance of the fulcrum from the load, determines the amount of advantage. Take the following steps to lift a load using a lever:

- Perform a load assessment (discussed in detail later in the chapter).
- Develop a lifting plan and conduct a briefing.
- Install initial stabilization.
- Clear enough area around and under the object to be lifted, to insert several pieces of 2 × 12-inch planks, which are used to create a solid base for the lift.
- Place a fulcrum(s) about 5 to 6-inches (13–15 cm) from the object when possible.
- Insert the lever under the object and on top of the fulcrum (class I lever). Place one rescuer per pry bar (lever). Two pry bar lifts performed simultaneously can provide a more controlled lift.
- Maintain ongoing stabilization. Lift an inch and crib an inch until the victim can be removed.

The **class II lever** is best applied for moving objects on a horizontal or near horizontal plane. In this application, the load is located between the fulcrum and the force **FIGURE 9-5**. A good example of this type of lever is the wheelbarrow. In this case, the force is applied in the lifting action of the handles. The load is in the hopper, with the fulcrum at the wheel. Using the formula, you can see that, up to a certain point, the longer the wheelbarrow handles are, the greater the advantage when moving objects.

FIGURE 9-6 An aerial ladder is a hydraulically powered class III lever.
© Muskoka Stock Photos/Shutterstock.

The **class III lever** is the hardest to understand because the primary example of it is the shovel, which can also be used as a class I or class II lever in some situations. This type of lever has the force located between the load and the fulcrum **FIGURE 9-6**. Imagine the action of moving a shovel full of sand. The sand is the load. The force is the hand that is used to lift and throw the sand, and the fulcrum is the hand on the end of the shovel that directs it to another location.

Pulleys

Pulleys can be used to develop a simple machine used for lifting. Understanding rope-based mechanical advantage systems and how they are used to raise and lower objects is a matter of understanding gravity and how pulleys can be used to make lifting easier. If you tie a rope to something to raise or lower it, you need to consider the object's entire weight. If the object weighs

200 pounds (91 kg), the system you create needs to be able to lift 200 pounds (91 kg). This is easier over short distances and becomes more challenging when the load must be moved over considerable distance. To make this job easier, pulleys can be used. The effect of anchoring one end of the rope and then having the pulley travel with the load divides the work in half, making it a 2:1 mechanical advantage. Be sure the rope is anchored above the load, is threaded through the pulley, and is then brought back up. In our example, the 200 pounds (91 kg) becomes 100 pounds (45 kg) because the anchored rope and pulley are supporting half the load.

Rope-Based Systems

Rope-based mechanical advantage systems work well for lifting and lowering objects that weigh hundreds of pounds. However, rope systems are not effective for many of the heavy objects (thousands of pounds) that are found on a construction (trench/excavation) site.

Basic Techniques

The concepts of mass, gravity, friction, center of gravity, **moment of force**, work, energy, inclined planes, levers, and pulleys can be applied in myriad ways to the trench rescue environment. Although you should always consider using modern equipment to effect positive rescue results, you should never forget that some tried and true basic techniques can be applied when the newer tools are not available.

Mechanics of load lifting include pushing and pulling. Some lifting tools push while others pull. Some lifting tools can be set up to either pull or push. An example is a mechanical jack. Although this kind of jack can pull, it is not rated by the manufacturer (Hi-Lift) to be used for overhead lifts.

Overhead lifting techniques are often useful for lifting loads that are in confined areas like a trench. In a confined area, it is usually difficult to find room under the object to set up both stabilizing and surface-based lifting tools and equipment. Using rigging (slings, chain, wire rope) to attach to the object can provide the room needed to conduct a safe lifting operation. Note that performing overhead lifts requires more knowledge and skills than using surface-based (pushing) tools.

Load Assessment

A load assessment is an essential part of creating a lifting and load stabilization plan. A load assessment for lifting or stabilization should include the object's weight, shape, center of gravity, type of material, friction, hazards, and whether any forces are acting on the object.

Weight

Several sources can be called upon to determine the weight of the load. Regardless of which source is utilized, the load should not be lifted until the weight is known or estimated. A basic formula can be utilized to determine the weight of common materials such as steel, concrete, brick, and wood. The formula used will depend on the shape of the object. Rescuers will also need to know the pounds per square foot (psf) and/or pounds per cubic foot (pcf) of the material. These formulas and weights can be found on any basic rigging card or guide such as the one seen in **FIGURE 9-7**.

While most passenger vehicles weigh between 2500 and 6000 pounds (1134 and 2722 kg), the weights of construction equipment and larger vehicles may be harder to estimate. Most times, the weight of a larger vehicle or piece of construction equipment can be determined by questioning the driver or operator or by locating the equipment tags and manufacturer's data (either onsite or by internet search).

Type of Material

The type of material the object is made of will help determine if rescuers can lift or move the object. Heavy objects that have soft exteriors or casings will bend and shift when lifted or spread. On the other hand, objects that have stiff and strong exteriors are better suited for lifting and spreading operations. However, brittle materials will fracture without giving a visual warning (such as bending). **Ductile materials** will bend (deform) before they fracture. Examples of brittle materials include cast iron, plastics, cast aluminum, and tungsten carbide. Mild steel, aluminum, copper, platinum, and lead are examples of ductile materials. Providing connecting points for placing lifting, spreading, and stabilizing tools under materials that are strong but will bend before they break (fracture) is a best practice.

Center of Gravity

The **center of gravity (COG)** of an object is the point at which the object will balance, and for lifting purposes, the entire weight of the object is considered to be concentrated at this point. During a lift, an object's COG may continue to move (or shift) as the lifting force is applied. Once the COG is moved beyond its base of

FIGURE 9-7 A rigging card.

Steel Plate	Reinforced Concrete	Material Weight	
1/8" - 5 lbs/sq ft	3" - 40 lbs/sq ft	Concrete (reinforced)	150 lbs/cu ft
1/4" - 10 lbs/sq ft	4" - 50 lbs/sq ft	Brick	120 lbs/cu ft
1/2" - 20 lbs/sq ft	6" - 75 lbs/sq ft	Steel	480 lbs/cu ft
3/4" - 30 lbs/sq ft	9" - 115 lbs/sq ft	Wood	45 lbs/cu ft
1" - 40 lbs/sq ft	12" - 150 lbs/sq ft	Soil	125 lbs/cu ft

Weight Formulas (all dimensions in feet)
Cube (Cu ft): Length × Width × Thickness × Weight/cu ft
Cube (Sq ft): Length × Width × Weight/sq ft
Pipe (Sq ft): Length × (Diameter x 3.2) × Weight/sq ft (based on thickness of pipe wall)
Pipe (Cu ft): Length × (Diameter x 3.2) × Thickness × Weight/cu ft
Cylinder (Cu ft): Diameter × Diameter × Thickness × 0.8 × Weight/cu ft

Inch to Feet Conversion Chart

1" = 0.08	4" = 0.33	7" = 0.58	10" = 0.83
2" = 0.17	5" = 0.42	8" = 0.67	11" = 0.92
3" = 0.25	6" = 0.5	9" = 0.75	

3/8" Chain Capacity

Grade	Vertical	2 Leg @ 60 Degrees
80	7,100 lbs	12,300 lbs
100	8,800 lbs	15,200 lbs
120	10,600 lbs	18,400 lbs

support, the object will topple unless it is supported by secondary stabilization methods.

Objects with a COG above their geometric center are considered top heavy. Top-heavy objects will be less stable during lifts and spreads. In general, the lower the COG of an object, the greater the angle (lean) will be before it falls over.

The geometric center of an object of **uniform shape** and composition will likely be the COG for that object **FIGURE 9-8**. Finding the COG of oddly shaped objects is not as simple as finding the center of uniformly shaped objects. One way of finding the COG of oddly shaped objects is to begin by dividing the object into geometrically shaped parts and then finding the COGs of each of the parts to estimate the COG of the entire object **FIGURE 9-9**.

Knowing the COG of an object is important. It will determine where the stabilizing equipment needs to be placed and will be critical for determining the position of the hook if the object is to be picked up (when the entire object is lifted off the ground).

Use the following steps to determine the COG of an oddly shaped object:

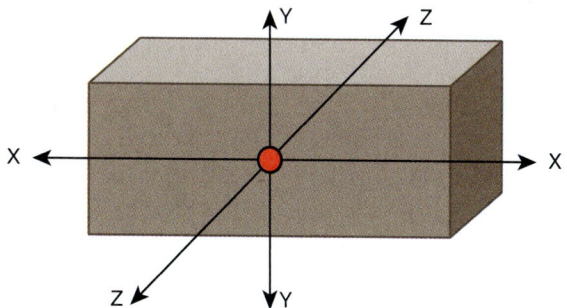

FIGURE 9-8 The center of gravity will be at the intersection of these three areas when the object in uniform in shape.

- Divide the object into uniformly shaped objects (in this case, a square and a rectangle).
- Find the COG of each object and then draw a line between the two COGs. The COG of the entire object is located on that line.
- Determine the ratio of the size/weight of the objects. In this case the rectangle is twice the size and weight of the square or 2 to 1 ratio.

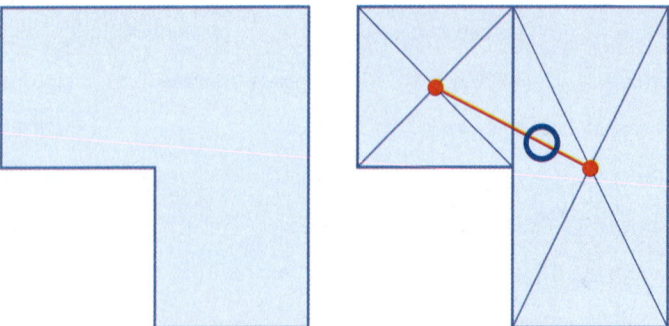

FIGURE 9-9 The rectangle is twice the size/weight of the square or a ratio of 2:1 of the total weight. That 2:1 ratio is used to determine the COG of the entire oddly shaped object.

- Measure the distance between the COG of the square and the COG of the rectangle. The COG of the entire object will be twice as far (2:1) from the small object (square) as it is from the large object (rectangle). In this case, with the distance between the square COG and the rectangle COG at 6 feet (1.8 m), the overall COG is 4 feet (1.2 m) from the square COG and 2 feet (0.6 m) from the rectangle COG (2:1).

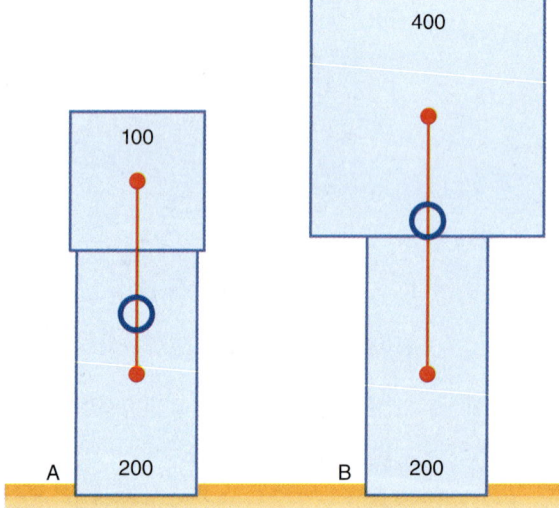

FIGURE 9-10 Using the same method described in Figure 9-9 we can determine the COG of these oddly shaped objects.

TIP

Safety Tip

Objects that are leaning create both horizontal and vertical forces. Both forces must be resolved during lifting and spreading. An object will lean without falling to a predictable angle In the example shown in **FIGURES 9-10** and **9-11**, a lever is used to lift both objects. When object A has reached its balance point (in this case at about a 65-degree angle) the COG is still within the base of support. Any further lifting will cause it to tip over or fall. With the top-heavy object (object B), the same lifting action (resulting at 65 degrees) would move the COG outside of the base of support and cause the object to tip over. As soon as the direction of the force of gravity (indicated by the red line in Figure 9-11) moves outside of the object's base, the object will fall over.

Forces Acting on the Load

The primary force that will impact a lifted load is gravity. The other forces acting on a load are any additional materials on the item to be lifted, mainly soil in the case of trenches, and the **friction** between the load and any adjacent materials/structures. Commonly, in a trench, the friction would come from collapsed soil resting against the load. If the soil is removed prior to the lift, then the friction force is reduced and rescuers have a greater chance at making a smooth lift.

Once the load is lifted outside the trench, wind load and how it will impact the object being lifted must be considered. This should be considered primarily when lifting items that have large surface areas. In these cases, tag lines can be secured to the load to help control it when it is lifted outside of the trench.

Hazards Associated with Heavy Objects

Lifting heavy objects at a trench rescue incident creates some specific hazards. Lifting and spreading

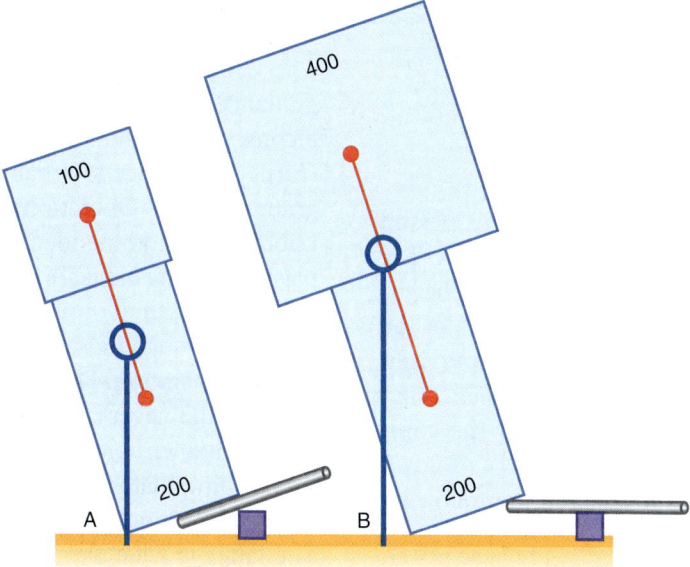

FIGURE 9-11 The object in Figure A has the COG at its base of support and is balanced. The object in Figure B has its COG outside of its base and will tip over.

operations place rescuers in the hot zone, close to the victim. Tight quarters, power tools, and the urgency of the situation require that rescuers give safety a high priority. Tools can slip, or hands can slip off tools, causing rescuers to strike an object that results in an injury. Rescuers must recognize the following lifting-specific hazards and must mitigate and manage the associated risks.

Rescuers should never place any part of their bodies between the load and a stationary object due to the danger of pinch points (or becoming trapped between the load and object). If the load is unstable enough to require support, then it is too unstable to risk reaching or crawling under it. Rescuers must identify pinch points and must avoid them.

Lifting and spreading can change the position of the object's center of gravity and cause it to slide, topple, or shift. Activating the force of a lifting tool on an object at rest causes a secondary movement that can injure or kill a victim or rescuer. Rescuers must identify and predict any potential secondary movement caused by lifting forces and must capture and control that movement with ongoing stabilization equipment and techniques.

Technician level rescuers performing lifting and stabilizing operations at a trench rescue will likely be in the trench during the operation. All rescuers assigned to lifting/stabilizing duties must wear a personal protective equipment (PPE) ensemble. This ensemble is described in Chapter 4, *Personal Protective Equipment and Equipment Basics*.

Just as rescuers need PPE, so does the victim. However, the mechanism of entrapment may not allow the use of victim PPE in the normal way. Always consider placing a helmet and eye protection on the victim, at a minimum. Entry team personnel must continuously monitor the safety of victims during lifting and extrication operations.

Load Stabilization

Load stabilization is divided into two phases, initial stabilization and ongoing stabilization. **Initial stabilization** stops the object from moving or prevents it from beginning to move. **Ongoing stabilization** is provided so that the object cannot shift or drop unexpectedly during the lifting operation. It is the safety net as the object is intentionally moved.

A stabilization plan must consider both vertical and lateral movement. If the object is to be moved or lifted, the stabilization plan should also be coordinated with the crew conducting the lift.

Mechanics of Load Stabilization

Load stabilization mechanics use tools and materials to prevent unwanted movement through compression or tension. Stabilization techniques that utilize compression to prevent or minimize movement place stabilizing equipment or tools between the load and a stationary object in the direction of movement. Techniques that use tension to prevent or minimize movement place stabilizing equipment or tools between the load and a stationary object on the side opposite the direction of movement.

Stabilization Plan

The stabilization plan is developed by the person in charge of stabilizing and/or lifting an object. For trench rescue, that person is the entry officer. The entry officer briefs the entry team on the details of the stabilization plan, assigns specific duties, supervises, and manages the plan. Chapter 3, *Initial Actions*, covers the incident management system and the command structure in detail.

A stabilization plan should include the following steps:

- Perform load assessment (determine the object's weight, COG, materials, and additional forces that will impact the lift)
- Stabilization equipment/tool selection
- Stabilization equipment/tool placement (location)
- Determine the sequence of equipment/tool placement
- Determine ongoing stabilization considerations and coordination with the lifting team if a lift is needed
- Provide a briefing (assignments, tools/equipment, safety, and communications)

Compression-Based Stabilization

Tools and equipment placed between an unstable object and a stable object are put into compression when the unstable object begins to move in that direction. Cribbing, wedges, shims, struts, and chocks are compression-based tools and equipment that are commonly used by firefighters and trench rescue team members. In this section, we will examine the fundamentals of the use of these compression-based stabilizers.

Cribbing

The U.S. Army Corp of Engineers provides excellent guidance for the capacity of wood cribbing. Rescue technicians should be familiar with the U.S. Army Corps of Engineers Urban Search and Rescue Program Shoring Operations Guide. Although wood cribbing is recommended, some fire departments use plastic cribbing blocks.

Cribbing can be used for both initial stabilization and for the ongoing stabilization that happens during a lifting operation. During each lift, the object being lifted should never be more than 1 inch (2.5 cm) from a substantial cribbing system.

Cribbing material for trench rescue is usually cut out of 2 × 4-inch and 4 × 4-inch pieces of lumber. The latter is what the boxes are usually constructed from, with the former used to take up odd spaces. 6 × 6 cribbing is less commonly carried on trench rescue apparatus but it is a viable alternative. Generally, cribbing is cut to standard 12-, 18-, 24-, or 36-inch (30-, 46-, 61-, or 91-cm) lengths.

Many types and sizes of synthetic cribbing are also available for rescue use. The manufacturer's recommendations should be followed when using these products.

Box Crib System

Three variations of the box crib systems are commonly used:

- 2x: The four-point box crib
- 3x: The nine-point box crib
- Solid box crib

The 2x and 3x systems are named based on the number of cribs in each tier (layer). Therefore, a 2x box crib system has two timbers for each layer. When the second and subsequent tiers are added, four contact points (places where lumber crosses) are created. A 3x box crib has nine contact points and three pieces of cribbing on each layer of the system **FIGURE 9-12**.

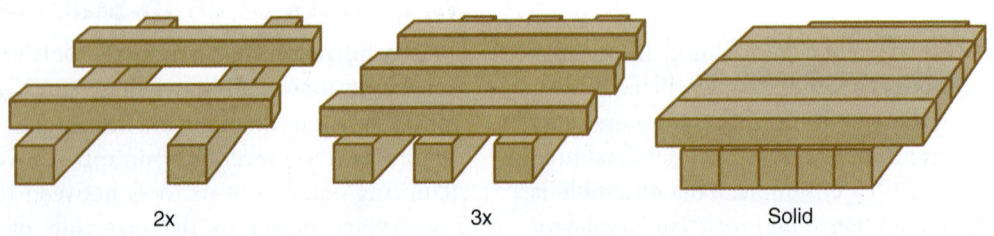

FIGURE 9-12 Box crib configurations: 2x, 3x, and solid.

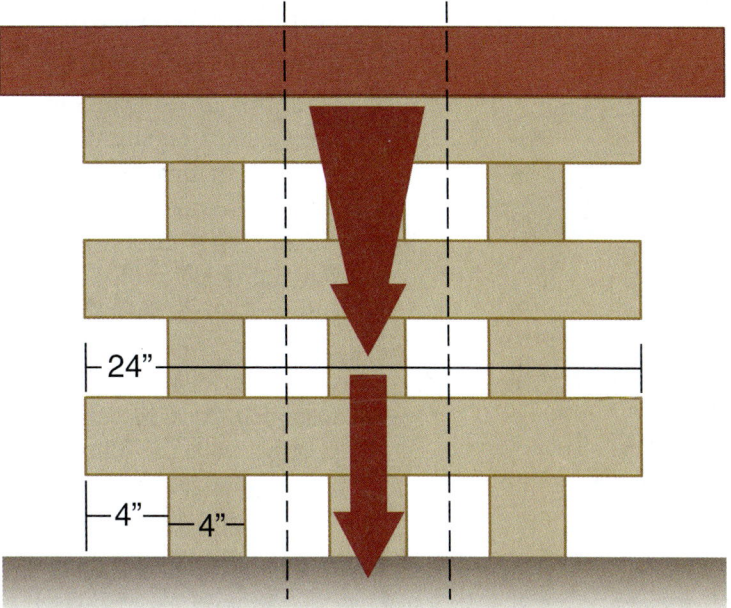

FIGURE 9-13 The object being stabilized should be as close to the center of the box crib as possible.

Cribbing Rules

The general rule for box cribbing is that each point of contact will support a standard amount of weight, depending on the type (wood species) and size of the lumber. A 4 × 4-inch crib (#1 Douglas fir and southern pine) can safely support 6000 pounds (2722 kg) per contact point, and a 6 × 6-inch crib approximately 15,000 pounds (6804 kg) per contact point **FIGURE 9-13**. By adding the number of points together, rescuers can determine the total capacity of the box crib. These capacities are only valid when all points of contact (4 points for a 2x box and 9 points for a 3x box) are loaded.

- **Design load for 4 × 4 Douglas fir and southern pine**
  2 × 2-member system = 24,000 pounds (10,886 kg)
  3 × 3-member system = 54,000 pounds (24,494 kg)

- **Design load for 6 × 6 Douglas fir and southern pine**
  2 × 2-member system - 60,000 pounds (27,216 kg)
  3 × 3-member system = 135,000 pounds (61,235 kg)

Because the weight supported may be more concentrated when on the box crib (due to less surface area), rescuers should also consider whether the ground on which the box crib rests is substantial enough to hold the load. This is especially true in the bottom of a wet trench. When the cribbing is placed on concrete or steel, there is no need to make a solid base layer. When placed on soil or asphalt, the bottom layer of the crib should be solid to fully spread out the load over the entire surface area. Generally speaking, allowable loads on soils and asphalts can range from 2,000 psf to 6,000 psf (95,760 to 287,281 pascals). If these values are exceeded, the cribbing can start to sink into the ground.

Cribbing Details

Cribbing is very stable and is very strong but it also has some limitations that need to be considered. The taller cribbing gets, the more unstable it becomes. The maximum height of cribbing used for stabilization should be less than three times the working length of the box. The **working length** is the length of the crib minus the required overlap (minimum 4 inches [10 cm] on both sides). Additionally, the size of the crib members affects the maximum height. The following maximum heights should be followed based on the cribbing member size:

- Recommended maximum height for 4 × 4 systems is 4 feet (1.2 m).
- Recommended maximum height for 6 × 6 systems is 6 feet (1.8 m).

All box crib systems are made by stacking the timbers in alternating rows (tiers). Never stack more than two tiers in the same direction. When using the box crib as a lifting platform to support a lifting tool, the top layer should always be solid. The height to which the box crib should be built as a lifting platform is no more than two times the working length of the box. It is also a good practice to use the 2x working length when using a box crib to support a lifting tool.

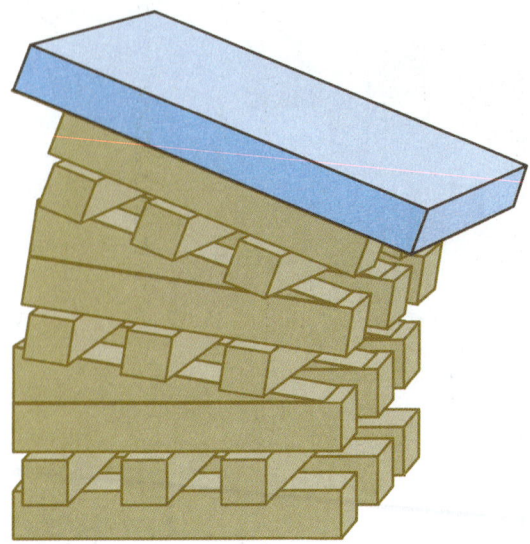

FIGURE 9-14 Angled box crib.

FIGURE 9-15 Not placing cribbing while on your knees but rather on the balls of your feet provides for a quick retreat if the load shifts during the stabilization process.
Courtesy of Ron Zawlocki.

Cribbing Angled Objects

Cribbing can be used to stabilize objects that are at angles (sloped) **FIGURE 9-14**. The maximum angle is 30 percent (about 15 degrees). The maximum height for angled crib boxes is 4 feet (1.2 m) when using 4 × 4-inch lumber and 6 feet (1.8 m) when using 6 × 6-inch lumber. Twenty-four-inch (61-cm) long shims are a good choice for changing the angle of a box crib **FIGURE 9-15**.

Wedges and Shims

Wedges are cut pieces of lumber that form an inclined plane. Wedges are used in matched sets and the two pieces oppose each other. Wedge sizes commonly used by rescuers include:

- 2 × 4-inch wedges at 12-inch (30 cm) lengths
- 4 × 4-inch wedges at 18-inch (46 cm) lengths
- 4 × 4-inch wedges at 24-inch (61 cm) lengths
- 4 × 6-inch wedges at 24-inch (61 cm) lengths

Shims are one piece of a wedge (set) and are commonly used during lifting operations to lift an inch and crib an inch. When used in this fashion, the shims take up the space between the object being lifted and the box crib, until a full piece of timber will fit under the object. While 12-inch (30-cm) and 18-inch (46-cm) shims and wedges are appropriate for taking up space, 24-inch (61-cm) shims are a much better option for changing (tilting) the angle of a box crib to match the angle of the object the box crib is supporting. Take the following steps to utilize cribbing:

- Complete a load assessment, develop a stabilization plan, and conduct a briefing.
- Determine the desired width and length dimensions of the crib.
- Determine the size of the members to be used, and the configuration of the crib layers.
  - Use 6 × 6-member configuration if the crib needs to be more than 4 feet (1.2) m high.
  - Note that the 3 × 3-member configuration is more than twice as strong as 2 × 2-member configuration.
- Decide whether the first layer needs to be a solid layer, depending on the type of bearing material (soil, asphalt, or another surface softer than a concrete slab).
- Carefully slide the members in for each layer, and keep the crib aligned and as square as possible ensuring the members are placed at least 4 inches (10 cm) from the end of the supporting member.
- When the crib reaches required height, add shims to make sure that all intersections of crib members are in solid contact with the supported object.
- Attach the crib to the supporting surface (or confine its movement), if practical.

Struts

Tabulated data exists for struts (or shores) between approximately 1 foot and 16 feet (0.3 m and 4.9 m) in length. Some can hold 20,000 pounds (9072 kg) with a 4:1 safety factor. All things being equal, shorter struts can safely hold larger loads than longer struts **FIGURES 9-16** and **9-17**. The capacities are based on manufacturer's tabulated data, which are based on tests conducted with the strut axially loaded, or perpendicular to the force. Forces are multiplied on struts when they are not perpendicular to the load. Rescuers must maintain at least a 2:1 safety factor.

FIGURE 9-16 Struts used to stabilize the boom of an excavator. A ratchet strap connecting the strut bases at the ground level is used to prevent "kick-out," or sliding.
Courtesy of Ron Zawlocki.

FIGURE 9-18 Wheel chock size should be properly matched to the tire diameter.
© Jones & Bartlett Learning. Photographed by Glen E. Ellman.

FIGURE 9-19 A rope-based tieback.
Courtesy of Ron Zawlocki.

FIGURE 9-17 Struts are available in a variety of adjustable lengths.
© FirePhoto/Alamy Stock Photo.

Wheel Chocks

Wheel chocks used for rescue should be made of aluminum or laminate construction **FIGURE 9-18**. It is important to use the right size chock for the tire diameter. To determine the right size chock, measure the physical height of the tire. The chock should be approximately 25 percent of the height (diameter) and should fit snugly into the tire. For example, a 24-inch (610-mm) diameter tire requires a 6-inch (150-mm) high chock.

If the vehicle is parked on a gradient, the safety chock needs to be large enough to hold the vehicle while in neutral gear and a minimum of two safety chocks will be needed. Check the vehicle owner's manual for specifics; some manufacturers may recommend safety chocks for both front and rear tires, or more, depending on the number of axles. When in doubt, choose a larger safety chock and chock all of the axles.

TIP

Safety Tip
Wedges, 2 × 4s, 4 × 4 cribs, and shims are not intended for chocking wheels from rolling.

Tension-Based Stabilization

Tieback is a term used to describe tension-based stabilization systems. Tiebacks are often used to control unwanted lateral movement; however, tiebacks can also be rigged to an overhead anchor to prevent or minimize unwanted downward movement. They can be constructed from synthetic rope, webbing, chain, and wire rope **FIGURE 9-19**. The system will likely incorporate connectors (such as shackles, carabiners, hooks, etc.) and a tightener (such as a come along, Griphoist, lever hoist, and rope-based mechanical advantage system). The capacity is based on the weakest item in the system, which could be the anchor. Tiebacks must be tightened enough to take slack out of the system but should not move the load. Anchor selection and angles are critical elements for tension-based stabilization systems.

Inspecting Stabilization Equipment

There are many different types of equipment that can be used for stabilization that are commonly available to rescuers. The most commonly available are wheel chocks, webbing, tiebacks, or lumber. Other equipment that may come with a trench rescue equipment cache includes cribbing, wedges, shims, and struts. It is critically important when selecting stabilization equipment to ensure that it has sufficient capacity and a factor of safety to support the load.

When performing a visual inspection of cribbing, wedges, or shims, examine for cracks, twisting, bowing, cupping, decay, crushed areas, and loose knots, and discard any cribbing and wedges with defects or damage. Visually inspect chocks for cracks and deformities.

A visual inspection for damage should be performed prior to every use. A designated person should perform a complete inspection for damage annually, and written records of this inspection should be maintained.

Inspecting Pneumatic Struts

Pneumatic struts must be checked before and during operation. Before turning on the equipment, check for loose hardware and cracked or deformed parts, then check for O-ring seal leakage while the system is pressurized. During operation, verify that the delivery pressure gauge(s) reads a relatively constant pressure regardless of the inlet pressure and flow rate, and also check for air leakage around any connection or a main housing fitting. Any leakage of air at these mating interfaces denotes either a loose connection or a defective O-ring seal that necessitates replacement.

If struts and ancillary equipment have not been used for training or collapse incidents during the last 3 months, they should be field tested to ensure they do not leak and are operational in preparation for their next use. Inspect them for damage, cracks, deformation, missing pins or parts, and components that do not move freely.

Inspecting Tiebacks

Because rope, webbing, lever hoists (cable and chain), and rigging equipment are typically available on most apparatus, they are commonly used to secure items at scenes. They can be readily used to stabilize materials that could fall or move in or around a trench. The materials to be used to stabilize the loads should be inspected with the following guidance.

Synthetic Rope

Synthetic rope should be inspected for abrasions, burns, glazing, discoloration, frays or cuts, deformation (such as appearing hour-glassed), stiffness (isolated areas), or mushiness (isolated areas).

Webbing Inspection

When inspecting webbing, look for missing or illegible sling identification; acid or caustic burns; melting or charring of any part of the sling; holes, tears, cuts, or snags; broken or worn stitching in load-bearing splices; excessive wear or abrasion; and knots in any part of the sling. Discoloration and brittle or stiff areas on any part of the sling may indicate chemical or ultraviolet (UV) damage. Fittings could be pitted, corroded, cracked, bent, twisted, gouged, or broken. Examine hooks in accordance with the removal criteria as stated in the American Society of Mechanical Engineers (ASME) B30.10 and examine rigging hardware in accordance with the removal criteria as stated in ASME B30.26. Check for other conditions, including visible damage, that cause doubt regarding the continued use of the sling.

Lever Hoist Inspection

For lever hoist inspection, check impact hooks and latches for deformation, cracks, and wear. An elongated or bent hook is an indication that the ratchet that was used to hoist was overloaded. Check the operation of latches. Safety latches on the hooks should not be removed or kept permanently open.

Inspect the cable or web strap for fraying, melting, charring, chemical damage, abrasive wear, cuts on the face or edges of webbing or cable, and/or any other damage such as holes, tears, or snags. Inspect web stitch patterns for broken or worn threads, and check ratchet teeth for gouges, burred edges, and/or other physical damage. Examine the equipment for bent or broken metal parts, rounded edges, and elongated holes, and inspect the metal parts for corrosion. Check pawls and levers for bends, cracks, and/or other damage and check the integrity of the springs.

Operate the U-frame, and verify that it does not rub against the main frame during operation. Test operating functions, such as lifting, lowering, and free release of the ratchet winch hoist. Verify that the U-frame is not missing the main frame spring, and check the hoist for broken stress links.

Inspecting Rigging Equipment

Rigging equipment includes wire rope, slings, chains, and hardware. This equipment is designed for heavy lifting and will most likely be part of the trench rescue equipment cache. Rigging equipment must meet Occupational Safety and Health Administration (OSHA) and ASME standards. The use of these items should consider the inspection procedures for wire rope, synthetic web slings and polyester round slings, alloy steel chain slings, and rigging hardware.

Wire Rope Inspection

Wire rope slings should be removed from service if they have missing or illegible sling identification, or broken wires. For running wire rope, look for 6 randomly distributed broken wires in one rope lay; for strand-laid, look for 10 broken wires in one strand in one rope lay. During inspection, review the slings for severe corrosion, localized abrasion or scraping of the rope, attachments, or fittings; evidence of kinking, crushing, bird caging or any other damage resulting in damage to the rope structure, or evidence of heat damage. Ensure that end attachments are not cracked, deformed, or worn to the extent that the strength of the sling is substantially affected. Look for any other condition that may cause a rescuer any doubt.

Synthetic Web Slings and Polyester Round Sling Inspection

Synthetic web slings should be removed from service if they have missing or illegible sling identification; acid damage, caustic burns, melting, or charring of any part of the sling; broken or worn stitching; or excessive abrasion, holes, tears, cuts, or snags. In addition, discoloration, or brittle or stiff areas of the sling may mean chemical or UV/sunlight damage. Look for knots in any part of the sling or evidence of knots, fittings that are pitted, corroded, cracked, bent, twisted, gouged, or broken, or any other condition that may cause the rescuer any doubt regarding the safety of the item.

Alloy Steel Chain Slings Inspection

Remove from service any alloy steel chain slings that have missing or illegible sling identification or excessive nicks, cracks, breaks, or gouges. Look for stretched, bent, twisted, or deformed chain links or components; evidence of heat damage; or excessive pitting or corrosion. Check for a lack of ability of the chain or components to hinge (articulate) freely, weld splatter, or any other condition that may cause the rescuer any doubt about the safety of the item.

Rigging Hardware: Hook, Shackle, Turnbuckle, and Carabiner Inspection

During inspection of the rigging hardware, it should be checked for missing or illegible manufacturer's name or trademark and/or rated load identification (or size as required); a 10 percent or more reduction of the original dimension; bent, twisted, distorted, stretched, elongated, cracked, or broken load bearing components; or any excessive nicks, gouges, pitting, and corrosion. Look for indications of heat damage, including weld spatter or arc strikes, evidence of unauthorized welding; loose or missing nuts, bolts, cotter pins, snap rings, or other fasteners and retaining devices; or unauthorized replacement components; or any other visible conditions that cause the rescuer any doubt as to the continued safe use of the sling.

Lifting Plan

A critical lift requires the development and use of a carefully considered lifting plan. The lifting plan is developed by the person in charge of the extrication (**entry team officer**) and must be shared with all members who will engage in the extrication operation. The plan is shared through a briefing, which is conducted by the entry team officer prior to beginning the lift. Before lifting, the team should do the following:

- Complete a load assessment (determine the object's weight, COG, and additional forces that will impact the lift)
- Install initial stabilization per the stabilization plan
- Select and inspect lifting tools (size, length, **working load limit (WLL)**, options)
- Place the lifting tool
- Determine the lifting sequence
- Coordinate the lift with ongoing stabilization (per the stabilization plan)
- Conduct the briefing, including:
  - Assignments
  - Sequence
  - Safety
  - Communications

Lifting Techniques

In this section, we will explore two lifting techniques and the tools and equipment that are associated with each of them. We will explain the use of surface-based lifting techniques that employ lifting tools that push (via compression) and the use of overhead lifting techniques that employ lifting tools that pull (via tension). Additionally, we will explain the use of the bridge lift technique that uses a combination of surface-based lifting tools (pushing) and overhead rigging (pulling).

Surface-Based Lifting Techniques

Surface-based lifting requires tools that push rather than pull. Surface-based lifts do not require the use of rigging equipment. However, when lifting with surface-based tools pushing from the trench floor, additional problems are often encountered. The biggest problem is finding enough room under the object to place and operate the surface-based tool. The trapped victim and the stabilizing equipment will occupy a sizeable area of prime real estate under the object, so it is often difficult to position and operate surface-based tools in the trench. When room is available, surface-based lifting tools can be safe and efficient. When using surface-based lifting tools, a solid platform, cribbing, or doubled pieces of ¾-inch (19-mm) plywood should be placed under the tool to distribute the weight of the load over a larger surface area in order to minimize penetration into the trench floor as the tool becomes loaded.

In this section we will discuss the use of pry bars, air bags, and mechanical jacks to conduct surface-based lifting operations. **Air bags** and mechanical jacks can be used as part of a bridge lifting operation with the addition of rigging equipment that connects them to the heavy object in the trench.

Types of Surface-Based Lifting Tools

Manual lifting tools are typically levers, such as a pry bar, Halligan tool, or pinch bar. **Mechanical lifting tools** commonly used by rescuers include mechanical (ratchet style) jacks. **Pneumatic lifting tools** typically used by rescuers are air (lifting) bags. **Hydraulic lifting tools** include jacks, wedges, and rams.

Lifting Tool Capacities

Manual. Pry bars can be used as a class I lever **FIGURE 9-20**. A 200-pound (91-kg) person using a 60-inch (152-cm) long pinch bar can lift about 2200 pounds (998 kg) by placing the fulcrum (5 inches [13 cm] from the load) to achieve a 11:1 mechanical advantage. If the load you are lifting weighs more than 2200 pounds

FIGURE 9-20 Pinch point pry bars (48–60 inches [122–152 cm]) are standard lifting tools in the rescue community.
Courtesy of AMES Compaines, Inc.

(998 kg), simply add more pry bars and more lifters (rescuers). Check with the manufacturer of the pry bars and use them in accordance with the manufacturer's recommendations for the specific model being used.

Mechanical. Most mechanical ratchet jacks have three ratings. For example, the Hi-Lift brand has a WLL of 5000 pounds (2268 kg) when used as a winch. When used as a clamp, it has a WLL of 750 pounds (340 kg). When used as a jack, it has a WLL of 4660 pounds (2114 kg). The rescuer must be familiar with and adhere to the manufacturer's recommendations for each brand and model of jack **FIGURE 9-21**.

Use the following steps to operate a mechanical jack **FIGURE 9-22**:

- Complete a load assessment, develop a lifting plan, and conduct a briefing.
- Clear enough space around and under the object to insert plywood (4 × 4-foot [1.2 × 1.2-m] minimum if possible), several pieces of 2 × 12-inch planks or 4 × 4-inch cribbing, which are used to create a solid platform for the jack(s) to distribute the load to a large surface area.
- Position the jack(s) under the lifting point.
- Extend the jack slowly.
- Lift an inch and crib an inch until the victim can be removed.

Pneumatic. Pneumatic lifting bags require contact surface area to conduct their lift **FIGURE 9-23**. The capacity of the bag will depend on the style of bag, the

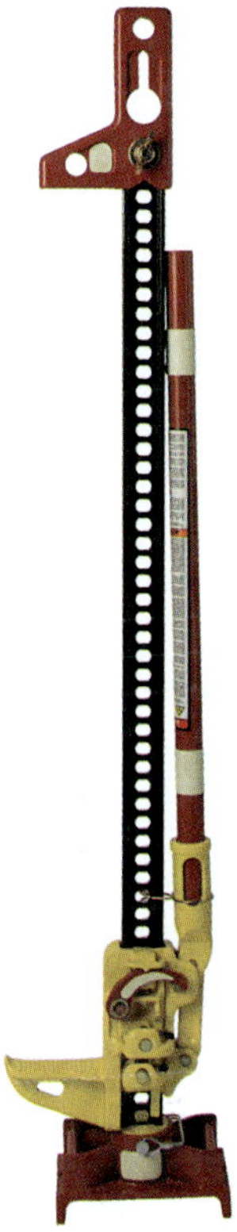

CHAPTER 9 Lifting and Load Stabilization **221**

FIGURE 9-23 High pressure air bags set up to lift an excavator.
Courtesy of Cecil V. "Buddy" Martinette, Jr.

FIGURE 9-21 Mechanical jacks are a good option for lifts that have a lot of distance between the ground and the object to be lifted.
Courtesy of Hi-Lift Jack Company.

FIGURE 9-24 Some compact hydraulic spreaders can generate 10,000 pounds (4536 kg) of force.
© aapsky/Shutterstock.

FIGURE 9-22 Operating a mechanical rachet jack.
Courtesy of Ron Zawlocki.

air pressure (pounds per square inch), and the contact surface area. Most air lifting bags have high load capacities but have limited lifting distances.

Hydraulic. Hydraulic lifting tools (jacks, wedges, and rams) use cylinders, pumps, and fluid to develop the forces needed to lift or push heavy objects **FIGURE 9-24**. The capacity of the tool will depend on the amount of pressure the pump can generate and the size (surface area) of the cylinder.

Surface-Based Lifting Tool Inspection

Generic inspection procedures are provided here. Be sure to follow your manufacturer's specific inspection

TIP

A mechanical jack can be used from a bridge, provided that the jack is positioned on top of the bridge so that it is in compression during the lift. A mechanical jack should not be suspended below the bridge so that it is in tension (pulling) the load.

procedures. Pry bars and lever equipment must be visually inspected before and after each use for damage, such as areas that are corroded, cracked, bent, twisted, gouged, or broken.

Inspect the mechanical jack carefully before each use. Ensure the jack is not damaged, worn, or missing parts. Check the climbing pins to make sure that they are not worn or damaged, and check the steel standard bar to make sure that it is straight and that nothing is blocking the steel standard bar holes. Do not use the jack unless it is in good clean working condition and properly lubricated. Using a jack that is not in good clean working condition or properly lubricated may cause serious injury.

A generic air bag inspection should be performed before each use. Check the manufacturer's inspection criteria for additional inspection details. Visually inspect air bag equipment for damage such as scuffs, kinks, tears, and ply separation. Audibly check for the leakage of air and visually check for leaks with the use of a soap/water solution to see if bubbles form.

Overhead Lifting Techniques

Overhead lifting requires tools that work in tension (or pull) rather than in compression (or push). Overhead lifts require the use of rated rigging equipment. In this section, we discuss using bridges, bipods (constructed by rescuers), and heavy equipment (such as excavators or backhoes) to conduct overhead lifting operations.

Critical Angles

Overhead lifts incorporate the use of slings (wire rope, chain, and synthetic slings) to connect the load to the lifting tool. Slings are commonly used in vertical, choker, or basket configurations. When two sling legs connect at one point, an angle is created. As the angle (shown in degrees) decreases, the sling capacity also decreases. For instance, if a sling has a WLL of 500 pounds (227 kg) in a vertical hitch, used in a basket hitch with the legs at a 90 degrees angle to the load, it would have a WLL of 1000 pounds (454 kg) **FIGURE 9-25**. The angle of the legs in a basket hitch has a critical effect on the sling's capacity. The same sling seen in **FIGURE 9-25** but used in a basket hitch with the legs at a 60 degree angle to the load has a WLL of only 866 pounds (393 kg) **FIGURE 9-26**. The sling angle factor is a geometric factor and can be found in load angle charts **FIGURE 9-27**.

When a sling is used in a vertical hitch, the full lifting capacity of the sling material can be utilized

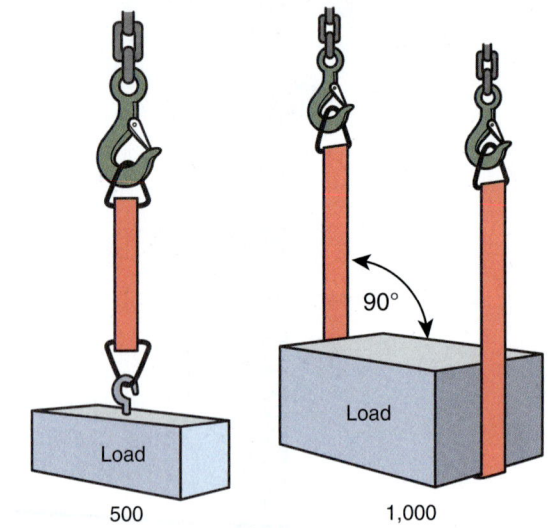

FIGURE 9-25 A basket hitch (90 degree legs) has twice the capacity of a vertical hitch.

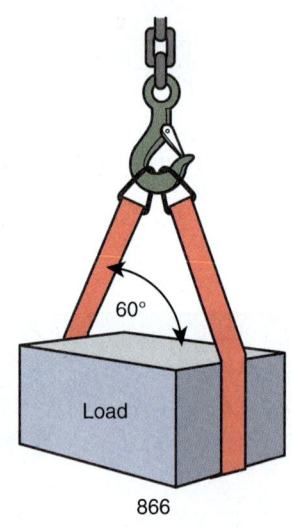

FIGURE 9-26 As the angle decreases so does the sling capacity.

FIGURE 9-28. Due to the stress created at the choke point, slings rigged with a choker hitch achieve only about 75 percent of their potential capacity. Always pull a choker hitch tight before a lift is made—never during the lift **FIGURE 9-29**.

The cradle configuration of a basket hitch (90 degrees) allows the two extending ends (legs) of the sling to function as if they were two separate slings. The capacity of the sling in this hitch is twice that of the same sling in a vertical hitch, but only if the sling angle of each leg is 90 degrees Lifting with both legs at 90 degrees would normally require two lifting devices or a spreader bar **FIGURE 9-30**.

For a basket hitch (less than 90 degrees), when slings or sling legs are used at an angle during a lift,

CHAPTER 9 Lifting and Load Stabilization   223

Load Angle Chart

Angle factor must be applied to calculate the reduced sling capacity when lifting force is not at 90° to the plane load.

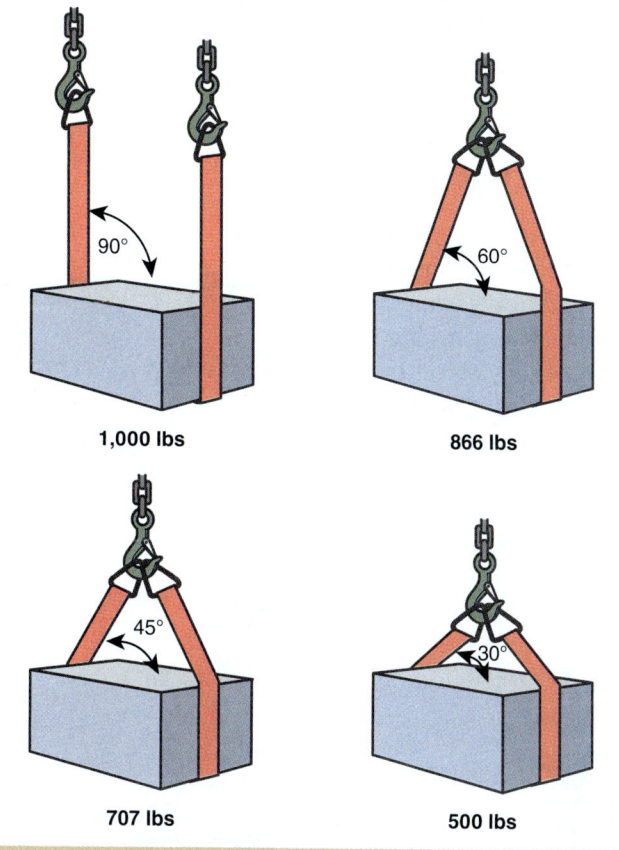

Multiply angle factor x sling's vertical rated load to calculate the reduced capacity at the angle.

Angle	Factor	Angle	Factor	Angle	Factor	Angle	Factor
90°	1.0000	70°	0.9397	55°	0.8192	40°	0.6428
80°	0.9848	65°	0.9063	50°	0.7660	35°	0.5736
75°	0.9659	60°	0.8660	45°	0.7071	30°	0.5000

FIGURE 9-27 This chart is used to calculate the reduction in sling capacity when the 90 degree (basket) capacity is known.

FIGURE 9-28 Vertical hitch.

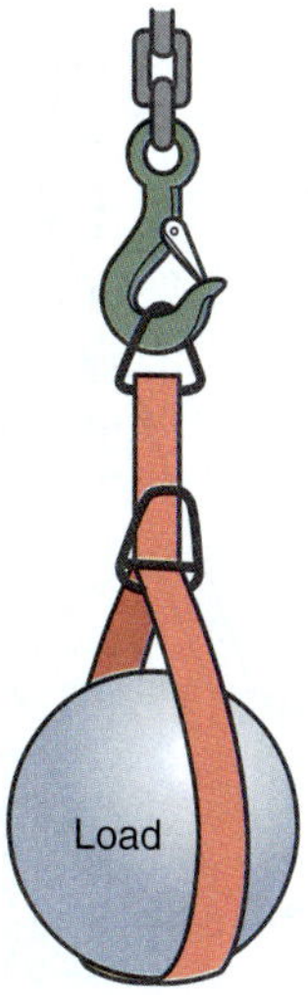

FIGURE 9-29 Choker hitch.

the sling capacity is reduced. How much it is reduced depends on the sling angle. Note that the rated capacity of a 30 degree basket is only one-half that of a 90 degree basket. Sling angles below 30 degrees are strongly discouraged. A sling angle of 60 degrees or more is preferred **FIGURE 9-31**.

Types of Overhead Lifts Used for Trench Rescues

There are three types of overhead lifts used for trench rescues: bridge lifts, bipod lifts, and heavy equipment lifts. A bridge that is made of timbers, or a **bridge lift**, can be positioned directly above a heavy object in the trench. Rigging is attached to the lifting tool positioned on or connected to the bridge and the heavy object **FIGURE 9-32**. Rigging can be attached to the head of the bipod system to raise a load in a **bipod lift**. Manufacturers such as Paratech Rescue Inc. have developed accessories for their rescue support struts that allow the rescuer to build a bipod for overhead lifts. Finally, the lifting eye of an excavator can be positioned directly above a heavy object in the trench to perform a **heavy equipment lift**. Rigging is attached from the lifting eye on the excavator arm or bucket to the heavy object. The hydraulic cylinders on the excavator or backhoe provide the force needed to lift the object.

FIGURE 9-32 Bridge lift.
Courtesy of Ron Zawlocki.

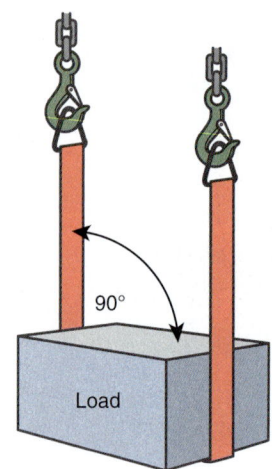

FIGURE 9-30 Basket hitch (90 degrees).

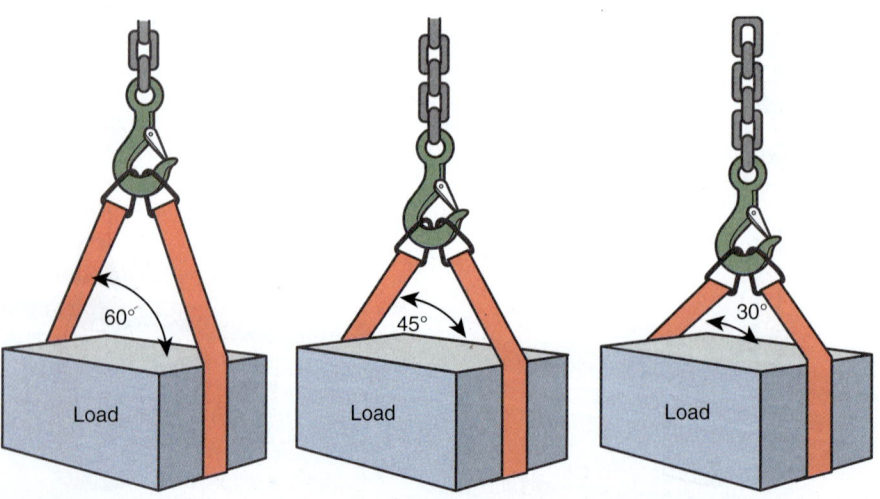

FIGURE 9-31 Basket hitch (less than 90 degrees).

> **TIP**
>
> **SAFETY TIP**
>
> Air bags tend to shoot out or roll round objects, such as pipes and boulders, when placed underneath them. Therefore, the preferred method uses the air bags in an overhead lifting configuration placed on a lifting bridge. The objects can be connected to the air bags with slings (chains, wire rope, or synthetic slings) in a choker or basket configuration.

Performing Bridge Lifts

As a technician level rescuer performing overhead bridge lifts, you need to know terminology, equipment inspection procedures, lifting capacities, hand signals, and lifting procedures. The components of the bridge lift are as follows:

- **Beam**: Many trench teams utilize long sections of wood as shoring or wales. These can also be used to span the top of the trench from lip to lip to create an overhead anchor.
- **Lip protection**: Keeps loads distributed near the trench; 4 × 4-foot (1.2 × 1.2-m) or 4 × 8-foot (1.2 × 2.4-m) sections of ¾-inch (19-mm) plywood are commonly used for lip protection.
- **Bridge supports**: Cribbing can be used to raise the bridge up over panels or to create a longer throw with the lifting tool when the bridge was not set at grade level. Depending on its construction, this support can also distribute the load across a larger footprint and/or keep it away from the lip of the trench.

Lifting Bridge Capacities

The chart below can be used for #1 Douglas fir. Pine and other fir timbers in first-class condition have 75 percent of these capacities. Safe working loads with the load concentrated at the center are as follows:

<u>4 × 4-inch Douglas Fir Beams</u>

6 foot (1.8 m) span – 700 pounds (318 kg)

8 foot (2.4 m) span – 500 pounds (227 kg)

<u>6 × 6-inch Douglas Fir Beams</u>

6 foot (1.8 m) span – 2300 pounds (1043 kg)

8 foot (2.4 m) span – 1700 pounds (771 kg)

10 foot (3.0 m) span – 1300 pounds (590 kg)

13 foot (3.7 m) span – 1100 pounds (499 kg)

<u>8 × 8-inch Douglas Fir Beams</u>

6 foot (1.8 m) span – 5800 pounds (2631 kg)

8 foot (2.4 m) span – 4300 pounds (1950 kg)

10 foot (3.0 m) span – 3500 pounds (1588 kg)

13 foot (3.7 m) span – 2900 pounds (1315 kg)

<u>7 × 7-inch LVL Beams (used in strong direction - individual plies in line with loads)</u>

6 foot (1.8 m) span – 13,200 pounds (5987 kg)

8 foot (2.4 m) span – 9900 pounds (4491 kg)

10 foot (3.0 m) span – 7900 pounds (3583 kg)

13 foot (3.7 m) span – 6600 pounds (2994 kg)

Air Bag Bridge Lift

To conduct a bridge lift with an air bag, follow these steps:

- Complete a load assessment, develop a lifting plan and stabilization plan, and then conduct a briefing and make specific assignments.
- Install initial stabilization per the stabilization plan.
- Position lip protection on each side of the trench, directly in line with the lifting point.
- Span the trench with appropriate beams placed on top of the lip protection.
- Select the lifting tool and center it on top of the beam directly over the object to be lifted.
- Connect a rigging over the bag or the lifting nose of the jack and extend the rigging into the trench (observe critical angle limitations).
- Connect the rigging from the object to the sling (around bag) with a rated chain that can be tightened.
- Direct the lifting tool equipment operator to slowly raise the object.
- Coordinate the lift with ongoing stabilization (per the stabilization plan).

Performing Bipod Lifts

Paratech Rescue Incorporated has developed rated accessories that provide for the use of a bipod for overhead lifts. Rigging can be attached to the head of the bipod system to raise a load and position the head/lifting point over the trench. A bipod can offer flexibility for the lifting team, and allow rescuers to use equipment with which they are familiar and have inspected, rather than relying on an excavator for which they do not have maintenance records being run by an operator who may or may not have the steady hand necessary to safely lift a load.

A lifting tool such as a Griphoist can be attached to the head of the bipod and operated from within the trench. Alternately, a change of direction (COD) pulley can be attached to the head of the bipod, and the lift can be controlled from outside the trench, because the lifting tool is secured to an object exterior to the trench and the wire rope runs over the COD and down to the object being lifted. A bat-wing rigging configuration uses two remote anchors and minimizes the risk of toppling the bipod because horizontal forces on the head are equalized. A bat-wing system eliminates the need for a back-tie on the bipod, but requires long lengths of wire rope/rope. Regardless of the lifting tool or configuration, all forces should be considered, and sound rigging practices should be used throughout.

Bipod Components

The components of a bipod include an engineered designed head (**bipod head assembly**) with a rotational capability that provides attachment points for lifting equipment and system stabilizing, struts/extensions that are used to gain elevation and reach, and finally, **bipod bases** that are high-strength, hinged rotating anchor plates.

Bipod Capacities

The safe working load of the bipod will depend on the manufacturer's tabulated data and the angle at which the bipod is set. For example, the Paratech bipod system is rated for over 4500 pounds (2041 kg) at an angle not less than 60 degrees (4:1 factor of safety).

Hand Signals for Bridge Lifts and Bipod Lifts

The hand signals for bridge lifts and biopod lifts are illustrated in **FIGURE 9-33**.

- Load up: Raise the load.
- Stop: Stop all actions.
- Set: Set stabilizers to capture the progress of the lift.
- Release: Release the lifting tool and place the load on stabilizers.

Overhead Lifting Tools

Most manufacturers of lifting tools do not approve their products when lifting loads over human beings. An example is the mechanical jack, which can be used as a jack (in compression) with a WLL of over 4500 pounds (2041 kg). When used as a pulling

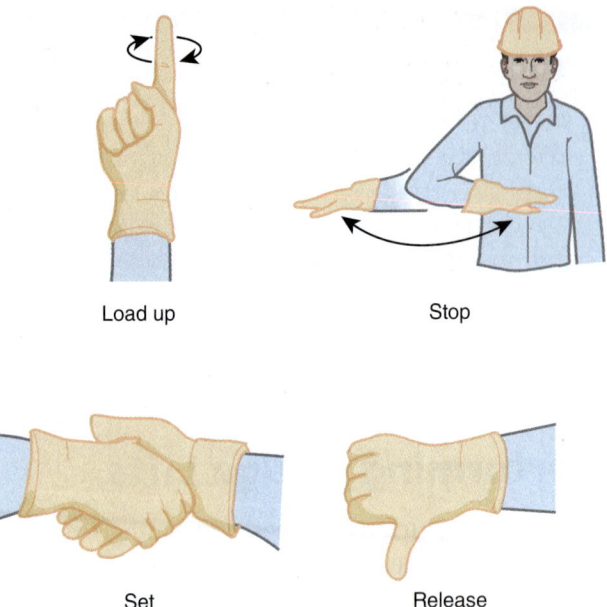

FIGURE 9-33 Hand signals.

tool (in tension) many mechanical jacks have a WLL of 5000 pounds (2268 kg), but they are not endorsed by manufacturers for use in lifting (pulling) an object from overhead at rescue incidents.

Griphoist produces pulling tools (tension) with models that are rated and approved for rescue application. Griphoists come in different sizes and are versatile lifting and pulling tools for rescue operations. They should only be used in accordance with the manufacturer's recommendations for the specific model being used.

Coordinating the Use of Heavy Equipment

Technician level responders may work in conjunction with heavy equipment and skilled operators to conduct a lift and should be familiar with the personnel at these sites and the roles they play in executing a lift. Coordination and clear lines of communications must be established between the entry team officer, signal person, cribbers, and the equipment operator. The entry team officer manages and supervises the lifting and stabilizing operation in order to extricate the victim. A signal person is needed when an equipment operator is at a distance from the victim such as a heavy equipment (excavator) operator. The signal person should be a technician level rescuer who has training and experience working with heavy equipment operators. Cribbers are personnel who place and operate stabilizing tools and equipment.

Equipment Operator

A heavy equipment operator involved in a trench accident will likely be emotionally charged by the incident. Remember that unlike firefighters and emergency medical personnel, they are not used to seeing traumatic injuries. Also remember that at the scene of a trench emergency where the operator is working, the victim is someone they know and is likely a friend or even a family member. Critical lifts are performed by experienced rescuers using their own rescue lifting equipment. If, however, heavy equipment must be used to lift an object that has trapped a victim, it is a best practice to get a professional equipment operator who was not involved in the accident. If time does not permit that, then care must be taken to carefully interview an operator to determine their emotional stability as well as their operator skill level.

The entry team officer will begin by talking to the operator and asking what happened. The officer will listen for signs of levels of excitement, sorrow, and anger. If the operator seems reasonably calm and collected, the operator will be told that the officer might want to use them to assist in the rescue and would like to check the team's rigging equipment compatibility with their equipment and make sure that hand signals are agreed upon. A quick practice lift will be run away from the trench to assess the operator's ability to precisely lift, hold, and move a load under the entry team officer's direction.

Communications

Utilizing heavy equipment during a rescue incident does not come without risk to both trapped victims and rescuers. Those risks can be reduced and managed through planning and good communications. Technician level rescuers must be familiar with heavy equipment terminology, lifting capacities, and methods of communication. Methods of communications include face-to-face, hand signals, and radio communications.

- Face-to-face: The first line of communications must be a face-to-face meeting with the entry team officer, the signal person, and the equipment operator. During this meeting, the lifting and stabilizing plan will be established and communications agreed on.
- Hand signals: Hand signals are a common means of communication when working with heavy equipment. When running, heavy equipment (excavators, backhoes, front end loaders, etc.) is noisy, making verbal communications difficult and sometimes impossible. Additionally, the equipment operator is at a significant distance from the object being lifted. As a result, the operator's view is limited. Having a signal person close to the object being lifted gives the operator another set of eyes on the target. Hand signals can effectively communicate and direct required movements of the heavy equipment from the signal person to the operator. Although universal hand signals exist for equipment operators, not all operators use them. Determining the exact hand signals that will be used must be established during the face-to-face meeting. A clear line of sight between the operator and the signal person must be established and maintained during the operation.
- Radio communications: The signal person and the equipment operator should be provided with hand-held radios. Hand-held (portable, two-way radio transceiver) radios should be used during the use of heavy equipment at a rescue operation. It is difficult for equipment operators to hold and push buttons on a hand-held radio while running their machines but a radio in the cab can allow the operator to listen to the signal person when verbal directions are needed.

Heavy Equipment Terminology

There is a difference between a backhoe and an excavator. An excavator's cab is mounted on a rotating platform. A backhoe's cab is fixed to the frame. A **backhoe** is a piece of excavating equipment consisting of a digging bucket on the end of a two-part articulated arm (boom and stick). They are typically mounted on the back of a tractor or front loader. The boom is attached to the vehicle through a pivot known as the kingpost, which allows the arm to slew left and right, usually through a total of approximately 200 degrees. A backhoe uses outriggers to transfer weight off of the rear tires and provide stabilization **FIGURE 9-34**. An excavator is an engineering vehicle consisting of an articulated arm (boom, stick), bucket, and cab mounted on a pivot (a rotating platform) atop an undercarriage with tracks or wheels **FIGURE 9-35**.

From a technical level rescuer point of view, the important parts of an excavator or backhoe include the following:

- **Boom**: The section of the arm closest to the vehicle.
- **Stick** (Arm): The section of arm that carries the bucket.
- **Bucket**: A specialized container attachment designed with teeth to facilitate digging.

FIGURE 9-34 Backhoe.

FIGURE 9-35 Excavator.
© bogdanhoda/Shutterstock.

FIGURE 9-36 A heavy equipment lift.
Courtesy of Ron Zawlocki.

Buckets are attached to the second arm (stick) of excavators and backhoes. Buckets come in a variety of sizes and volumes. Backhoes commonly have loader buckets and digging buckets.

- **Lifting eye**: The factory-installed rigging point on a bucket.
- **Outriggers**: Hydraulically operated stabilizers found on tractor-style backhoes.
- **Center pin**: The center of the rotating platform on an excavator.
- **Swing area**: The dangerous area or radius around which a rotating cab can reach.

Lifting Capacities

An essential component in a successful lifting plan is the determination of the lifting tool's (excavator/backhoe) capacity. To determine how much the machine can lift, check with the operator, look at the chart in the cab, and review the operator's manual. The lifting capacity will vary based on the orientation to the tracks or wheels as well as the extension and elevation **FIGURE 9-36**.

Hand Signals Needed for Heavy Equipment Lifting

Establish verbal communications with the operating engineer and agree on hand signals that will facilitate rigging, lifting, and moving. The essential hand signals include the following:

- Boom (up/down)
- Stick or arm (in/out)
- Bucket (in/out)
- Swing (left/right)

In order to lift and move an object, you need to know the following signals **FIGURE 9-37**:

- Load (in/out)
- Load (up/down)
- Slow
- Travel (ahead/back)
- Stop
- Emergency stop

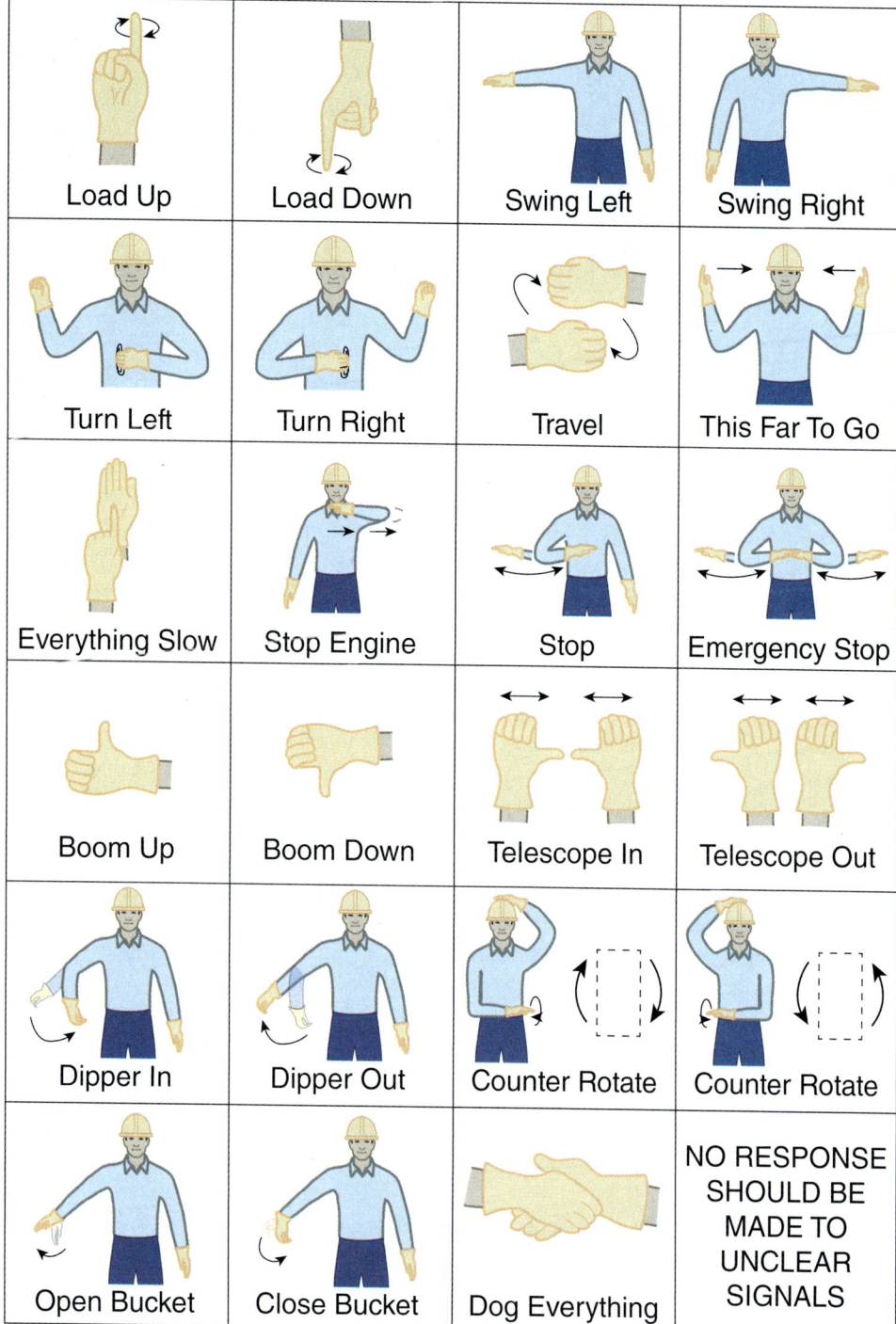

FIGURE 9-37 Excavator hand signals.

TIP

Safety Tip

The hand signals in Figure 9-37 are national standard signals taught and used by professional operators. Some operators will have their own versions of hand signals, just like various regions of the United States have their own dialects and accents as they speak the English language. Agree on the signals with the equipment operator before you begin the operation, and be prepared to use their version of hand signals.

Voices of Experience

A call came in for a possible trench rescue in the south end of the city. The soil types there are known to be a mix of sand, clay, and rock as it is closer to Lake Ontario. The first-in apparatus confirmed that it was a trench call and that a mini excavator had fallen into a trench, pinning the operator against the foundation of the house. There was no safety cage on this excavator as it may have been removed.

Our system uses the swap loader bin system and we keep our trench components inside a bin that sits on the ground because when it is on the truck, it will not fit into the apparatus bay floor. We loaded the bin onto the truck and made our way south to the address; this added time to the response.

Once our trench team arrived, we walked into the backyard where we were directed by crews already on scene. A male victim was pinned against the foundation of the house and a mini excavator had tipped into the trench with one of its tracks on the lip and the rest of the machine pinning the victim against the wall. The workers were waterproofing the basement and had a 9-foot (2.7-m) deep by 4-feet (1.2-m) wide trench around the entire back of the house.

Our trench bin had trouble maneuvering through all of the fire trucks and police cars already on scene and needed a large enough area to drop the container.

I walked up to the incident commander and asked if the victim was "code 5" due to traumatic asphyxia (our term for obviously dead). He replied, "He was making respiratory effort when we first arrived on scene." This directed our crews to go into rescue mode, and we started by rigging the machine back to a large tree that we used for an anchor and called in a city excavator and a heavy tow. The excavator weighed approximately 8000 pounds (3629 kg) and the come-alongs that we had were only rated for 4000 pounds (1814 kg), so we set up two of them, but the machine was wedged too tightly to budge.

Observing that the victim was now extremely cyanotic, we wanted to reassess the situation. We cantilevered a ladder over the trench, and we sent a rescuer out and he could not feel a pulse on the victim. We attached a cardiac monitor to the victim and observed that there was no electrical activity in the heart. At this point I asked the paramedics to call online medical control to obtain a field pronouncement and we would switch to recovery mode. The paramedics reported to the doctor that they were not capable of doing a full assessment on the victim so the physician would not grant them a pronouncement. This kept us in rescue mode for longer than we should have been.

There was a delay in the arrival of the heavy equipment, and the effort was looking futile. More time had passed, and we were putting the lives of rescuers in potential danger for a person that we knew would not survive. I walked up to the police sergeant and asked him if he could get the coroner on the line. The sergeant took out his phone and called the coroner directly. Because I also work as a flight paramedic, I was comfortable talking to this physician directly. I explained to the coroner that the victim had no cardiac activity, had been trapped for over 45 minutes, and there was a high likelihood of a rescuer being hurt or killed trying to move this excavator. The police coroner gave us a pronouncement of death immediately, which put us into a safer recovery mode, and we waited for the heavy excavator and tow to assist.

Lessons Learned

- We should have all of our trench equipment on a bin instead of on the apparatus that could get to the call faster. If we had had a few ground pads and a

Voices of Experience

couple panel sets on the rescue squad, it would have saved quite a bit of time. First-in crews need to keep an area open for the bin to get dropped. Our new rescue trucks are being designed with an area to keep panels and timbers on a vehicle that can go directly to the scene of an emergency to start shoring operations while the support vehicle makes its way there.

- The paramedic's attempt to paint good picture to their online medical control did not get them a pronouncement of death. This kept us in rescue mode much longer than we should have been. It needs to be stressed to the doctor that an attempt to rescue has a high probability of injury or death to rescuers.

- Heavy rigging equipment would make rescues or recoveries much safer for rescuers. It would not have made a difference in the outcome of this call, but it opened our eyes to our limitations. Our department has since invested in a full complement of rigging equipment, including Griphoists, slings, and associated hardware.

Mike Tesarski
Mississauga Fire

After-Action

IN SUMMARY

- A critical lift is any lift with a human being below the object, and/or any lift where the object exceeds 75 percent of the lifting or stabilizing tool's rated capacity.
- Critical lifts require a lifting plan that includes a load assessment, equipment inspection, ongoing stabilization, lifting procedures, and coordination between stabilization and lifting team members.
- Initial stabilization prevents or limits unwanted movement and increases the safety of the victim and rescuers in the trench.
- Ongoing stabilization is a technique that prevents the object from shifting or dropping during the lifting operation. It is the "safety net" as the object is intentionally moved.
- The concepts of mass, gravity, friction, center of gravity, moment of force, work, energy, inclined planes, levers, and pulleys can be applied in myriad ways to the trench rescue environment.
- The theory of lifting mechanics is subdivided into two areas: energy and work. The effective use of rescue tools is often determined by a thorough understanding of these concepts and their application in a given situation.
- Surface-based lifting techniques push the load.
- Overhead lifting techniques pull the load.
- During a critical lift, the object being lifted should not be more than 1 inch (2.5 cm) from an ongoing stabilization system.

KEY TERMS

Air bags Inflatable bags used for rescue operations available in low pressure (7 psi), medium pressure (14 psi), or high-pressure (100-150 psi) varieties. These systems include pressure regulators, hoses, controllers, and bags.

Backhoe A piece of excavating equipment consisting of a digging bucket on the end of a two-part articulated arm (boom and stick).

Beam Long section of wood or metal used by trench teams as shoring or a wale. It can also be used to span the top of the trench from lip to lip to create an overhead anchor point.

Bipod bases High-strength hinged rotating anchor plates for specialized lifting equipment.

Bipod head assembly An engineered, designed head with a rotational capability that provides attachment points for lifting equipment and system stabilizing.

Bipod lift Rescue support struts such as those manufactured by Paratech Rescue Inc. that allow rescuers to build a two-legged structure for overhead lifts. Rigging can be attached to the head of the system to raise a load.

Boom The section of the arm closest to the vehicle.

Bridge lift A lifting tool positioned on or connected to a bridge made of timbers that can be positioned directly above a heavy object in the trench. Rigging is attached to the bridge and the heavy object.

Bridge supports Cribbing used to raise a bridge up over panels or to create a longer throw with the lifting tool when the bridge is not set at grade level. This support can also distribute the load across a larger footprint and/or keep it away from the lip of the trench.

Bucket A specialized container attachment designed with teeth to facilitate digging. They are attached to the second arm (stick) of excavators and backhoes, and they come in a variety of sizes and volumes.

Center pin The middle of the rotating platform on an excavator.

Center of gravity (COG) The point on a body around which the body's mass is evenly distributed or balanced.

Class I lever Type of lever in which the fulcrum is located between the force and the load.

Class II lever Type of lever in which the load is located between the fulcrum and the force.

Class III lever Type of lever in which the force is located between the load and the fulcrum.

Critical lift Any lift in which a human being is below the object, and/or any lift where the object exceeds 75 percent of the lifting or stabilizing tool's rated capacity.

Ductile materials Materials that bend before they fracture.

Energy The capacity for doing work and overcoming resistance.

Entry team officer The individual who manages and supervises the extrication functions.

Excavator An engineering vehicle consisting of an articulated arm (boom, stick), bucket, and cab mounted on a pivot (a rotating platform) atop an undercarriage with tracks or wheels.

Friction Measure of the amount of force it takes to move an object across the surface of another object.

Heavy equipment lift Type of lift that employs the hydraulic cylinders on an excavator or backhoe to provide the force needed to lift an object. Rigging is attached from the lifting eye of an excavator bucket to the object.

Hydraulic lifting tools Tools that use cylinders, pumps, and fluid to develop the forces needed to list or push heavy objects.

Inclined plane Angled plane that gains efficiency by reducing required force over time.

Initial stabilization A technique used to prevent or minimize the movement of a load.

Lever A simple machine used to move a load. It includes a fulcrum (pivot point), a force, and a load.

Lifting eye The factory installed rigging point on a bucket.

Lip protection Ground pads, usually made of plywood that keep loads distributed near the trench edge.

Manual lifting tools Used in trench rescue, these are typically levers, such as a pry bar, Halligan bar, or pinch bar.

Mechanical advantage The force achieved using a tool, mechanical device, or other equipment.

Mechanical lifting tools Used in trench rescue, these are typically a mechanical ratchet-style jack (Hi-Lift brand is common).

Mechanics A branch of physics that deals with energy and forces in relation to bodies.

Moment of force The amount of force rotating around the fulcrum times the distance from the fulcrum.

Ongoing stabilization A technique that prevents the object from shifting or dropping during the lifting operation. It is the "safety net" as the object is intentionally moved.

Outriggers Hydraulically operated stabilizers found on tractor-style backhoes.

Pneumatic lifting tools Used in trench rescue, these are typically high-pressure air lifting bags.

Stick The section of a backhoe or excavator arm that carries the bucket.

Swing area The dangerous area or radius around which a rotating cab can reach.

Tieback Term used to describe tension-based stabilization systems.

Uniform shape A single, rigid symmetrical object with constant density. The balance point (center of mass) is located at the object's geometrical center (centroid).

Work Distance times force.

Working length The length of the crib minus the required overlap (minimum 4 inches [10 cm] on both sides).

Working load limit (WLL) The maximum allowed weight (load). It is a percentage of the "break strength" that is often expressed as a ratio. Example: A chain that breaks at 10,000 pounds has a value of 2,500 (25%) at a 4 to 1 factor of safety.

On Scene

1. If you are first to arrive at a trench rescue incident, what stabilization equipment will be available to you and how will you utilize it?

2. Can you quickly estimate the center of gravity of oddly shaped loads, such as an excavator bucket with soil in it?

3. What equipment do you have available on your first due units and on your trench rescue response to lift loads?

4. Do you have trusted local operators that you can call to operate heavy equipment at an incident?

5. Are you confident in your ability to signal to a heavy equipment operator?

CHAPTER 10

Technician Level

Technician Level Trench Rescue Shoring

KNOWLEDGE OBJECTIVES

After studying this chapter, you should be able to:

- Recognize conditions that indicate expertise beyond the trench technician level rescuer may be required.
- Create a shoring plan for an intersecting trench. (**NFPA 1006: 12.3.2**, p. 232)
- Create a shoring plan for a trench more than 8 feet (2.4 m) deep. (**NFPA 1006: 12.3.3**, p. 232)
- Describe common collapse conditions in intersecting trenches. (**NFPA 1006: 12.3.1**, p. 233)
- Identify differences between shoring a straight trench and shoring intersecting trenches. (**NFPA 1006: 12.3.1**, pp. 232–235)
- Describe considerations when developing a shoring plan for an intersecting trench. (**NFPA 1006: 12.3.1**, pp. 233–253)
- Describe considerations when developing a shoring plan for a trench deeper than 8 feet (2.4 m). (**NFPA 1006: 12.3.3**, pp. 253–256)
- Identify shoring equipment for intersecting trenches. (**NFPA 1006: 12.3.1, 13.3.2, 12.3.5**, p. 236)
- Describe the procedure for primary and secondary shoring of an intersecting trench. (**NFPA 1006: 12.3.1, 13.3.2, 12.3.5**, pp. 233–253, 256–262)
- Describe the procedure for corner shoring. (**NFPA 1006: 12.3.1, 13.3.2, 12.3.5**, pp. 233–234)
- Describe differences in shoring L and T intersecting trenches. (**NFPA 1006: 12.3.1, 13.3.2, 12.3.5**, p. 233)
- Describe the procedure for primary and secondary shoring of a deep trench. (**NFPA 1006: 12.3.3, 12.3.4**, pp. 253– 262)
- Explain when supplemental shoring is required. (**NFPA 1006: 12.3.3**, pp. 260–261)
- Describe the procedure for installing supplemental shoring. (**NFPA 1006: 12.3.6**, pp. 256–262)
- Explain safety considerations when considering spot shoring. (**NFPA 1006: 12.3.7**, pp. 262–263)
- Describe the failure potentials associated with spot shoring. (**NFPA 1006: 12.3.7**, pp. 262–263)

SKILLS OBJECTIVES

After studying this chapter, you should be able to:

- Release a victim from soil entrapment. (**NFPA 1006: 12.3.11**, p. 232)
- Disentangle a victim from objects in a trench. (**NFPA 1006: 12.3.11**, p. 232)
- Utilize a shoring plan to shore an intersecting trench. (**NFPA 1006: 12.3.2, 12.3.5**, pp. 143–145)
- Utilize a shoring plan to shore a trench more than 8 feet (2.4 m) deep. (**NFPA 1006: 12.3.3, 12.3.4**, pp. 257–260)
- Install supplemental shoring at 2-foot (0.6-m) intervals. (**NFPA 1006: 12.3.6**, p. 262)

You Are the Rescuer

Your department recently completed a trench rescue training program. During the technician level rescuer course, you shored three trenches: a 12-foot (3.7-m) deep trench, an 8-foot (2.4-m) deep T-trench, and an 8-foot (2.4-m) deep L-trench. All trenches were dug in stable soil, and the lips and walls remained intact during the shoring lesson. During the training, your chief brought the media out to cover the events. In a television interview, the chief explained the purpose of the trench rescue team and described how it will serve the city and the neighboring communities.

Only a few weeks after the training, your department receives a mutual aid request for a man buried in a trench. When you arrive at the scene, you see a trench that looks nothing like your training experiences. This 17-foot (5.2-m) deep trench is 8 feet (2.4 m) wide where it intersects with another trench. At the intersection, a partially built concrete catch basin will clearly create an obstruction for conventional shoring. The inside corner of the L-trench has a wedge collapse and has resulted in a large void. In your training, the instructor had mentioned that the course met the requirements found in the NFPA standards. Now you are wondering if that 3-day class really provided the knowledge, skills, and abilities needed to conduct rescue operations at the types of trenches you see being dug in your response district.

Today, your chief is on the scene along with the chief of the neighboring fire department. The TV cameras are on the ground and are circling above in helicopters. The spotlight is on you and your team of "trench rescue experts." If your previous training provided you with a clear understanding of shoring fundamentals, you should be able to confidently answer these technician level questions.

1. Which shoring practices could rapidly protect a victim and be easily expanded to create a safe working area for rescuers to enter this trench?
2. Which type of equipment and procedures are needed to install rescue shoring in a 17-foot (5.2-m) deep trench?
3. Which equipment and techniques can be used to shore an area with an obstruction that cannot be removed?

 Access Navigate for more practice activities.

Introduction

The job performance requirements (JPRs) in NFPA 1006, *Standard for Technical Rescue Personnel Professional Qualifications*, for a technician level rescuer are an indication of the higher sets of knowledge, skills, and abilities that are required to hold this title. Whereas operations level JPRs for trench rescuers are limited to a straight 8-foot (2.4-m) deep trench or less, the technician level expands those limits to include developing and implementing a trench shoring plan to support an intersecting trench and a trench more than 8 feet (2.4 m) deep. Additional JPRs for the technician level rescuer include installing supplemental sheeting and shoring for each 2 feet (0.6 m) of depth dug below an existing approved shoring system and the use of spot shoring in appropriate soil conditions. These JPRs are addressed in this chapter.

Technician level rescuers also must possess the following requisite knowledge and skills (these were addressed earlier in this text):

- Chapter 2: Soil and collapse mechanics
  - Recognition of potential collapse
  - Types of collapses and techniques to stabilize, emergency procedures
- Chapter 3: Initial actions
  - Scene safety and security measures
  - Establishment of zones
- Chapter 4: Personal protective equipment (PPE) and equipment basics
  - Selection of PPE
- Chapter 5: Hazard mitigation
  - Consulting related tabulated data
  - Designating a cut station location
  - Additional material and equipment needs
- Chapter 6: Managing the trench incident
  - Criteria for a safe zone within the trench
  - Use of shoring strategies and tactics and prebriefing of team members
  - Addressing areas of the trench that are blown out or undercut

- Chapter 8: Victim care and extrication
  - Consideration of the weights and hazards associated with the soils
  - Identification of mechanisms of entrapment
  - The location of the victim and projected path for removal are incorporated.
  - Consideration of selected stabilization tactics for extrication and victim safety.
  - Victim release techniques and projected path for removal.

Based on the authors' experience, we believe that the majority of trench rescue incidents are in trenches that go beyond the conditions that operations level rescuers can manage. Additionally, some trench rescue incidents may require skills and knowledge beyond the operations and technician levels. With that in mind, the authors have crafted Appendix A, *Advanced Trench Rescue Shoring*.

This chapter will focus on the shoring tactics that are described in the trench rescue technician sections of the NFPA 1006 and 1670 standards. Technician level victim care, handling, and removal techniques are covered in Chapter 8, *Victim Care and Extrication*, and extrication tactics are covered in Chapter 9, *Lifting and Load Stabilization*.

Intersecting Trenches

An **intersecting trench** is one where a second trench intersects the first trench, creating corners within the trench intersection. Typically, the intersections are at near 90-degree angles but could be practically any angle. The most common intersecting trenches are the **T-shaped trench** and **L-shaped trench**. An L-shaped trench is where one trench merges into a second trench that runs approximately perpendicular to the first, creating a "L" shape when viewed from above **FIGURE 10-1A**. Quite simply, a T-shaped trench is where two trenches meet at a near 90-degree angle and create a "T" shape when viewed from above **FIGURE 10-1B**. Both types have similar complexities and unique shoring challenges that must be resolved.

On the chance that you are called for a rescue where the victim is within the straight portion of an intersecting trench and not near the intersection (corner), the rescue may be considered a straight trench rescue as long as a failure of the corner would not threaten the victim or the rescuers. If the completed straight wall shoring (primary shoring expanded to create an appropriate safe zone) is greater than either 4 feet (1.2 m) from an inside corner or beyond any cracking or soil disturbance related to the inside corner wall, then shoring the corner typically would not be required. If there is any question regarding the stability of the inside corner wall, or the soil near it to support the opposite trench wall, then the corner should be shored as discussed later in this chapter.

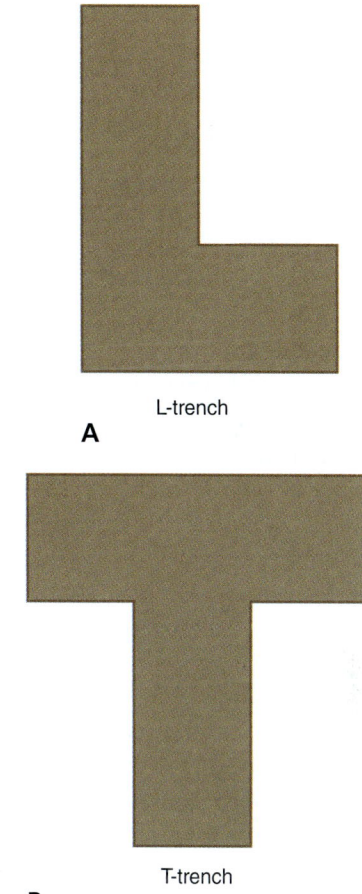

FIGURE 10-1 Top view of the two most common types of intersecting trench. **A.** L-shaped trench. **B.** T-shaped trench.

Deep Trenches

Technician level rescuers are expected to conduct rescue and recovery operations in trenches that are deeper than 8 feet (2.4 m). Although a maximum trench depth is not identified in the NFPA job performance requirements, the authors of this text have set a maximum depth at 20 feet (6.1 m). That limit is based on the practical limitations of the kinds of shoring equipment that can be carried in fire department apparatus and on the fact that the overwhelming majority of trench rescue incidents occur in trenches deeper than 8 feet (4.2 m) but less than 20 feet (6.1 m) deep. Similarly, the Occupational Safety and Health Administration (OSHA) allows construction workers to shore trenches up to 20 feet (6.1 m) deep provided they follow either manufacturer's tabulated data or a shoring plan designed by a registered professional engineer (RPE). In this chapter, we offer an engineered shoring plan for rescue operations in trenches that are up to 20 feet (6.1 m) deep.

Sheeting and Shoring for the Technician Level

Supplemental Sheeting and Shoring

Supplemental sheeting and shoring include operations that involve the use and placement of sheeting and shoring when greater than 2 feet (0.6 m) of exposed trench wall exists below the bottom of the initially installed shoring panels. Traditional sheeting and shoring involve the use of 4 × 8-foot (1.2 × 2.4-m) panels, with a strongback attachment, supplemented by a variety of conventional shoring (strut) options. Supplemental sheeting and shoring involve the use of 2 × 4-foot (0.6 × 1.2-m) panels with a strongback attachment supplemented by a variety of conventional shoring (strut) options. Installing supplemental sheeting and shoring concurrently with victim extrication (digging) operations requires additional training beyond that of the traditional sheeting and shoring taught at the operations level.

Spot Shoring

Spot shoring is shoring that does not use traditional strongbacks or panels, but instead uses specially designed rail bases installed directly on the trench walls **FIGURE 10-2**. Spot shoring can be a dangerous shoring technique at most trench rescue incidents. In this chapter, we will provide an engineered approach to this concept to help make it safer for rescuers and victims.

Rescue Plan

At the technician level, the shapes, depths, configurations, and conditions of the trenches become more complex. Accordingly, the tactical options required for safe and efficient rescue operations become more technical, more time consuming, and more personnel and equipment intensive, all of which require more planning. The rescue plan is the tactical portion of the incident action plan (this plan is covered in detail in Chapter 3, *Initial Actions*). The rescue plan includes the shoring plan, victim care and extrication (release and removal) tactics, and specific assignments. The tactical options are based on the hazard identification and mitigation (risk versus benefit analysis) efforts and resource availability.

FIGURE 10-2 The authors believe that spot shoring and skip shoring should not be used at trench collapse incidents.
Courtesy of Speed Shore Corporation.

Risk versus Benefit Analysis

When crafting a rescue plan, an analysis and subsequent decision must be made at the command level that is based on a hazard identification and situation assessment, which weighs the risks likely to be taken against the benefits to be gained for taking those risks. The choice of tactics used in the rescue plan must be based on the results of a thorough risk–benefit analysis. Calculated risks are appropriate for live victims; however, risks are not appropriate for body recovery operations.

Shoring Plan

The shoring plan section of the rescue plan includes specific duties assigned to the panel team and shoring team. The trench rescue shoring plan must always start with providing immediate protection for the victim(s). This is detailed in Chapter 7, *Operations Level Trench Rescue Shoring*. The tactics used to support this plan must be easily expanded from protecting the victim to creating a safe working area for rescuers. The shoring system plan is then completed with a continuation of efforts that are necessary to maximize rescuer and victim safety during the extrication process. The shoring plan includes the following:

- Shoring size-up: An assessment and evaluation of the trench and victim conditions
- Lip protection: Providing a safe area on the trench lip

- Primary shoring: Rapidly protecting victims from additional collapse
- Secondary shoring: Creating a safe area in the trench for rescuers
- Complete shoring: Expanding, enhancing, and supplementing shoring
- Shoring performance assessment: Physical inspection of all components every hour

Shoring Performance Assessment

The purpose of the **shoring performance assessment** is to evaluate and adjust the shoring to maintain safety. All shored areas, spoil piles, and trench walls adjacent to the shored area must be monitored throughout the rescue or recovery operation. To begin, physically inspect all shores twice during the first hour and then once per hour. Check for the following:

- Struts loosening: Check collars for tightness. If the collars are loose, reshoot, tighten collars, and release air pressure.
- Increased soil force: This may be indicated by panels/wales bending or deflecting more than 1 inch (2.5 cm) between struts. Wood in the shoring system provides warning signs (such as cracking, creaking, groaning, or popping noises). When signs of loading are present, add struts halfway between existing struts to reduce the unsupported surface area on the bending wood members.
- Signs of moving soil: Signs include widening fissures, bulging, leaning, sloughing, and raveling. Visually check the walls (on both sides of the trench) that are next to the shored area. Check under the lip protection for new or widening cracks. Expand the shoring to support adjacent trench walls that are showing movement.

TIP

Each of the shoring plans discussed in this chapter must include a shoring performance assessment.

Victim Release

This portion of the rescue plan includes specific methods for the extrication of the victim from entrapment and the medical treatment prior to and during the extrication process. Extrication is an entry team duty. Patient assessment and treatment of the victim may be provided by basic life support (BLS) or advanced life support (ALS) personnel working under the supervision of trench rescue personnel. The victim release section of the rescue plan includes the following:

- Patient assessment and treatment
- Extrication method and equipment needs
- Patient packaging for removal
- Removal technique and equipment and personnel needs

Path for Removal

This portion of the rescue plan includes the following steps (developing a plan is covered in detail in Chapter 5, *Hazard Mitigation*):

- Ensuring that the path selected is safe and clear of obstructions
- Positioning of stretcher(s), personnel, and ambulance(s)
- Communications with the receiving hospital and a predetermined patient transport route

Pre-entry Briefing

Every person involved in the rescue operation must understand the plan and their specific role in the execution of that plan. The pre-entry briefing of the rescue plan should include, but not be limited to, information regarding the following:

- Hazards and mitigation methods in place
- Anticipated environmental concerns
- Rescue or recovery mode
- Rescue plan tactical assignments and specific equipment needs include:
  - Shoring
  - Victim release
  - Path for removal
- Communications protocols, procedures, and details
- Emergency procedures
- Anticipated logistical needs
- Time frames for operations
- Debriefing procedures

Shoring Intersecting Trenches

The most common types of intersecting trenches are the T-trench and the L-trench. The most common failure and collapse location at intersecting trenches is the inside corner. The process of digging this type of trench typically disturbs a wedge of soil on the inside corner. This disturbance may result in that wedge shearing and falling into the trench. Most times the wedge will fall in the direction that the corner is pointing (diagonally); however, the distribution of the collapsed soil seldom results in equal accumulations on the trench floor. Most often, one side of the corner will have a taller pile of dirt. The side with the taller pile is called the **pile side**. This definition will be referenced throughout this chapter.

> **TIP**
>
> If a joined corner panel is going to be used to shore an inside corner, it is important to build and install the corner panels from the pile side.

Intersecting trenches have complexities that are that are unique and must be thoroughly understood for a successful shoring **FIGURE 10-3**. One of those complexities is that intersecting trenches have unsupported areas of trench wall that do not have opposing trench walls that can be used for support. Figure 10-3 only shows the primary shoring in place at a victim's location, and expanding the safe zone with secondary shoring must include supporting the areas shown in red. In a T-trench, this can be accomplished relatively quickly by placing a panel to cover the red area and using a wale that is supported with struts on both sides of the force (two-point supported beam), with the force transferred to both inside corner walls as seen in **FIGURE 10-4**. In a L-trench, such a two-point supported beam system cannot be achieved.

Historically, firefighters have been taught to shore the outside corner by using thrust blocks to support struts running diagonally from the outside corner to the panels on the inside corners **FIGURE 10-5**. Unfortunately, this method does not have sufficient capacity to resolve the soil forces generated at the red-highlighted areas seen in Figure 10-3. With angled walls, when a strut meets a panel at anything other than a right angle, a sliding force is created (as discussed in Chapter 9, *Lifting and Load Stabilization*). That force needs to be resolved to ensure safety and stability. If it is not resolved, the shoring system will become unstable and the system will fail.

The thrust block method uses angled struts that create sliding forces but does not resolve the sliding forces that are transferred to the inside corner panels. Testing has shown that even low strut activation forces will cause sliding forces that will push the inside corner panels and struts down the trench walls, resulting in complete and catastrophic failure **FIGURE 10-6**. If a shoring system shows signs of failing from normal strut activation forces, that system does not provide the stability required to resist the forces of active soil conditions at the outside corner walls.

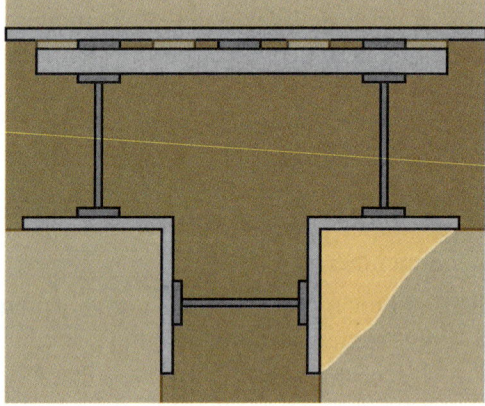

FIGURE 10-4 The unsupported area of soil in a T-trench can be collected with panels and wales, and the lateral soil forces can be transferred to the corner walls across the trench.

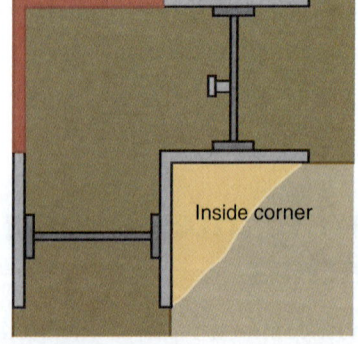 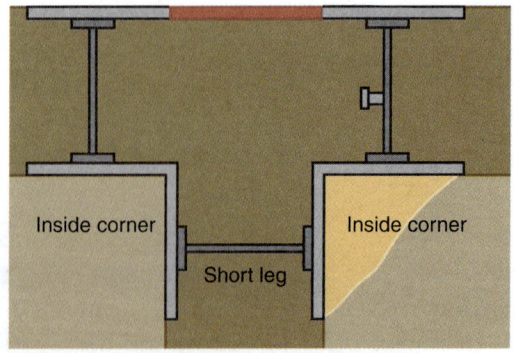

FIGURE 10-3 One of the complexities of intersecting trenches is that they have unsupported areas of trench wall (highlighted in red) that do not have opposing trench walls that can be used for support.

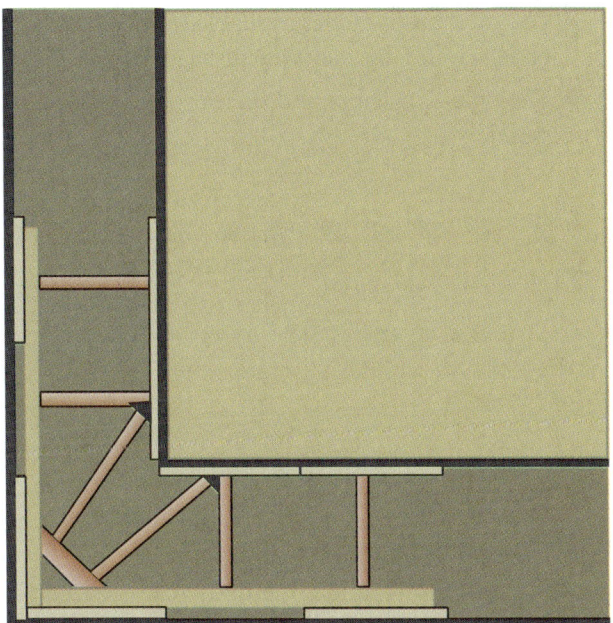

FIGURE 10-5 Previous editions of this text have included the use of the thrust block shoring design. Recent testing and engineering reviews no longer consider the use of the thrust block shoring as a tactical option for trench emergency shoring.

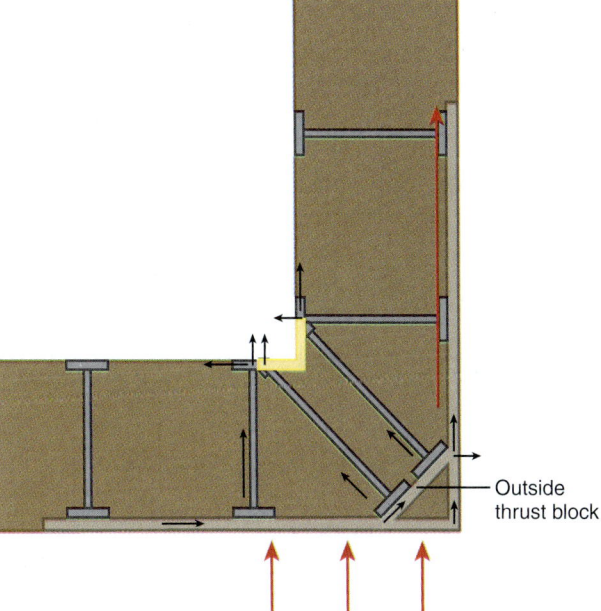

FIGURE 10-7 The vectors (direction and magnitude of force) from dynamic soil conditions at the outside wall cannot be resisted by the thrust block shoring designs.

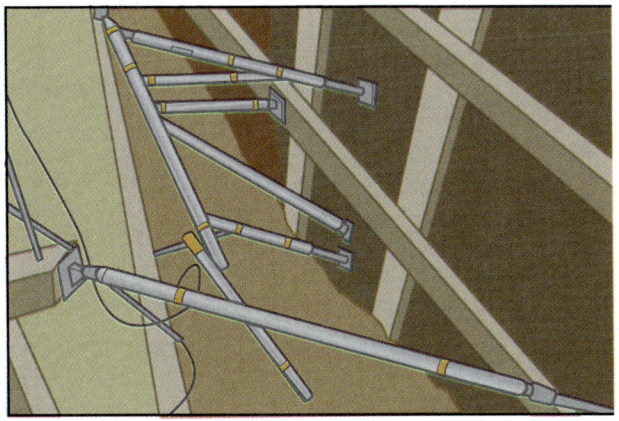

FIGURE 10-6 This illustration shows the struts dislodging and sliding down the wall as a result of minimal pressure on the angle struts used in thrust block systems.

More recent attempts at resolving the thrust block sliding problem have included the use of ratchet straps, ropes, and brackets attached to the two inside corner panels to prevent them from being pushed (slid) down the wall **FIGURE 10-7**. Although the sliding of the panels at the inside corner was reduced or eliminated, continued testing revealed another critical design problem. The lateral force found a new point of least resistance at the wales on the outside corner walls. These wales were not designed to resist forces applied at their ends and running in line with the length of the wale. Those forces create sliding that occurs between the wale and the panel strongback point of contact. The only resistance to that sliding is the activation force of the struts on the wale that cause friction between the wale and the panel strongbacks. The small amount of frictional resistance to sliding that the strut activation force can create is further reduced by the minimal friction coefficient developed between the wale and the strongback and between the panel and the trench wall. Testing has shown that when the inside corner panels are connected, the force finds the point of least resistance and pushes (slides) the wales and the attached struts down the outside wall, once again resulting in complete and catastrophic failure.

For these reasons, the authors recommend that rescuers never use the thrust block method to support an L-trench. It provides so little protection that it is not worth the time spent building it. The L-trench shoring methods described in this chapter resolve the sliding issues by using a **cantilever wale (beam)**. With this system, it is important that the wales not touch or be connected in the corner (via outside thrust blocks). This keeps each wale isolated from the other wales and prevents sliding forces from being transferred to the wales.

Even if the inside corner has not failed, the presence of the corner still complicates the trench rescue shoring. Figure 10-3 shows primary shoring (protecting the victim) in place in both a L-trench and a T-trench. Only one strut is visible in the top view of those primary shores. In reality, there are at least two and possibly three struts on those primary panel sets. Activating the struts to high activation forces on one side of an inside corner without supporting the other side of the corner could fracture the corner and push it into the opposite leg of the trench leading to the primary panel set becoming unstable/failing. Rescuers must recognize this collapse potential and install primary struts at low pressures (<500 pounds [227 kg] of strut activation force). Then they must install panels and struts at low pressures on the opposite corner before bringing all corner struts (simultaneously) up to recommended strut pressures (1000–1250 pounds [454–567 kg] of strut activation force). On an L-shaped trench, rescuers should address this once, by placing two sets of panels in one corner. On a T-shaped trench, rescuers should address this twice, once on each corner, by placing three sets of panels **FIGURE 10-8**.

Intersecting trenches require significantly more equipment than straight trenches of the same depth. Consequently, more rescuers will be needed to install this equipment within a time frame that will prevent a rescue from becoming a recovery. Agencies must recognize that emergencies that require technician level rescuers will often require more equipment and more personnel than a typical trench rescue team can provide. Provisions for the deployment of multiple trench rescue teams should be preplanned and practiced to ensure that enough personnel and equipment are available for intersecting trench rescue situations.

Plan Application for an Intersecting Trench

The application of a trench rescue shoring plan for an intersecting trench should be determined by the following:

- The type of trench to shore (T- or L-shaped)
- Any obstructions in the desired shoring area
- The method or tactics available (joined corner panels, wales, backshores, air bags, etc.)
- The type of collapse and the existing collapse pattern

TIP

Inside corners on intersecting trenches are highly likely to result in a wedge collapse.

Precautions for an Intersecting Trench

All the hazard mitigation methods (see Chapter 5, *Hazard Mitigation*) apply here, with close attention to the following:

- Limiting the activity and surcharged loads on trench corner areas
- Providing coordinated and simultaneous operations from the panel team, shoring team, and the entry team

Preparation to Stabilize an Intersecting Trench

Take the following steps to prepare to stabilize an intersection trench:

- Assign shoring team, panel teams, and wall number designations.
- Mitigate hazards.
- Provide a clear shoring plan.
- Place lip protection.
- Measure trench/place escape ladder(s).
- Prepare struts and backfill.
- Prepare panels and wales.

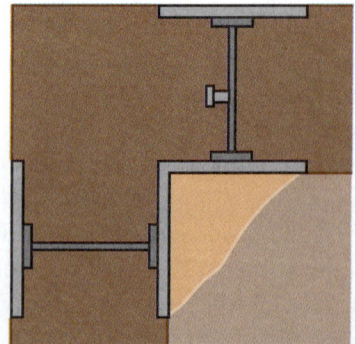
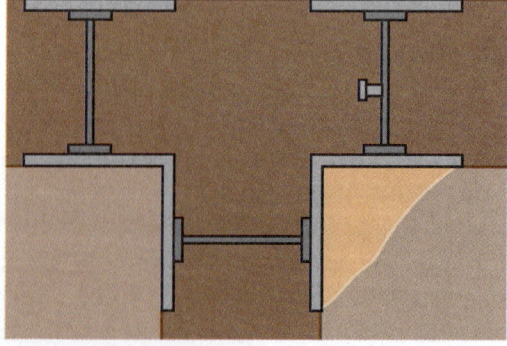

FIGURE 10-8 Shoring the inside corners of intersecting trenches must be done carefully and systematically.

Intersecting Trench–Specific Shoring Equipment

Special equipment has been developed to successfully shore intersecting trenches. Rescuers need to develop an understanding of this equipment and obtain skills in how to use it. The following equipment is needed or is extremely helpful in shoring T- and L-trenches.

- **Wales**: 7.25 × 7-inch laminated veneer lumber (LVL) wales will support the high loads required for the cantilevered spans needed to shore the outside corner of L-trenches and potential long spans of T-trenches.
- **Wale hangers**: Pipe clamps can be preassembled and hung on the top of a panel and support a wale at a specific height for placing struts. This provides a much more stable method of supports rather than counting on rescuers to hold ropes to steady the wales.
- **Joined corner panels**: These panels use two aluminum or steel brackets placed in tiers, secured with ½-inch (13-mm) diameter bolts and nuts, this allows two trench rescue panels to be bolted together at a 90-degree angle. This prefabricated corner panel is easy to slide into the trench corner, get struts installed to opposing panels, to quickly provide primary shoring at an inside corner.
- **Cantilever lip bridges**: With an inside corner trench failure, the shoring construction requires reaching out over the tops of the panels to install struts. With the earth missing from the corner, a lip bridge has to be cantilevered out in order to drop in and position the struts. This is accomplished by placing a ladder on the ground, extending over the missing lip, placing 2 × 10-inch boards over the rungs, and anchoring the far side of the ladder with ladder rung brackets with pickets driven into the ground **FIGURE 10-9**. This allows a single rescuer to stand on the overhanging lip bridge to install the shoring.

FIGURE 10-9 Cantilever lip bridge.
Courtesy of Ron Zawlocki.

Standard Shoring Equipment for Intersecting Trenches

Standard shoring equipment includes the following (see Chapter 4, *Personal Protective Equipment and Equipment Basics*, for further detail):

- Tape measure
- Ladders
- Lip bridges
- Joined corner panels (or corner brackets)
- Trench rescue shoring panels
- Ropes
- Pneumatic struts, bases, extension, ropes, and air supply system
- Hammers, nails, and remote strut collar locking equipment
- 7 × 7-inch LVL wales, hangers, and fillers
- Backfill and installation equipment

Personnel for Intersecting Trenches

The required personnel include the following:
- Panel team 1
- Panel team 2
- Panel team 3 (L-trench)
- Shoring team
- Entry team

TIP
Joined corner panels are not required for shoring intersecting trenches; however, with proper installation, they offer several tactical advantages and a reduction in the time needed to complete shoring.

L-Trench Shoring Plan

This L-trench shoring plan is based on a tested and engineered design specific to collapse conditions **FIGURE 10-10**. This design is also appropriate for shoring L-shaped trenches that have not collapsed.

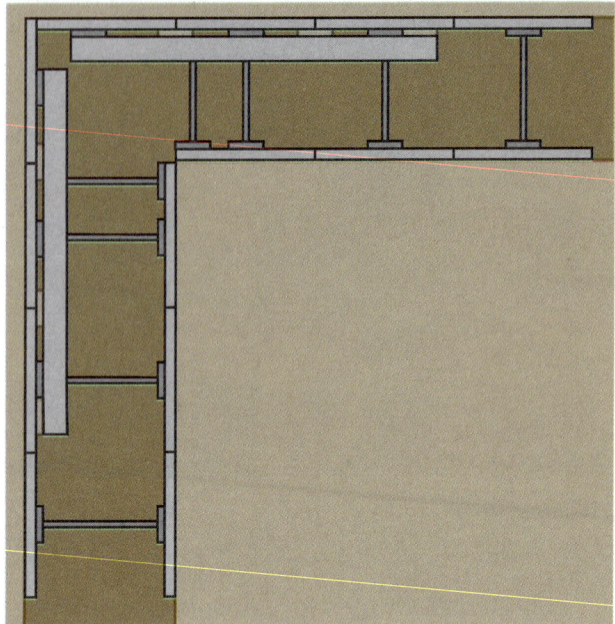

FIGURE 10-10 Top view of the L-trench shoring design.

Like all of the shoring plans supplied by the authors, this plan includes a shoring size-up, lip protection, primary shoring, secondary shoring, complete shoring, and ongoing shoring performance assessments.

This plan assumes the most likely L-trench collapse, which is an inside corner (most likely a wedge failure) collapse. Primary shoring must cover the area above the victim's head and chest. If the victim is not trapped in the corner area, straight wall trench primary shoring procedures should be used. The following procedure is used for shoring the corner area.

Shoring Size-Up for an Intersecting Trench

Before shoring the trench, first size-up the trench. Assess the lip of the trench. Lip protection must be installed before shoring and panel team members can begin shoring operations. Given a collapsed trench incident, the best practice is the use of lip bridges.

Next, measure the depth from the lip to the debris (collapse) pile to determine the number of panels needed. Measure the width of the trench to determine the length of struts needed. Measure the "L" for the trench and any surcharge loads to determine the Total L. Estimate the wall angles.

Assess the collapse potential of the trench. Identify signs of impending collapse. Look for fissures (cracks), cornices (undercut walls), bulging walls, surcharge loads within the L area, layered soil, and flowing water in the trench walls.

Assess the voids. Voids are created by sections of trench walls that have collapsed. The voids must be filled to both minimize the movement of the back wall of the void and to create an adequate point of resistance to any movement (load) being transferred from the opposing trench wall. In order to determine the best backfill technique for each void, rescuers must know the void size (height, width, and length), the angle of the void back wall, and whether it is an open or closed lip void.

Next, assess the victim(s). The victim's condition will dictate the benefit and subsequent risk levels that will be acceptable. Calculated and reasonable risks are appropriate for live victims. Minimal risks are appropriate for clearly deceased victims. Finally, use tabulated data to determine the possibility of collapse (see Chapter 5, *Hazard Mitigation*).

Walkthrough of Shoring a L-Trench

Primary Shoring

Primary shoring must cover the area above the victim's head and chest. If the victim is not trapped in the corner area, straight wall trench primary shoring procedures should be utilized. Shoring placed to support walls at the inside corner has the potential to cause failure at the other leg of the inside corner. Both sides of inside corners *must* be supported before strut pressures are increased to the recommended operating pressures. The following procedure is used for shoring the corner area **FIGURE 10-11**.

Panel Team 1. Panel team 1 installs Wall #1.
1. If using a joined corner panel, it should be built 4 feet (1.2 m) back from the trench. The joined corner panel must be built and installed from the outside wall (Wall #1 or Wall #2) that has the highest debris (cave-in) pile of dirt on the trench floor.
2. Panel team 1 inserts 4 × 4-inch rails into the trench and cuts them flush with the trench lip.
3. Panel team 1 places two 6-foot (1.8-m) long 6 × 6-inch runners on the lip protection. The runners reduce friction when the joined panels are pushed toward the lip and help get the panels over the lip and on the rails.
4. Panel team 1 slides the corner panels into the trench on the rails.
5. Panel team 1 sets wales on the trench floor unless imminent signs of collapse are present. In that case, the focus should be on protecting the victim.

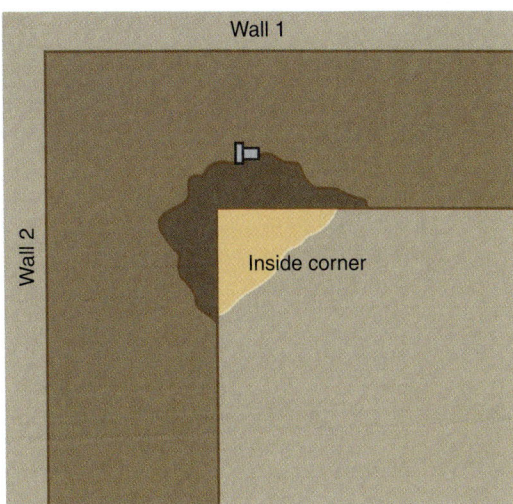

FIGURE 10-11 Wall designations. For consistency, Wall #1 is always designated to the area where the victim is located.

Panel Team 3. Panel team 3 stabilizes the collapsed corner.
1. Panel team 3 installs a lip bridge at the collapsed corner.
2. Panel team 3 prepares the backfill. An inside corner void can be backfilled using air bag, buttress, soil, or back shore techniques depending on the void's size and its back-wall angle.
3. Panel team 3 receives the corner panels from panel team 1 and stands them up at the inside corner.
4. Panel team 3 begins installing backfill after a minimum of one strut is installed from the corner panels to the opposing panels on Wall #1 and Wall #2.

TIP

Corner brackets can be bolted onto panels to create a joined corner panel. Corner brackets take about 10 minutes to bolt on but typically speed up the overall shoring process. They should not be used if an impending collapse is threatening a victim in the corner area. In that case, simply install individual primary panels to protect the victim.

Panel Team 2. Panel team 2 installs Wall #2.
1. Panel team 2 sets wales on the trench floor (collapse assessment permitting).
2. Panel team 2 sets the primary panel on Wall #2.

Shoring Team. The shoring team then completes the primary shoring.

1. The shoring team measures the trench and places the ladder.
2. The shoring team installs the positioning strut from the corner panel to the primary panel on Wall #1 (initial activation ≤0.5 kips or ≤100 psi on 2.5-inch [6.4-cm] diameter pneumatic struts).
3. The shoring team installs the positioning strut from the corner panel to panel on Wall #2 (initial activation ≤0.5 kips or ≤100 psi on 2.5-inch [6.4-cm] diameter pneumatic struts).
4. The shoring team installs backfill struts from Wall #1 and Wall #2 to the corner panels (initial activation ≤0.5 kips or ≤100 psi on 2.5-inch [6.4-cm] diameter pneumatic struts).
5. The shoring team installs compliance struts from Wall #1 panels and Wall #2 panels to the corner panels.
6. The shoring team installs the struts on the panel set at the opposite side of the corner.
7. Using a Y connector, the shoring team simultaneously increases all struts to 1.0 kips (200 psi on 2.5-inch [6.4-cm] diameter pneumatic struts).

TIP

Place the positioning struts at the strongest section of the remaining trench wall on the inside corner.

Secondary Shoring

Secondary shoring expands the size of the safe zone in order to accommodate rescue and medical personnel. Secondary shoring typically expands the primary shoring area (one panel set wide) to a 12-foot (3.7-m) wide area (three panel sets with compliant strut spacing). The following procedures provide for the installation of secondary shoring in a L-shaped trench with a victim trapped near the inside corner.

Panel Teams 1 and 2. Panel teams 1 and 2 install Walls #1 and #2.
1. Panel teams 1 and 2 set panels into outside corner Walls #1 and #2.
2. Panel teams 1 and 2 place a third panel on Walls #1 and #2.
3. Panel teams 1 and 2 position the first tier wales on Walls #1 and #2, **FIGURE 10-12**. The sequence of wale installation (first tier, second tier, etc.) is determined by the collapse assessment. The most likely area to collapse must be shored (first tier) first.

4. Panel teams 1 and 2 install wale spacers between wale and panels between the strongbacks.
5. Panel teams 1 and 2 install backfill as needed.

> **TIP**
> The end of the wale should be approximately 12 inches (30 cm) from the outside corner of the trench to avoid contact with the opposite wale.

Panel Team 3. Panel team 3 stabilizes the corner.
1. Panel team 3 completes the backfill for the inside corner.
2. Panel team 3 installs the second panel sets on both corner walls.
3. Panel team 3 backfills behind the other panels as needed.

Shoring Team. The shoring team completes the installation.
1. The shoring team installs a minimum of two struts (first tier) from wales on outside walls to panels on inside walls. A wye is used to simultaneously pressurize struts from the wale to the inside wall panels on each tier.
2. The shoring team has the hose from the positioning strut connected to the controller and pressurizes it whenever the struts on the wale are shot. This will prevent it from loosening and falling out during secondary strut installation.
3. The shoring team initially pressurizes all struts to ≤0.5kips (≤100 psi for Paratech Gray struts).
4. The shoring team repeats the previous steps on the opposite side of the corner and repeats this procedure for each additional tier of shores. The number of tiers of shores and wales is dependent on the depth of the trench and the total L calculation.
5. After backfill has been completed, the shoring team increases the pressure in all struts (one tier at a time) to between 1.0 and 1.25 kips (200–250 psi on Paratech Gray struts).

Expand the Safe Zone
1. The safe zone is expanded with panels and struts by adding an additional panel set to each wall **FIGURE 10-13**.
2. All strut bases are nailed.
3. Supplemental shoring is installed as needed during extrication operations.

FIGURE 10-12 The first tier wale (in this case the bottom) is supported with struts before the second (top) wales are installed.
Courtesy of Ron Zawlocki.

Putting It All Together: Shoring the L-Trench

SKILL DRILL 10-1 puts the entire skill of shoring the L-trench together. After the shoring size-up and the preparation steps are complete, this shoring skill drill may begin.

T-Trench Shoring Plan

This T-trench shoring plan is based on a tested and engineered design specific to collapse conditions **FIGURE 10-14**. Many of the techniques and skills from an L-shaped trench apply to this situation with some minor differences. This design is also appropriate for shoring T-shaped trenches that have not collapsed. Like all of the authors' shoring plans, this plan includes a shoring size-up, lip protection, primary shoring, secondary shoring, complete shoring, and ongoing shoring performance assessments.

CHAPTER 10 Technician Level Trench Rescue Shoring   247

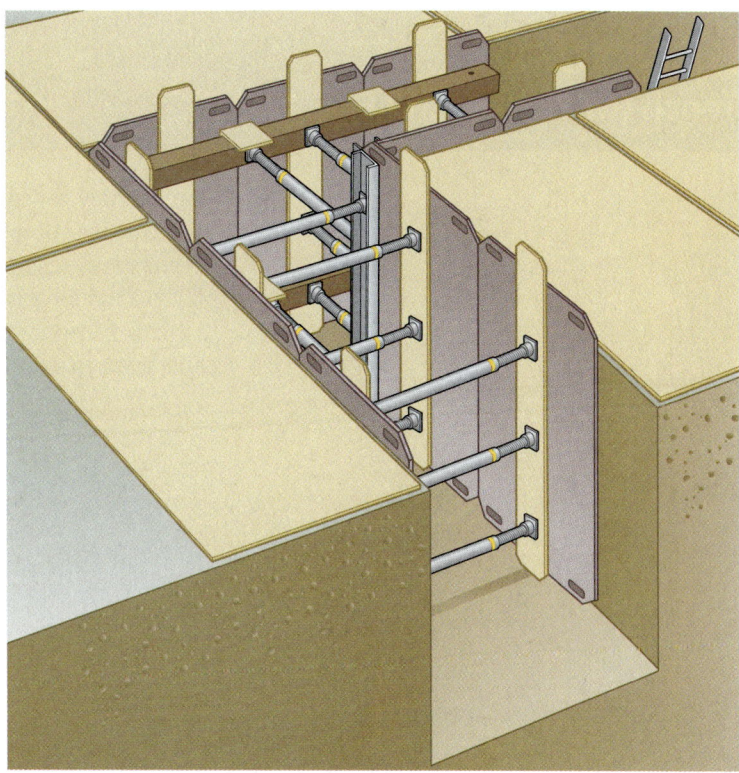

FIGURE 10-13 A completed two tiered (8 feet [4.2 m] deep or less) L-trench shoring design.

SKILL DRILL 10-1
Shoring the L-Trench (NFPA 1006: 12.3.2, 12.3.5)

This skill drill assumes the victim to be in the corner area of the L-trench. This shoring plan directs the efforts to protect that area first. If the victim is trapped in another section of the trench, then begin shoring to protect the victim area first.

Primary Shoring

1. Lower a wale to the trench floor of Wall #1 and Wall #2. If signs of a secondary collapse are present, this wale placement must wait until the victim is protected with the installation of primary panels and struts.

(continues)

SKILL DRILL 10-1
Shoring the L-Trench (NFPA 1006: 12.3.2, 12.3.5)

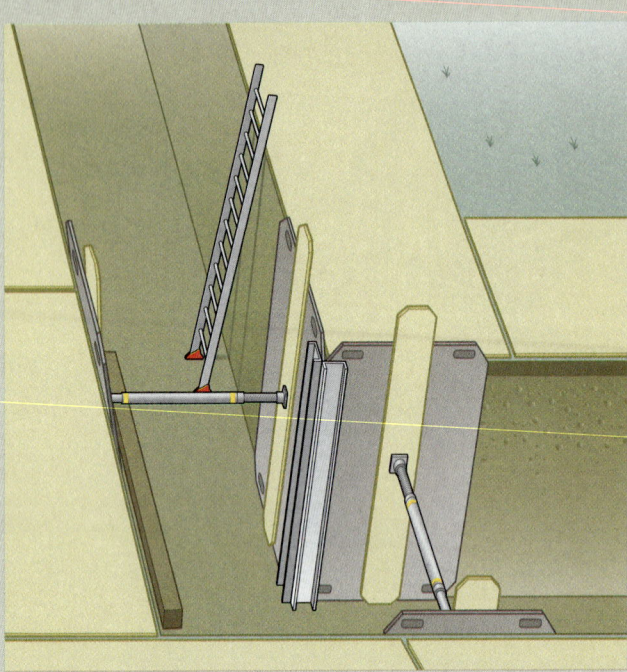

2 Place primary panels (inside corner/Wall #1 and inside corner/Wall#2) and install positioning struts at 500 pounds (227 kg) of activation force (100 psi on 2.5-inch [6.4-cm] diameter strut) to position the panel and protect the victim.

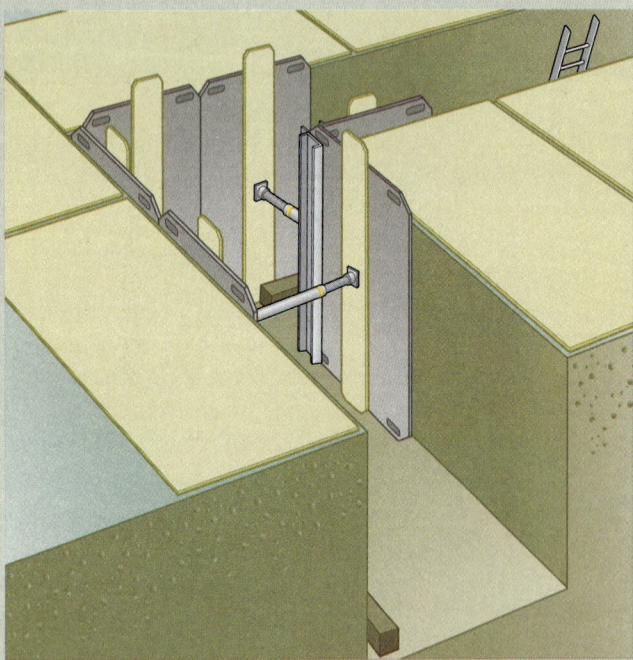

Secondary Shoring

3 Place panels at both outside corners (Wall #1 and Wall #2).

SKILL DRILL 10-1
Shoring the L-Trench (NFPA 1006: 12.3.2, 12.3.5)

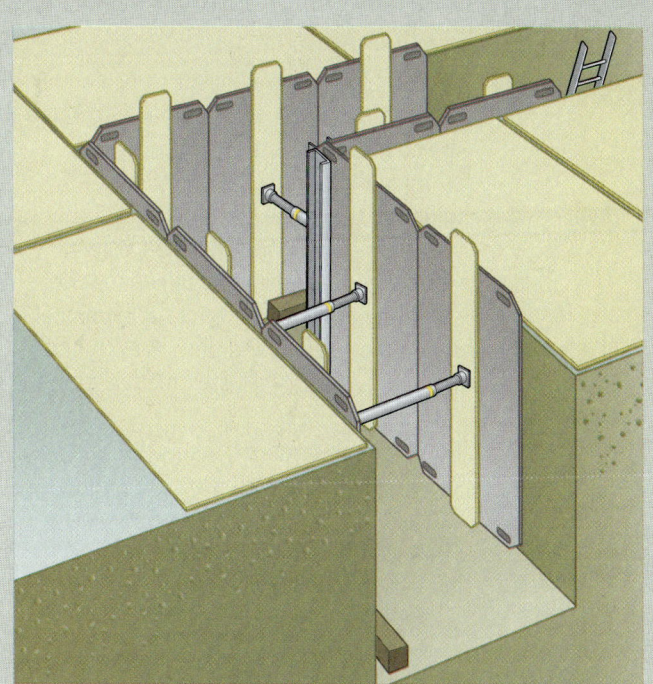

4 Place additional panel sets (inside corner wall to Wall #1 and inside corner wall to Wall #2) and install a positioning strut (as close as possible to half-way between the lip and floor) at 500 pounds (227 kg) of activation force (100 psi on 2.5-inch [6.4-cm] diameter strut).

5 Position the first tier of wales on Wall #1 and Wall #2. The first wales must be positioned to prevent the most likely area of secondary collapse. If the most likely area of collapse is near the top of the trench wall, then the wale must be between 1 to 2 feet (0.3–0.6 m) below the lip. If the most likely area of collapse is near the bottom of the trench wall, then the wale must be between 1 to 2 feet (0.3–0.6 m) above the floor or as close to that as the debris pile will permit. Install wale support struts at 500 pounds (227 kg) of activation force (100 psi on 2.5-inch [6.4-cm] diameter strut). Finish backfill installation and bring all struts activation forces to 1000 (454 kg) pounds of activation force (200 psi on 2.5-inch [6.4-cm] diameter strut).

(continues)

SKILL DRILL 10-1
Shoring the L-Trench (NFPA 1006: 12.3.2, 12.3.5)

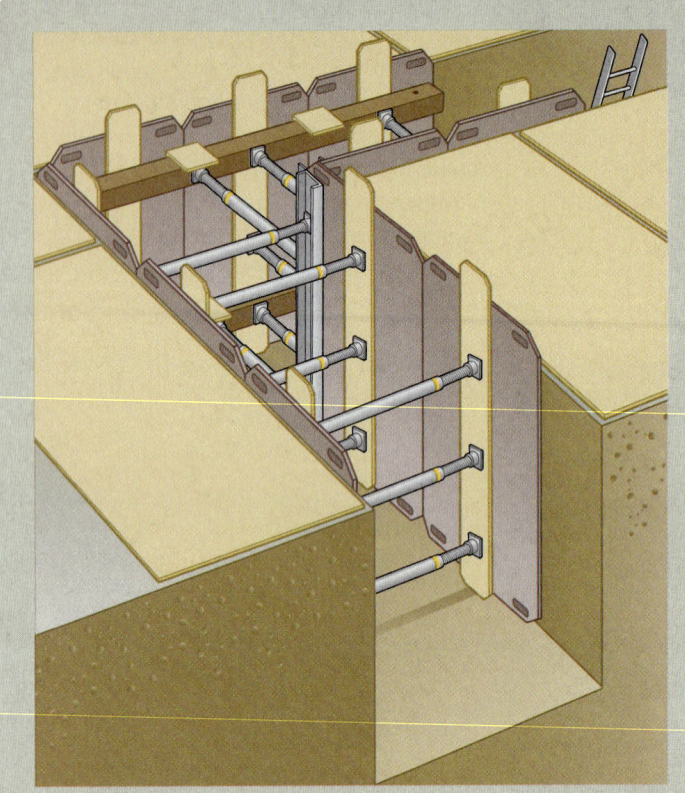

Complete Shoring

6 Enter the shored area of the trench and nail all strut bases. While remaining within the confines of the shored panels, add struts to the uprights at the edges of the panels at the inside corner (inside corner to wales on Wall #1 and wales on Wall #2). Add additional panel sets to expand the safe zone as needed. Install supplemental shoring as needed and provided ongoing shoring performance assessments.

NOTE: The illustrations in this skill drill show aluminum wales used as uprights on the edges of the inside corner panels. These can be substituted with 2″ × 12″ lumber bolted onto the panels.

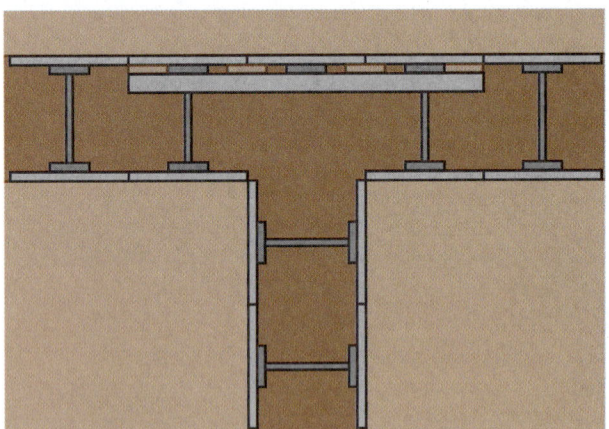

FIGURE 10-14 Top view of T-trench shoring design.

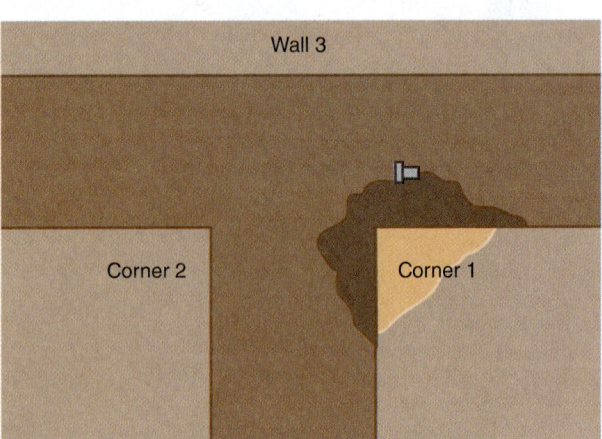

FIGURE 10-15 Wall/corner designations.

Walkthrough of Shoring a T-Trench

Primary Shoring

This plan assumes the most likely T-trench collapse, which is an inside corner (most likely a wedge failure) collapse **FIGURE 10-15**. Primary shoring must cover the area above the victim's head and chest. If the victim is not trapped in the corner area, straight wall trench primary shoring procedures should be used. The following procedure is used for shoring the corner area with a trapped victim. This walkthrough describes the duties of the three panel teams and the shoring team during each phase of the shoring process.

Panel Team 1. Panel team 1 is assigned to Corner #1 and performs the following tasks:
1. Panel team 1 installs the lip bridge at the collapsed corner.
2. Panel team 1 prepares backfill for the inside corner void.
3. Panel team 1 receives a corner panel from panel team 3 and stands it up at the inside corner.
4. Panel team 1 receives a corner panel from panel team 2 and stands it up at the inside corner.
5. Panel team 1 installs backfill in the inside corner void.

> **TIP**
> Joined corner panels can be used on one or both inside corners of a T-trench. See the L-trench shoring plan for joined corner panel details

Panel Team 2. Panel team 2 is assigned to Corner #2 and performs the following tasks:
1. Panel team 2 installs lip protection in the Corner #2 area.
2. Panel team 2 inserts 4 × 4 rails into the trench from Corner #2 to Corner #1
3. Panel team 2 uses the one-side panel set and slides a panel on the rails from Corner #2 to Corner #1.
4. Panel team 2 places a primary panel directly across from that panel on Corner #2.
5. Panel team 2 installs a second panel on Corner #2 panel facing Wall #3.

Panel Team 3. Panel team 3 is assigned to Wall #3 and the following tasks are performed:
1. Panel team 3 installs lip protection in the Corner #2 area.
2. Panel team 3 inserts 4 × 4 rails into the trench from Corner #2 to Corner #1.
3. Panel team 3 uses the one-side panel set and slides a panel on the rails from Wall #3 to Corner #1.
4. Panel team 3 sets a wale on the trench floor (based on collapse assessment). If a collapse is imminent, the wale will need to be fished in between struts afterwards.
5. Panel team 3 places a primary panel directly across from that panel on Wall #3.
6. Panel team 3 installs a second panel on Wall #3 directly across from the panel on Corner #2.
7. Panel team 3 places a wale on the floor of Wall #3.

Shoring Team. The shoring team performs the following tasks:
1. The shoring team measures the trench and places a ladder.
2. The shoring team installs a positioning strut from the Corner #1 panel to the panel on Wall #3.
3. The shoring team installs a positioning strut from the Corner #1 panel to panel on Corner #2.
4. The shoring team installs backfill struts from Corner #1 to Corner #2 panels.
5. The shoring team installs compliance struts from Corner #1 to Corner #2.
6. After backfill is in place, the shoring team uses a Y connector to simultaneously increase all struts to 1000 pounds (454 kg) of force (200 psi on 2.5-inch [6.4-cm] diameter pneumatic struts) **FIGURE 10-16**.

> **TIP**
> The initial strut activation force is 500 pounds (227 kg) of force.

Secondary Shoring

Panel Team 1. Panel team 1 is assigned to Corner #1 and the following tasks are performed:
1. Panel team 1 installs additional panel(s) on the Corner #1 walls.
2. Panel team 1 assists the shoring team.

Panel Team 2. Panel team 2 is assigned to Corner #2 and the following tasks are performed:
1. Panel team 2 installs additional panels on Corner #2 walls.
2. Panel team 2 assists panel team 3 and/or the shoring team.

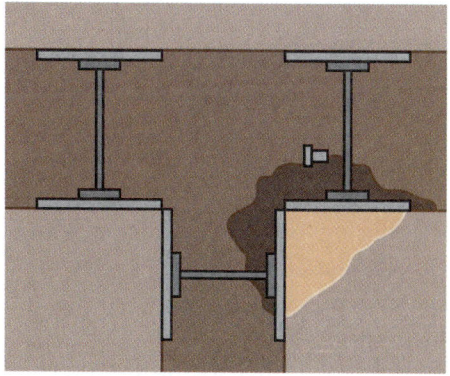

FIGURE 10-16 Primary shoring.

Panel Team 3. Panel team 3 is assigned to Wall #3 and the following tasks are performed:
1. Panel team 3 installs the middle panel on Wall #3. The panel is cut to fit or overlap panels as needed.
2. Panel team 3 hangs wales on Wall #3 (top or bottom first as dictated by secondary collapse potential).
3. Panel team 3 assists the shoring team as needed.

Shoring Team. The shoring team performs the following tasks:
1. The shoring team installs a minimum of two struts (first tier) from wales on Wall #3 to panels on the inside corners. The shoring team uses a wye to simultaneously pressurize struts from the wale to the inside wall panels on each tier.
2. The shoring team initially pressurizes all struts to ≤0.5 kips (≤100 psi for Paratech Gray struts). This procedure is repeated for each additional wale. The number of wales is dependent on the depth of the trench and the total L calculation.
3. After backfill has been completed, the shoring team increases the pressure in all struts (one tier at a time) to between 1000 and 1250 pounds (454 and 567 kg) of force (200–250 psi on 2.5-inch [6.4-cm] diameter struts).
4. The shoring team installs a minimum of two struts to the secondary shoring panel sets **FIGURE 10-17**.

> **TIP**
> Be sure to have the hose from the positioning strut connected to the controller and pressurize it whenever the struts on the wale are shot. This will prevent it from loosening and falling out during secondary strut installation.

Complete the Shoring

Shoring Team. The shoring team performs the following tasks:
1. The shoring team brings all struts up to operating pressures.
2. The shoring team enters the trench and nails all strut bases.
3. The shoring team enhances and expands the safe zone as needed.
4. The shoring team assists with supplemental shoring.

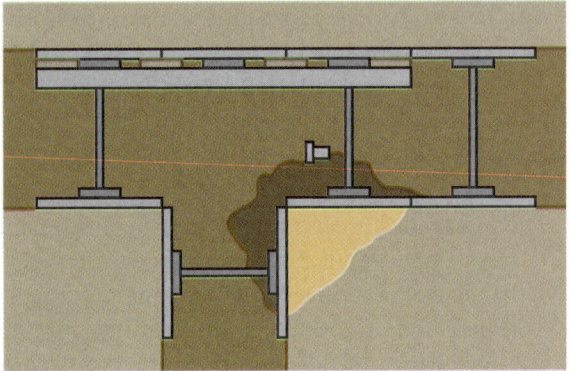

FIGURE 10-17 Secondary shoring.

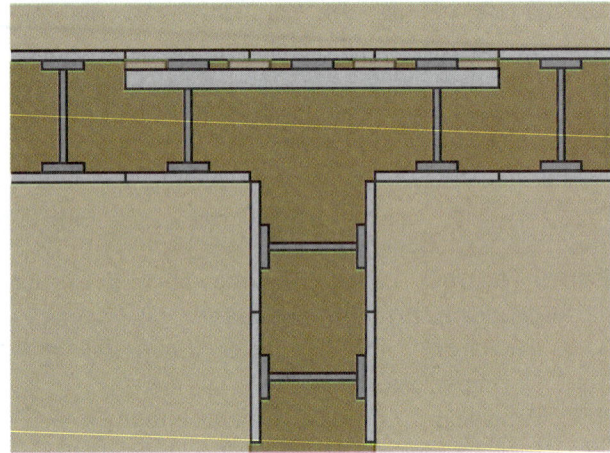

FIGURE 10-18 Complete shoring.

Entry Team. The entry team installs supplemental shoring as needed during excavation and extrication.

Panel Team. The panel team enhances and expands the safe zone as needed and assists with supplemental shoring. After backfill is in place, the panel team uses a Y connector to simultaneously increase all struts to 1000 pounds (454 kg) of force (200 psi on 2.5-inch [6.4-cm] diameter pneumatic struts or 150 psi on 3-inch [7.6-cm] diameter struts) **FIGURE 10-18**.

Putting It All Together: Shoring a T-Trench

SKILL DRILL 10-2 puts the entire skill of shoring the T-trench together. After the shoring size-up and the preparation steps are complete, this shoring skill drill may begin.

> **TIP**
> If the trench has signs of failure at the lip (top of the trench wall) then install the top wale first.

SKILL DRILL 10-2
Shoring the T-Trench (NFPA 1006: 12.3.2, 12.3.5)

This skill drill assumes the victim to be in one of the corner areas of the T-trench. This shoring plan directs the efforts to protect that area first. If the victim is trapped in another section of the trench, begin shoring to protect the victim area first.

Primary Shoring

1. Install lip protection, including lip bridges over voids and failed areas.

2. Install primary panels and the first strut at 500 pounds (227 kg) of activation force (100 psi on 2.5-inch [6.4-cm] diameter strut) to position the panel and protect the victim.

(continues)

SKILL DRILL 10-2
Shoring the T-Trench (NFPA 1006: 12.3.2, 12.3.5)

3 Lower a wale to the trench floor of the long wall.

4 Install a positioning strut on the panels (Corner #1 to Corner #2 and Corner #2 to Wall #3) at 500 pounds (227 kg) of activation force (100 psi on 2.5-inch [6.4-cm] diameter strut).

SKILL DRILL 10-2
Shoring the T-Trench (NFPA 1006: 12.3.2, 12.3.5)

Secondary Shoring

5. Install a compliance strut in the second (Corner #1 to Corner #2) panel set at 500 pounds (227 kg) of activation force (100 psi on 2.5-inch [6.4-cm] diameter strut). Place a panel on the Wall #3 between the two shored panels.

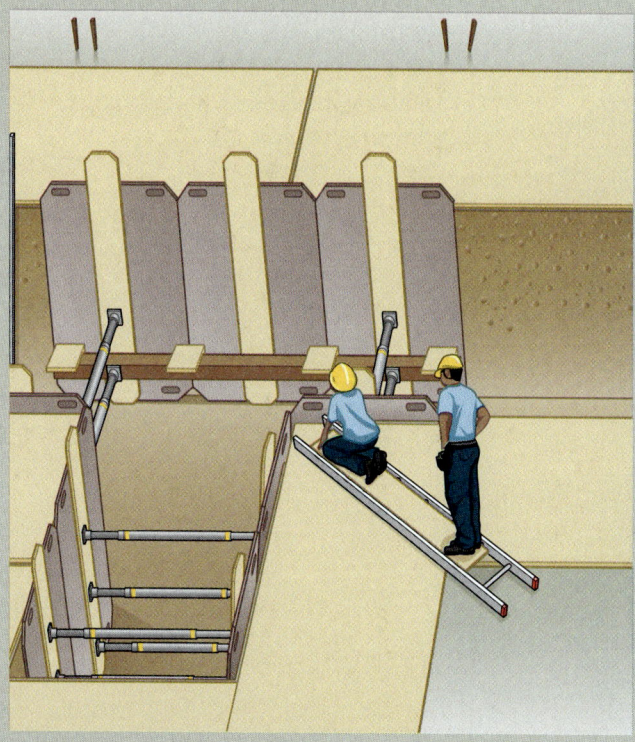

6. Raise the bottom wale (not more than 2 feet [0.6 m] above the trench floor) and install the wale support struts at 500 pounds (227 kg) of activation force (100 psi on 2.5-inch [6.4-cm] diameter strut).

(continues)

SKILL DRILL 10-2
Shoring the T-Trench (NFPA 1006: 12.3.2, 12.3.5)

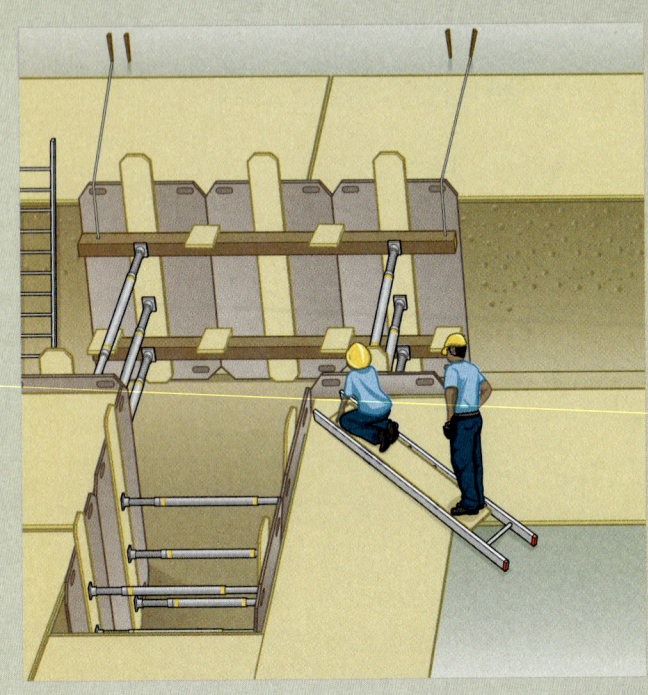

7 Position the top wale (not more than 2 feet [0.6 m] below the lip) and install the wale support struts at 500 pounds (227 kg) of activation force (100 psi on 2.5-inch [6.4-cm] diameter strut).

8 Complete backfill and bring all struts activation forces to 1000 pounds (454 kg) of activation force (200 psi on 2.5-inch [6.4-cm] diameter strut). Expand and enhance the safe zone (additional panels and struts as needed). Install two 16d nails in each strut base. Install supplemental shoring as needed.

Voices of Experience

I was assigned to the back end of a ladder truck. A beautiful lakeshore summer day was unfolding at our Fire Department (St. Clair Shores) headquarters. Morning fire department training gave way to rigorous physical exercise. We ran a vehicle entrapment on the expressway (I-94), which was followed by firehouse banter and camaraderie centered around our first chow. While we were just finishing getting cleaned up, my cell phone began to ring with a call from Macomb County Emergency Dispatch. As a rescue squad officer on the regional technical rescue team, I was part of an initial call group for a phone conference. The call for assistance came in from a neighboring community, about 10 minutes away. Two workers were trapped in a soil collapse at a construction site. During our group call, team leaders dispatched trench rescue resources based on proximity and equipment availability.

St. Clair Shores Fire Department (SCSFD) had five MCTRT (Macomb County Technical Rescue Team) members on duty that day. We all boarded our rescue and headed north to Chesterfield Township. While en route, the updated information from first-arriving TRT members (Lenox Township) on the scene was that two live victims were trapped in an intersecting (L trench). The crews were beginning initial actions, including hazard control, lip safety, and primary shoring to protect the victims.

We arrived on the scene and checked in with the incident commander. I was assigned to run shoring at the inside corner wall and immediately assigned a panel and shoring team. I had them begin pulling panel team and shoring team equipment from the rescues and trailers. Panels were rigged and pneumatic struts and air systems assembled and staged at the trench.

While the teams were gathering, preparing, and staging equipment, I began my size-up and made the following assessments: situation assessment, a deep intersecting trench collapse with two live victims; hazards assessment, unprotected trench wall with extremely unstable soil conditions, high water table due to the close proximity of a canal leading to Lake St. Clair, and a running backhoe near the intersecting point of the trenches; victim assessment, two workers pinned by not only soil, but by the lumber they were using to create concrete forms, which were being used to extend the footings for a basement wall, both men were conscious and did not have apparent life-threatening injuries; trench assessment, L-shaped trench, 12 feet (3.7 m) deep and 8 feet (4.2) wide, with near-vertical walls with one wall angled (sloped).

The shoring workload was separated into two divisions, each with their own panel and shoring teams. Division 1 was the inside corner with a victim (Victim 1) trapped in that area. The second victim (Victim 2) was trapped near the outside corner (Division 2), which included a sloped wall. My assignment was to plan and implement shoring in Division 1, make entry to treat, package, extricate, and remove Victim 1. Lip protection and primary shoring in Division 1 quickly went into place and were followed expediently by secondary shoring. The plan allowed rescuers (entry team) to enter the trench within a shored area and begin to dig the victim out. Our team was able to reach the victim and render medical aid and extrication from the soil and entangled lumber forms within 10 minutes of the assignment. Victim 1 was completely removed from the trench within 15 minutes.

While this was unfolding, primary shoring in Division 2 was taking place at the outside corner area of the L-shaped trench. As usually is the case with shoring L-trenches, the outside corner area is more difficult. In this case, the difficulty was compounded by a sloping wall near the intersection and the surcharge of the backhoe that was positioned near the outside corner wall.

Voices of Experience

The panel and shoring teams worked methodically and diligently, amid tactical challenges and setbacks, to contribute to the entry team being able to make a successful rescue of the victim at the apex. When our shoring system was all said and done, it certainly was not the most plumb, level, and square shoring job our team had ever completed, but overall, we felt that we won that day. Two victims were rescued and were able to return to their families, no fire/rescue personnel were injured, our team knowledge, skills, and abilities were enhanced, and lessons were shared within our fire and rescue community.

Lessons Learned

- Initial actions: Train with as many engine, truck, and squad companies as you can in your region to share information at a minimum of the awareness level. Trench rescue specialists arriving on a scene where initial actions have already been completed will expedite the time needed to effect a rescue. If your response starts off chaotically, it will likely remain chaotic. If your response starts out calm and organized, it will likely remain calm and organized.
- Assignments: We have consistently seen the value and impact in having a designated shoring team, panel team, and entry team. Each team has prescribed duties, responsibilities, and procedures. Each team has the expected and anticipated techniques and sequences we look for in our teammates. Together, this has a synergistic effect on how quickly and safely we are able to mitigate hazards and rescue trapped victims.
- A good plan now is better than a perfect plan later: We are in the rescue business. We draw upon our training and past experiences to make decisions related to strategies and tactics that will result in the most success possible. You may have a great plan. You may know a way that might work a little bit better. You may be a rescue genius in your own mind at the time. Whoever is the boss has been put there for a reason and a rescue scene is not the place for rescue by committee. At a rescue scene, you should simply ask yourself, "Will the boss's plan work and is it safe for the victim and rescuers?" If the answer is yes, support it. Work as hard as you can to make that plan succeed. If the answer is no, bring the specific concern to the boss's attention.
- Train hard: Being chronically exposed to the trench rescue environment is the only way to get good at it. Rather than training until you get it right, train until you cannot get it wrong. Once you achieve this, add new challenges to your training. Training in collapsed conditions with voids in the trench walls is essential. Of all the trench rescue incidents to which I have responded, none of them fit the NFPA classification for operations level. Create difficult training scenarios and resolve them. Utilize subject matter experts (SMEs) to assist in your training. SMEs include geotechnical engineers, Urban Search and Rescue structure specialists, and experienced rescuers who are involved in teaching the discipline. Seek reputable educators with testing and experience to backup what they share. There is a lot of trench shoring information available that has not been engineered and tested for use in failed (collapsed) soil conditions. Question it all.

CHAPTER 10 Technician Level Trench Rescue Shoring

Voices of Experience

- Just a note to bosses: A trench rescue is an intense experience and something that most firefighters will only experience once or twice in their career. As an officer (boss), you set the tone. Always retain your composure. Do not get dead-set on any plan, if plan A is not working, adapt and transition to plan B. Your planning needs to be three steps ahead of what your team is engaged in at any given moment. Do not get fixated on the situation inside of the trench. Look outside of the hole once in a while and see what your people are doing. Are there a bunch of people just standing around and looking in? Get them back from the lip and get them working toward the common goal. Use simple, clear communication, and make a call.

Michael DeCraene
Firefighter, City of St. Clair Shores
Rescue Team Manager, MCTRT
Rescue Squad Officer, MI-TF1

Deep Trench

Deep trench shoring procedures begin with shores that are placed at depths of more than 8 feet (2.4 m) below the trench lip. Shoring below the 8 foot (2.4 m) mark requires rescue panels to be stacked vertically. The width of the shored area should be equal to or exceed the depth of the trench. Large lateral forces resulting from deep vertical (or near vertical) walls can cause sudden cave-ins.

Deep trenches result in large spoil piles. Deep trenches are also likely to result in larger (higher L value) collapse zones, which results in more spoil pile within the simple L distance, which equates to large surcharge loads. When dealing with higher loads, stronger shoring equipment becomes necessary. Trench rescue shoring panels are used to support the potential loads of deep trenches (as discussed in Chapter 7, *Operations Level Trench Rescue Shoring*). These panels are built from ¾-inch (19-mm; 14-ply) premium arctic birch plywood (FinnForm or equivalent) with 2 × 12 LVL strongbacks screwed and glued to create a composite panel. This is the only wood panel system that can support the loads that can exist at a 20-foot (6.1-m) deep trench.

Deep trenches require extensive amounts of shoring equipment and materials. A 20-foot (6.1-m) deep trench will require 30 panels and at least 30 pneumatic struts (see **FIGURE 10-19**) to shore a safe zone for rescuers and more panels and struts to conduct victim extrication (digging) and supplemental sheeting and shoring. That is more equipment than most trench rescue teams carry and is why it is important for agencies to have working mutual aid agreements with several regional trench rescue resources.

With deep trenches comes the likelihood that the failure areas are longer and deeper. Backfilling large and deep voids also requires specialized equipment and a high skill level. One of the most serious hazards is the fall risk when working on a leading edge. While falling into a shallow trench is dangerous, falling into a deep trench could be fatal. For this reason, edge restriction devices should be used. By securing a rope to a rescuer working on the lip and the other end to a 1-inch (2.5 cm) diameter steel picket driven at least 36 inches (0.9 m) into the ground, the edge limiter system can prevent a rescuer from going over the edge.

FIGURE 10-19 Deep trenches require extensive amounts of shoring equipment and materials.
Courtesy of Ron Zawlocki.

Application for Deep Trenches

Deep trench shoring procedures are applicable whenever trench depths exceed 8 feet (4.2 m), resulting in the need to stack panels (vertically) on the walls. Deep trench shoring procedures must occur when struts are placed 8 feet (4.2 m) or more below the lip. These procedures are applicable to trenches up to 20 feet (6.1 m) deep.

Precautions for Deep Trenches

Limit the number of rescuers working on the trench lip and use edge limiter techniques to prevent those working the lip area from slipping or falling into the trench.

Preparation Tasks for Shoring Deep Trenches

All hazards shall be identified and mitigated, and lip protection installed. Lip bridges are a best practice at deep trench incidents. Use edge limiters for rescuers working on the lip whenever the lip is slippery or angled (not flat).

Shoring Equipment for Deep Trenches

The shoring equipment for deep trenches includes the following:

- Tape measure
- Ladders

- Lip bridges and ground pads
- Trench rescue shoring panels
- Panel bender
- Ropes
- Pneumatic struts, bases, extension, ropes, and air supply system
- Hammers, nails, and remote strut collar locking equipment
- Backfill and installation equipment
- Edge limiters and harnesses may be needed

Personnel for Deep Trenches

Personnel requirements for deep trenches include the following:

- Panel team 1
- Panel team 2
- Shoring team
- Entry team

Shoring Plan for a Deep Trench

Size-Up

Before shoring a deep trench, the following size-up tasks should be performed:

- Assess the lip: Lip protection must be installed before the shoring and panel team members can begin shoring operations. Remember that the best practice for lip protection is the use of lip bridges.
- Assess the trench: Measure the depth of the trench floor to determine the level (operations or technician) of the incident. Measure the depth from the lip to the debris (collapse) pile to determine the number of panels needed. Measure the width of the trench to determine the length of struts needed. Measure the L for the trench and any surcharge loads to determine the total L. Estimate the wall angles.
- Assess the collapse potential: Identify signs of impending collapse. Look for fissures (cracks), cornice (undercut walls), bulging walls, surcharge loads within the L area, layered soil, and flowing water in the trench walls.
- Assess the voids: Voids are created by sections of trench walls that have collapsed. The voids must be filled to both minimize the movement of the back wall of the void and to create an adequate point of resistance to any movement (load) being transferred from the opposing trench wall. To determine the best backfill technique for each void rescuers must know the void size (height, width, and length), the angle of the void back wall, and whether it is an open or closed lip void.
- Assess the victim(s): The victim's condition will dictate the benefit and subsequent risk levels that will be acceptable. Calculated and reasonable risks are appropriate for live victims. Minimal risks are appropriate for clearly deceased victims.
- Use tabulated data.

Walkthrough of Shoring a Deep Trench

Primary Shoring

Panel Team 1 and 2. Panel teams 1 and 2 perform the following tasks:

1. Panel teams 1 and 2 install lip protection.
2. Panel teams 1 and 2 mark the victim location on the ground pad or lip bridge.
3. Panel teams 1 and 2 install three 42-inch (1.1-m) long and 1-inch (2.5-cm) diameter steel pickets (driven 36 inches [0.9 m] into the ground at 8 feet [2.4 m] back from trench wall) on each side of the trench, the middle picket in line with the victim. The other pickets are spaced out 4 feet (1.2 m) to each side. Edge limiters (if needed) are attached, and the teams prepare for panel installation.
4. Panel teams 1 and 2 install backfill as needed.
5. Panel teams 1 and 2 install a panel set (both walls) at the bottom of the trench (panel should be in line with the victim's head and chest).

Shoring Team. The shoring team performs the following tasks:

1. The shoring team positions a ladder in the rescue area.
2. The shoring team installs a positioning strut near the center of the panels **FIGURE 10-20**.
3. If the wall behind the strut is solid, the shoring team sets the strut pressure to 200–250 psi.
4. If a void is present behind the strut, the shoring team sets the strut pressure to 100–125 psi.

Primary Shoring (Top Panel Set)

Panel Teams 1 and 2. Panel teams 1 and 2 install the top panels (half panels for trenches up to 12 feet [6.1 m] deep and full panels for trenches 14 feet [4.3 m] and deeper). The teams install backfill as needed.

FIGURE 10-20 Bottom panel positioning strut.

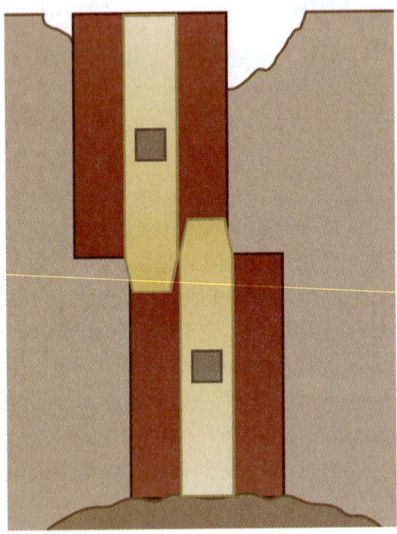

FIGURE 10-21 Top panel positioning strut.

Shoring Team. The shoring team installs a positioning strut on the top panels **FIGURE 10-21** and completes the strut installation (backfill struts and compliance struts) on the primary panels **FIGURE 10-22**.

Secondary Shoring (Bottom Panels)

Panel Teams 1 and 2. The panel teams install bottom panels on each side of the primary panels on both walls and install backfill as needed.

Shoring Team. The shoring team installs a positioning strut near the center of each bottom panel set **FIGURE 10-23**.

Secondary Shoring (Top Panels)

Panel Teams 1 and 2. The panel teams install the top panels on each side of the primary panels and install backfill as needed.

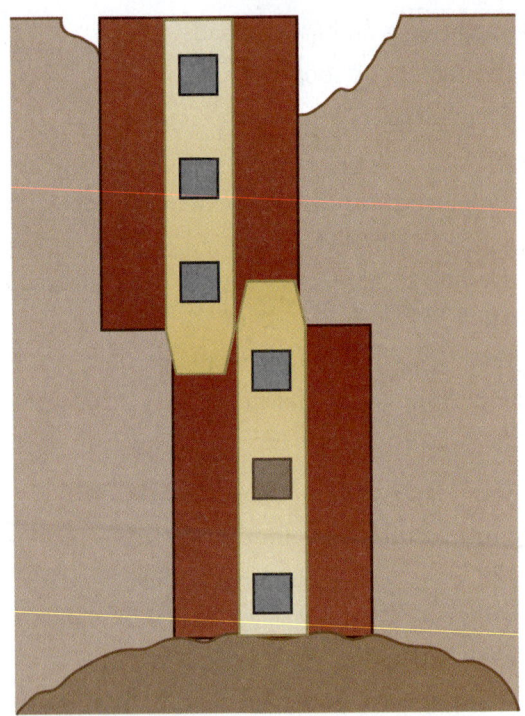

FIGURE 10-22 Primary panels and struts.

FIGURE 10-23 Bottom secondary panels with positioning struts.

TIP

Whenever installing more than two struts on a strongback or wale, always use wyes to position all struts simultaneously to prevent struts from loosening and falling out.

Shoring Team. The shoring team installs a positioning strut on each top panel set. The placement of these struts is dependent on the location and size of voids in the trench wall. In this example, the positioning struts on the top panels are placed at 6 feet (1.8 m) below the trench lip **FIGURE 10-24**. The shoring team installs additional struts (backfill struts and compliance struts) on all secondary shoring panels and ensures that backfill is complete and all struts in the secondary shoring area are set at operating pressures **FIGURE 10-25**. The shoring team enters the trench and nails strut bases to strongbacks in the secondary shoring area. Finally, the shoring team ensures that all trip hazards (panel/shore ropes, etc.) are removed.

Complete Shoring

To install complete shoring, additional panels and struts are installed to create a safe zone that is at least as wide as it is deep. Supplemental shoring is installed as needed **FIGURE 10-26** and **FIGURE 10-27**. The team ensures that all strut bases are nailed, backfill is complete, and struts are set at operating pressures.

Putting It All Together: Shoring a Deep Trench

SKILL DRILL 10-3 puts the entire skill of shoring a deep trench together. After the shoring size-up and the preparation steps are complete, this shoring skill drill may begin.

Utilizing Supplemental Shoring

Per NFPA 1670, supplemental sheeting and shoring includes operations that involve the use of commercial sheeting/shoring systems and/or isolation devices or cutting and placement of sheeting and shoring when greater than 2 feet (0.6 m) of exposed trench wall exists below the bottom of the panel. Supplemental sheeting and shoring require additional training beyond that of traditional sheeting and shoring. Traditional sheeting and shoring involves the use of 4 × 8-foot (1.2 × 2.4-m) panels with a strongback attachment supplemented by a variety of conventional shoring options such as hydraulic, pneumatic, and/or screw shores.

Supplemental shoring is used when rescuers must remove soil that has caved in from the trench. Soil removal techniques during rescue operations

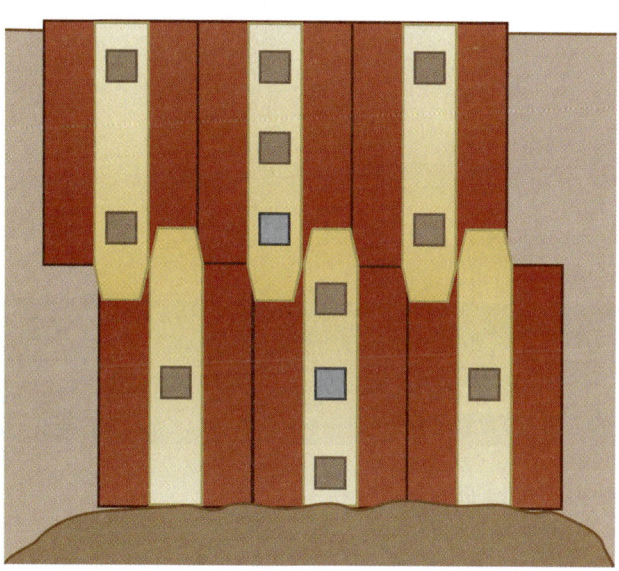

FIGURE 10-24 Secondary panels with positioning struts.

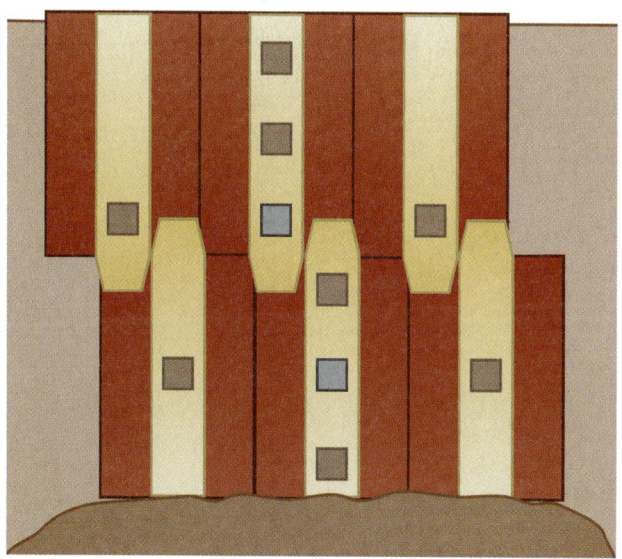

FIGURE 10-25 Secondary shoring.

FIGURE 10-26 Complete the shoring.

264   *Trench Rescue: Principles and Practice to NFPA 1006 and 1670*

FIGURE 10-27 Install supplemental shoring as needed.

SKILL DRILL 10-3
Shoring a Deep Trench (NFPA 1006: 12.3.3, 12.3.4)

Primary Shoring

1. Install lip bridges and position escape ladder. Place the first set of primary panels at the bottom of the trench wall to protect the victim's head and chest. Install backfill as needed. Install the first strut at 500 pounds (227 kg) of activation force (100 psi on 2.5-inch [6.4-cm] diameter strut) in the center of the panels to position the panel and protect the victim.

SKILL DRILL 10-3
Shoring a Deep Trench (NFPA 1006: 12.3.3, 12.3.4)

2 Install the next set of panels directly above the first set making sure to interlock the panels by using the overlapping strongbacks to hold the upper panels in place. Install backfill as needed.

3 The placement of the positioning strut on the second (top) set of panels is dependent on the location of the void(s) but it is usually placed about 4-6 feet (1.8 m) below the trench lip. Install this strut at between 500 to 1000 pounds (227–454 kg) of activation force (100–200 psi on 2.5-inch [6.4-cm] diameter strut) in the center of the panels to position the panel and protect the victim. Install struts 2 feet (0.6 m) from the top and 2 feet (0.6 m) from the bottom of the first panel set. Install backfill and compliance struts as needed on the bottom panel set. Install these struts at 1000 pounds (454 kg) of activation force while bringing all struts on the panel set up to 1000 pounds (454 kg) by using a wye (Y) adapter.

(continues)

SKILL DRILL 10-3
Shoring a Deep Trench (NFPA 1006: 12.3.3, 12.3.4)

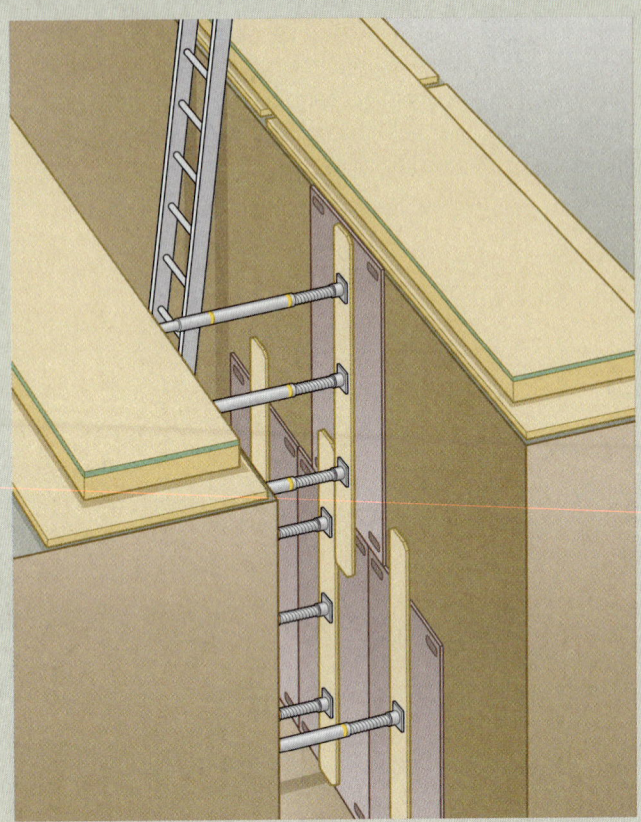

4 Place panel sets on both sides of the bottom primary panels. Install backfill as needed. Install struts at between 500–1000 pounds (227–454 kg) of activation force (100 psi on 2.5-inch [6.4-cm] diameter strut) in the center of the panels.

Secondary Shoring

5 Install the next sets of panels directly above the (bottom) secondary panel sets making sure to interlock the panels by using the overlapping strongbacks to hold the upper panels in place. Install backfill as needed.

SKILL DRILL 10-3
Shoring a Deep Trench (NFPA 1006: 12.3.3, 12.3.4)

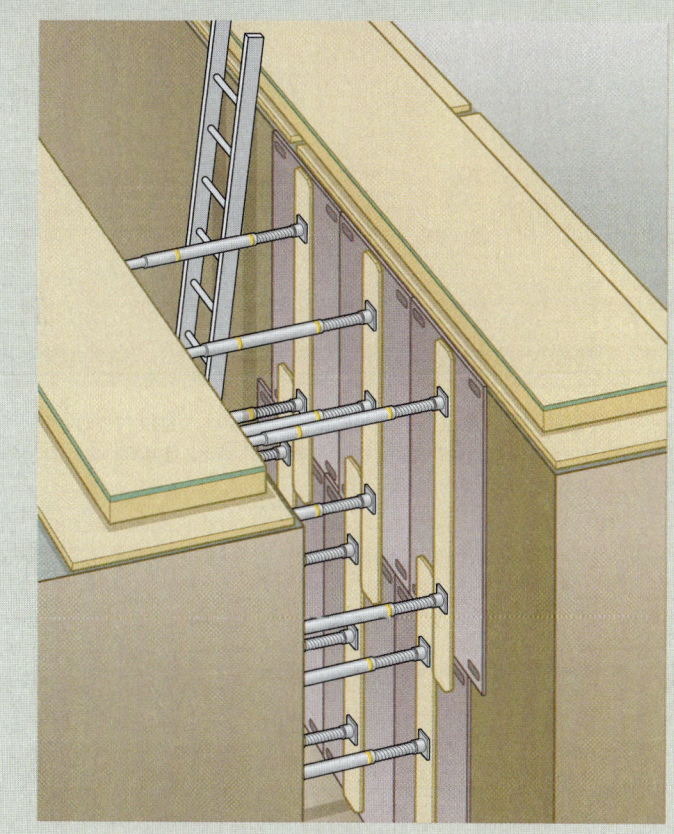

6. Install backfill and compliance struts as needed on the bottom secondary panel sets. After completing the backfill of any voids, install these struts at 1000 pounds (454 kg) of activation force while bringing all struts on the bottom panel sets up to 1000 pounds (454 kg) by using a wye (Y) adapter. Install backfill and compliance struts as needed on the top secondary panel sets. After completing the backfill of any voids, install these struts at 1000 pounds (454 kg) of activation force while bringing all struts on the bottom panel sets up to 1000 pounds (454 kg) by using a wye (Y) adapter. Expand the safe zone (horizontal shored area equal to or greater than the depth of the trench) by repeating the panel and strut installation steps found in the secondary shoring steps listed earlier. Install two 16d nails in each strut base. Install supplemental shoring as needed. Perform an ongoing shoring performance assessment.

commonly involve entrenching shovels, buckets, air knives, and vacuum systems. Equipment used for supplemental shoring includes 2 × 4-foot (0.6 × 1.2 m) sections of shoring panels, sections of 2 × 12-inch strongbacks and (rescue) struts **FIGURE 10-28**. Supplemental shoring is a technician level skill that is needed to enhance safety during advanced trench rescue operations.

Application for Supplemental Shoring

Initial panels and struts (primary and secondary shoring) should be installed as close as possible to the soil pile (cave in) on the trench floor. Rescuers must remove soil to extricate victims from cave-in incidents. Whenever these digging operations expose 2 feet (0.6 m) of trench wall, supplemental shoring must be implemented. Supplemental panels (¾-inch [19-mm] 2 × 4-foot [0.6 × 1.2-m] FinnForm) help collect the

FIGURE 10-28 Supplemental shoring.
Courtesy of Ron Zawlocki.

soil forces, distribute the strut pressure, and keep active soil from falling into the rescue area. *Supplemental sheeting/shoring is required whenever 2 feet (0.6 m) of trench wall is exposed below a shored panel.*

Precautions for Supplemental Shoring

Problems and hazards associated with digging and supplemental shoring may include the following:

- Creating a secondary cave-in by exposing unprotected portions of the trench wall
- Dislodging rescue struts during digging and supplemental shoring operations
- Injuring the patient with improper digging

Preparation for Supplemental Shoring

Take the following steps to prepare to install supplemental shoring:

- Obtain supplemental panels: 2 × 4-foot (0.6 × 1.2-m) ¾-inch (19-mm) FinnForm panels (prebuilt).
- Ensure that all hazards are mitigated and rescuer entry has been approved by command.
- Because of the reduced size (2 × 4-foot [0.6 × 1.2-m]) of the supplemental panels, aluminum spot shore bases can be used with 14- or 16-ply engineered arctic birch plywood instead of lumber (strongbacks).

Walkthrough of Installing Supplemental Shoring

1. The panel team stages supplemental sheeting/strongback and the shoring teams place struts near the trench prior to beginning soil removal operations. Steps 2 through 7 are performed by the entry team.
2. The entry team stops soil removal (digging) when 2 feet (0.6 m) of trench wall is exposed below the lowest strut **FIGURE 10-29**.
3. The entry team inserts supplemental (2 × 4-foot [0.6 × 1.2-m]) panel sections directly below the existing panel.
4. The entry team installs a strut on the supplemental panel. The strut can be placed up to 2 feet (0.6 m) below a shored panel **FIGURE 10-30**.
5. The entry team nails strut bases to strongbacks.
6. The entry team continues removing soil and adding supplemental shoring as needed.

FIGURE 10-29 The entry team stops soil removal (digging) when 2 (0.6 m) feet of trench wall is exposed below the lowest strut.

FIGURE 10-30 One supplemental shoring panel installed.

7. If loading begins to occur on the supplemental shoring, the entry team adds struts to supplemental panels.
8. If it is necessary to dig deeper to extricate the victim, repeat steps 2 through 7 **FIGURE 10-31**.

Putting It All Together: Supplemental Shoring

SKILL DRILL 10-4 puts the entire skill of supplemental shoring together. After the shoring size-up and the preparation steps are complete, this shoring skill drill may begin.

CHAPTER 10 Technician Level Trench Rescue Shoring

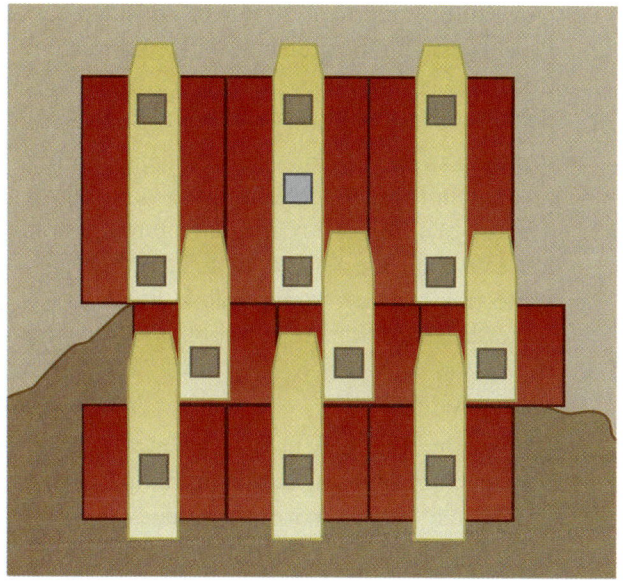

FIGURE 10-31 Additional supplemental shoring installed.

Spot Shoring

Spot shoring can be a high-risk endeavor. In some soil conditions, spot shoring is not capable of preventing collapse. In other soil conditions, spot shoring can cause the soil to collapse. The safe use of spot shoring requires onsite soil analysis and an understanding of soil mechanics that is beyond the training and experience of rescuers. Shoring without incorporating uprights or panels at a rescue incident requires a list of conditions and limitations. *The use of spot shores or skip shores in the unstable or dynamic soil conditions that are common to trench rescue incidents will most likely result in additional collapse.*

Because spot shoring is addressed in NFPA 1006, the authors offer the following engineered approach to the concept of spot shoring to help ensure the safety of rescuers and victims. From an engineering standpoint, the term *spot shoring* is incorrect. A more accurate

SKILL DRILL 10-4
Supplemental Shoring (NFPA 1006: 12.3.3, 12.3.4)

1 Stage supplemental sheeting/strongback and struts near the trench tip. Stop soil removal (digging) when 2 feet (0.6 m) of trench wall is exposed below a strongback. Install backfill if a void is present. Insert supplemental (2 × 4-foot [0.6 × 1.2-m]) sheeting sections directly below the existing panels on both sides of the trench. Install a strut on the supplemental panel/strongbacks.

term is single-point shore. Therefore, the authors will use this more accurate term in our discussion.

Single-Point Shoring

Single-point shores should not be used as a principal shoring technique; they should only be used when conventional trench rescue shoring cannot be properly installed. The use of single-point shoring requires a soil analysis conducted by a competent person with specific training on the implementation, applicability, and construction of alternative shoring systems to make a determination of the soil's compatibility to the technique. The implementation, application, and construction of a nonstandard shoring system should be implemented under the direction of a competent person. The use of single-point shores to supplement a conventional trench rescue shoring system (i.e., back shore) should include the use of supplemental panel sections with a minimum size of 2 × 2 feet (0.6 × 0.6 m).

Severe Environmental Conditions

Rescuers may be subjected to a variety of **severe environmental conditions** while working at trench or excavation incidents. In fact, severe environmental conditions often precipitate or cause accidents at trench sites. These conditions create hazards that must be mitigated to enhance both victim and rescuer safety. Severe environmental conditions are beyond the scope of awareness and operations level personnel but must be addressed by technician level rescuers.

Extreme environmental conditions may cause hazards that cannot be eliminated. Carefully selected mitigation techniques may improve, but not guarantee rescuer safety. A risk and benefit analysis must always be conducted when severe environmental conditions are encountered.

Severe environmental conditions include the following:

- Frozen soil: Frozen soil can be shored with traditional sheeting and shoring equipment and techniques. If the frozen soil is thawing (melting), space the panels about 1 inch (2.5 cm) apart, leaving room for the melting water to escape and reducing the buildup of hydrostatic pressure behind the panels. Heavy equipment can slope or bench trench walls with frozen soil. Large chunks of frozen soil that have collapsed and trapped a victim can be hard to remove using shovels and buckets. Vacuum trucks, in combination with hydro excavation or pneumatic cutting (Air Knife) equipment, should be considered. Rescuers should be protected with cold weather (insulated coverall–type) ensembles, hoods, helmets, insulated gloves, and boots.

- Running soil: Traditional sheeting and shoring are not appropriate for running soil conditions. The use of a trench box and/or sloping the walls to a safe angle are a more suitable method of protecting rescuers from collapse. Running (granular) soils that are entrapping victims in a trench can be removed with shovels, buckets, or Vactor systems.

- Heavy rain: Rain may be diverted away from the trench by building dikes and ramps. A tarp or tent may be set up over the rescue area of the trench. Water may be removed from the trench by using trash pumps and dewatering equipment. Flowing water commonly carries soil with it, which can destabilize trench walls. Rescuers should be protected with Gore-Tex–type rain gear, helmets, eye protection, gloves, and water-resistant footwear. Live victims must also be provided with appropriate PPE.

- High winds: If possible, create a wind break by positioning the rescue truck or trailer in a safe spot upwind of the trench. Cut off trench panels that stick up more than 2 feet (0.6 m) above the trench lip to avoid the additional pressure created by wind hitting the large panels. Carry panels and ground pads as low to the ground as possible. Rescuers should be protected with Gore-Tex–type wind breakers, helmets, eye protection, gloves, and safety footwear.

- Cold temperatures: Warm air from heated ventilation fans should be ducted into the trench. Rescuers need to be rotated out of cold environments and periodically placed into warming huts. Rescuers should be protected with cold weather (insulated coverall–type) ensembles, hoods, helmets, insulated gloves, and boots. Victims should be covered with blankets and warmed IV solutions should be considered (contact medical control). Live victims must also be provided with appropriate PPE.

- Hot temperatures: Cooling air can be circulated into the trench with ventilation equipment. Rescuers need to be rotated out of hot environments and periodically placed into air-conditioned areas. Rescuers should be protected with light-colored cotton shirts, cotton pants, vented helmets, lightweight gloves, and breathable boots.

- Snow/ice: The rescue area in the trench can be covered by tarps or tents. Snow and ice should be periodically removed (shoveled) from the trench and trench lip areas. Rescuers should be protected by water-repellent cold-weather ensembles, helmets, water repellent and insulated gloves, and boots with cleats. Victims should be covered with blankets and warmed IV solutions should be considered (per emergency medical services [EMS] and administered by EMS).
- Lightning: Anytime that lightning strikes are in the area and the incident has been declared a recovery rather than a rescue, rescuers should leave the area and return when the lightning has stopped.
- Hazardous atmospheres: Hazardous atmospheres should be controlled with confined space ventilation techniques. Hazardous atmospheres should be avoided in all recovery (nonrescue) situations. After the area had been purged and monitored, recovery can resume. If possible, remove the source of the hazardous atmosphere and control the residual air with ventilation techniques. Rescuers should be protected by respirators, appropriate ensembles, helmets, gloves, and boots. Live victims must also be provided with appropriate PPE.
- Night operations: Illuminate the scene with scene lighting equipment and use helmet-mounted lights.

Victim Removal Review

Once the trench has been shored and any severe environmental conditions accounted for, victim removal may proceed. Chapter 8, *Victim Care and Extrication*, covers this topic in detail, including the use of vacuum trucks to rapidly remove soil and victim considerations. Remember to consider the possibility that the victim may be experiencing crush or compartment syndrome and to have emergency medical personnel on scene to assess and treat the victim.

If a victim is trapped under heavy objects, then technician level rescuers will need to work with heavy machinery operators to disentangle trapped victims. Chapter 9, *Lifting and Load Stabilization*, covers this topic in detail.

After-Action

IN SUMMARY

- Trench rescuers trained to the technician level must have knowledge, skills, and abilities to shore trenches that are beyond the operations level limitations (straight trench, 8 feet (4.2 m) deep or less that can be supported with traditional sheeting and shoring).
- Trench rescue technicians must be able to shore trenches that are deeper than 8 feet (4.2 m), shore intersecting trenches, install supplemental shoring when digging exposes more than 2 feet (0.6 m) of unsupported trench wall, and safely use spot shoring (single-point shoring) within clearly defined engineered limitations.
- Additionally, trench rescuers at the technician level must have prerequisite knowledge, skills, and abilities to perform the following:
  - Recognize potential for collapse.
- Types of collapses and techniques to stabilize, emergency procedures.
  - Know protocols on making the general area safe.
  - Establish of safe zones.
  - Select appropriate PPE.
  - Consult related tabulated data.
  - Designate cut station location and material and equipment needs.
  - Know criteria for a safe zone within the trench.
  - Use shoring strategies and tactics and prebriefing of team members.
  - Address areas of the trench that are blown out or undercut.
  - Consider of the weights and hazards associated with the soils.

- Identify of mechanisms of entrapment.
- Consider selected stabilization tactics for extrication and victim safety.
- Employ victim release techniques and projected path for removal.
- All shoring plans must include a shoring performance assessment to evaluate and adjust shoring to maintain safety. All shored areas, spoil piles, and trench walls adjacent to the shored area must be monitored throughout the rescue/recovery operation. Check for:
  - Struts loosening
  - Increased soil force
  - Signs of moving soil
- A rescue plan should include:
  - Pre-entry briefing
  - Risk versus benefit analysis
  - Shoring plan
  - Victim release
  - Path for removal
- The most common failure and collapse location at intersecting trenches is the inside corner.
- A shoring size-up should include assessment of the lip, trench, collapse potential, voids, victims, and the use of tabulated data.
- Primary shoring must cover the area above the victim's head and chest. If the victim is not trapped in the corner area of a T-trench or L-trench, normal straight wall trench primary shoring procedures should be used.
- Deep trench shoring procedures begin with shores that are placed at depths of more than 8 feet (4.2 m) below the trench lip and require vertically stacked rescue panels, stronger shoring equipment, and edge restriction devices.
- Supplemental shoring is a trench rescue technician skill that is used when rescuers must remove soil that has caved in from the trench. Equipment used for supplemental shoring includes 2 × 4-foot (0.6 × 1.2-m) sections of shoring panels, sections of 2 × 12-inch strongbacks and (rescue) struts.
- Single-point shoring (spot shoring) should not be used as a principal shoring technique and should only be used when conventional trench rescue shoring cannot be properly installed. It requires:
  - A soil analysis conducted by a competent person
  - Implementation, application, and construction under the direction of the competent person
  - The use of supplemental panel sections (minimum size: 2 × 2-foot [0.6 × 0.6 m])
- Hazard mitigation techniques include control, avoidance, removal, and personal protective equipment.
- Severe environmental conditions include: frozen and running soils, heavy rain, high winds, cold and hot temperatures, snow/ice, lightning, and hazardous atmospheres.

KEY TERMS

Cantilever lip bridges Lip bridges are created by placing a ladder on the ground, extending it over the missing lip, placing 2 × 10s over the rungs, and anchoring the far side of the ladder with ladder rung brackets with pickets driven into the ground to allow a rescuer to stand on an overhanging lip bridge.

Cantilever wales (beams) Beams that have two or more supports but extend beyond one end support and end in clear space.

Deep trench A trench that is deeper than 8 feet (4.2 m).

Intersecting trench A trench where multiple trench cuts or legs converge at a single point.

Joined corner panels A prefabricated corner panel constructed from two aluminum or steel brackets placed at tiers, secured with ½-inch (13-mm) diameter bolts and nuts to allow two trench rescue panels to be bolted together at a 90-degree angle.

L-shaped trench An intersecting trench where one cut or leg converges with another cut or leg resulting in the shape of the letter "L."

Pile side In a T-trench or L-trench, the side of the trench with the taller pile of dirt that resulted from a trench collapse.

Rescue plan The tactical portion of the incident action plan.

Severe environmental conditions Operations involving frozen soil, running soil (e.g., gravel, sand, liquid), severe weather (e.g., heavy rain, wind, lightning, or flooding), or nighttime (dark) operations.

Shoring performance assessment An inspection of in-place shoring that verifies continuous safety by regularly verifying the shoring integrity.

Single-point shores A nonstandard shoring type that should be used only as a secondary, not primary, means of trench shoring and should only be used when conventional trench rescue shoring cannot be properly installed.

Spot shoring Shoring without the use of panels or strongbacks.

Supplemental shoring Sheeting and shoring operations that involve the use of commercial sheeting/shoring systems and/or isolation devices or that involve cutting and placement of sheeting and shoring when greater than 2 feet (0.6 m) of shoring exists below the bottom of the strongback.

T-shaped trench An intersecting trench where one cut or leg converges with another cut or leg resulting in the shape of the letter "T."

Traditional sheeting and shoring The use of 4 × 8-foot (1.2 × 2.4-m) sheet panels, with a strongback attachment, supplemented by a variety of conventional shoring (strut) options such as hydraulic, screw, and/or pneumatic shores.

Wale hangers Pipe clamps that are preassembled that can hang on the top of a panel and support a wale at a specific height for placing struts.

Wales Beams used horizontally in a trench to collect loads where it is not practical to collect the loads with struts due to lack of trench wall to push against or to leave a clear opening for victim removal operations.

On Scene

1. What are the trench shapes and depths that differentiate the operation level and the technician level job performance requirements?

2. What pre-built equipment is needed for the implementation of supplemental shoring during the removal of soil and victim extrication process?

3. The use of spot shoring at most trench emergency incidents is not supported by engineering principles and not endorsed by rescue strut manufacturers. What soil analysis equipment is needed to assure that alternative techniques are safe for a trench emergency where the trench has not collapsed?

4. What severe environmental conditions are most likely in your response district and what equipment and procedures are needed to resolve them?

Appendix A
Advanced Trench Rescue Shoring

KNOWLEDGE OBJECTIVES

After studying this chapter, you should be able to:
- Identify advanced trench rescue shoring topics.
- Identify methods for shoring trench end walls.
- Identify a method for replacing inadequate shoring without increasing the risk of cave-in during the replacement operation.
- Identify methods for shoring angled walls.
- Identify methods for providing protection from cave-ins at trenches and excavations with extreme conditions.

SKILLS OBJECTIVES

After studying this chapter, you should be able to:
- Install cross shoring.
- Install excavator shoring.
- Install an end wall raker shoring.
- Install replacement shoring.
- Install angled wall shoring.

Introduction

Trench rescue incidents present conditions and challenges to each level of trench rescuers. NFPA 1006, *Standard for Technical Rescue Personnel Professional Qualifications,* speaks to shoring straight trenches, intersecting trenches, and adding supplemental shoring that results from digging (extrication) operations. These shoring job performance requirements are clearly stated in NFPA 1006. However, some emergencies may pose an additional level of complexity and require advanced shoring techniques, above and beyond technician level training.

It would be impossible to address every trench condition variable in a single volume textbook, so the authors will address some variations that occur rather frequently at trench rescue incidents. Here, the authors will explore end wall shoring, excavation shoring, replacement shoring, shoring angled walls, and protective systems for extreme soil conditions.

This appendix was written to introduce rescuers to possible solutions to conditions that are not included in NFPA 1006 and 1670: *Standard on Operations and Training for Technical Search and Rescue Incidents.* These solutions have been reviewed by professional engineers and subjected to practical testing procedures; however, they should not be attempted without learning the practical (hands-on) application from an experienced advanced trench rescue shoring instructor. The application of the solutions is also dependent on the local authority having jurisdiction (AHJ).

Rescuers must also understand that while installing shoring, following engineering shoring principles and practices is essential for victim and rescuer safety. In addition, assessing the performance of a shoring system after it is installed is of equal importance. A shoring performance assessment is required after the installation of trench rescue shoring panels and struts. As previously discussed, the purpose of the shoring performance assessment is to evaluate and adjust shoring to maintain safety. All shored areas, spoil piles, and trench walls adjacent to the shored area must be monitored throughout the operation. Review the steps of the shoring performance assessment in Chapter 7, *Operations Level Trench Rescue Shoring.*

End Wall Rescue Shoring

End walls can become a serious hazard, and rescuers must be protected from collapse in the following situations:
- There are signs of movement, such as bulging, fissures, sloughing, raveling, or collapse.
- The walls have not been properly benched or sloped.
- The walls are wider than 4 feet (1.2 m) and/or deeper than 8 feet (4.2 m).

Having several proven tactical options is the key to success in the rescue business. In this section, we will explore three options for shoring end walls. These options include the following:
- Cross-shoring
- Excavator shoring
- End wall raker shoring

Please note that the wall designations and corresponding team assignments for the panel and shoring teams must be made and discussed during the pre-entry briefing.

Cross-Shore Design

The cross-shore design is applicable for end wall protection when an opposing end wall is within the reach of the longest strut and when the opposing (end) wall is suitable to resist the potential lateral forces. Trenches that are dug to repair utilities are often relatively small holes that have short distances between their end walls. Cross-shores can be installed to protect the victim and create a safe working area for rescuers. Primary shoring to protect the victim from collapse of four walls with two tiers (levels) of struts shores in trenches 8 feet (4.2 m) deep or less can be installed by a seasoned trench rescue squad in 30 minutes or less. Secondary shoring, or walls shored to create a 12-foot (3.7-m) horizontal safe zone for rescuers, can be completely installed in an additional 30 minutes **FIGURE A-1**.

Typically, primary shoring starts with the long walls as the area of greatest risk. The long walls are shored first because the larger area of exposed wall usually has a greater collapse potential. However, trench rescuers must always use the shoring size-up criteria to determine the area that is most likely to collapse next and install the primary shoring there. See Chapter 7, *Operations Level Trench Rescue Shoring,* and Chapter 10, *Technician Level Trench Rescue Shoring,* for a detailed review of shoring size-up criteria.

Application and Precautions

Trench rescue teams often carry struts (timbers or pneumatic) which are 16 feet (4.9 m) or less in length. When the walls of a trench are farther apart than the struts can span, cross-shoring is not feasible. Problems

FIGURE A-1 Primary shoring in a "repair hole" can be installed by a seasoned trench rescue squad in 30 minutes or less. Secondary shoring should be completed in an additional 30 minutes or less.
Courtesy of Ron Zawlocki.

and hazards associated with the use of cross-shores at rescue incidents include the following:

- Lip protection will be needed over all four walls. Lip bridges should always be used when working on the lips of unstable walls.
- Simultaneously shoring from two directions increases the probability of victim injury resulting from rescuers losing control and dropping the shoring during the installation process and therefore must be planned and coordinated.
- Shoring in two directions will result in more overhead obstructions, and will require advance planning to ensure there is a clear window for victim extrication.
- Shoring capacity decreases substantially at the long lengths.

Preparation

Remember, hands-on instruction by qualified instructors and practice are required for cross-shoring proficiency. The first step in preparation to place cross-shoring is to mitigate all hazards. Hazard recognition and mitigation techniques are detailed in Chapter 5, *Hazard Mitigation*. Typically, non-collapse hazard mitigation is assigned to entry team personnel.

The next step is to conduct a shoring size-up. Chapter 7, *Operations Level Trench Rescue Shoring*, details the process of completing a shoring size-up. The information gathered during the size-up becomes a critical element of the shoring plan, and it allows rescuers to determine what type of shoring is needed and whether additional resources (person power or equipment) or additional expertise (like an engineer) are needed at the incident.

Upon completion of the size-up, the available personnel onsite are assigned to teams. Each team must have a designated leader (or officer) who will coordinate operations with the operations officer and other team leaders. Assigning personnel to specific functions and designated areas provides an organized approach that can improve the safety and efficiency of a trench rescue operation. For trenches that require end wall shoring, the teams and minimum number of team members recommended by the authors are as follows:

- Long wall panel team (6 members)
- Long wall shoring team (4 members)
- End wall panel team (6 members)
- End wall shoring team (4 members)
- Entry team (4 members)

After team assignments, a clear shoring plan must be developed and briefed to the team. The shoring plan and briefing must include the techniques and assignments needed to conduct a shoring size-up, protect the victim(s) (primary shoring), create a safe area for rescuers (secondary shoring), enhance the shoring (complete shoring), and conduct shoring performance assessments. It will be up to the teams to carry out the steps to complete all five items. The fifth step is always performing a physical inspection of the shoring with a strut performance assessment.

Walkthrough

Step 1: Trench Size-Up

The purpose of the trench size-up is to gather the information needed to shore the trench. That information includes the following assessments (review

Chapter 10, *Technician Level Trench Rescue Shoring*, for a detailed explanation of each step):

- Assess the lip.
- Assess the trench.
- Assess the collapse potential.
- Assess the voids.
- Assess the victim(s).
- Use tabulated data.

Step 2: Primary Shoring

Long Wall Panel Team. This team does the following:

- Places lip protection over the victim area on both long walls
- Installs primary panels to protect the victim's head and chest from additional collapse
- Completes backfill (if required) after the backfill shore is in place

Long Wall Shoring Team. This team does the following:

- Measures the trench width and places the ladder
- Installs the positioning strut on the primary panel set (long wall)
- Installs backfill and compliance (if needed) struts

End Wall Panel Team. This team does the following:

- Places lip protection over the victim area on both end walls
- Installs primary end wall panels to protect the victim's head and chest from additional collapse

End Wall Shoring Team. This team does the following:

- Places an end wall strut on the trench floor before any strut placement occurs
- After the positioning strut is secured on the long wall, installs the positioning strut on the primary panel set (end wall) **FIGURE A-2**
- Installs the backfill and compliance (if needed) strut

Step 3: Secondary Shoring

Long Wall Panel Team. This team installs the secondary panels on the long wall.

Long Wall Shoring Team. This team installs struts on the secondary panel sets **FIGURE A-3**.

FIGURE A-2 Positioning strut installed on the long wall primary panels.
Courtesy of Ron Zawlocki.

FIGURE A-3 Secondary shoring in place.
Courtesy of Ron Zawlocki.

End Wall Panel Team. This team installs secondary panels on the end wall as needed.

End Wall Shoring Team. This team installs struts on the end wall secondary panel sets as needed.

Step 4: Complete Shoring

All panel and shoring teams do the following:

- Install fillers in gaps between wales and panels if wales are used.
- Expand the safe zone with panels and struts as needed.
- Install supplemental shoring as needed during extrication operations.
- Expand the safe zone with panels and struts as needed **FIGURE A-4**.
- Nail all strut bases.

Finally, perform a shoring performance assessment. Physically inspect all struts twice during the first hour and once per hour thereafter.

Excavator Shore Design

The excavator shore design is applicable for end wall protection when an opposing end wall is beyond the reach or capacity of the longest strut or when the opposing (end) wall is not suitable to resist the potential lateral forces. Excavator shoring can be installed to protect the victim and create a safe working area near a vulnerable end wall. Primary shoring for a two-tiered excavator shore can be installed by a seasoned trench rescue squad in 30 minutes or less. Secondary shoring can be completely installed in an additional 30 minutes **FIGURE A-5**.

Application

When an opposing end wall is beyond the reach of cross-shore designs or when the opposing (end) wall is not suitable to resist the potential lateral forces using an excavator arm and bucket, the excavator shore design is a potential solution. The application of the excavator shore design includes the following:

- End walls that are not sloped or benched (nearly vertical) and are wider than 4 feet (1.2 m).
- Any end wall with unstable soil that is nearly vertical and is showing signs of movement.
- The excavator/backhoe being used has flat areas on the bucket and/or arm that can support struts and wales.

FIGURE A-4 Complete the shoring.
Courtesy of Ron Zawlocki.

FIGURE A-5 Excavator shore design.
Courtesy of Ron Zawlocki.

Precautions

Problems and hazards associated with the use of excavator shores at rescue incidents include the following:

- Lip protection will be needed over all four walls. Lip bridges should always be used when working on the lips of unstable walls.
- Victims may be injured if rescuers lose control and drop the heavy struts during the installation operation.
- The excavator must be in good mechanical condition and be able to support loads without hydraulic leak-down.
- Operation of an excavator with a bucket in the trench in the area of a victim is inherently dangerous.
- Must have room to place the excavator bucket/arm into the trench to support shoring without being over the victim.
- The front of the track or the outriggers must be no closer than 4 feet (1.2 m) from the L (farthest point of soil failure).
- The surcharge load and vibrations associated with positioning an excavator on an unstable trench lip may cause soil failure and collapse.

Preparation

The first step when preparing to place excavator shoring, like all other shoring, is to mitigate hazards. The next step is to conduct a shoring size-up. Upon completion of the size-up, the available personnel onsite are assigned into teams with leaders.

Excavator Shore Walkthrough

Step 1: Trench Size-Up

The purpose of the trench size-up is to gather the information needed to shore the trench. That information includes the standard six assessments listed previously.

Step 2: Primary Shoring

Entry Team. This team does the following:

- Determines the skill level and the emotional stability of the heavy equipment operator (see Chapter 7, *Operations Level Trench Rescue Shoring*).
- Determines the suitability of equipment and mechanical soundness. The heavy equipment operator should load the bucket with dirt—as far away from the trench as possible without moving the equipment—and then shut the equipment off with the bucket suspended to verify that there is no hydraulic leak-down the arm/bucket over 5 minutes.
- If the heavy equipment operator is suitable for the task, develop a communication plan with the heavy equipment operator.
- Directs the positioning of the excavator or backhoe in relation to the trench to ensure a safe area. This point can be either between the victim and the end wall or beyond the victim's position.
- Cuts two wales to fit between the long walls with panels on them.
- Measures the depth of the trench to determine the placement of the wales that will be attached to the excavator arm and bucket.
- Uses ratchet straps to attach the bottom wale to the bucket so that it will be within 2 feet (0.6 m) of the trench floor and the top wale to the excavator arm so that it will be within 2 feet (0.6 m) from the lip **FIGURE A-6**.

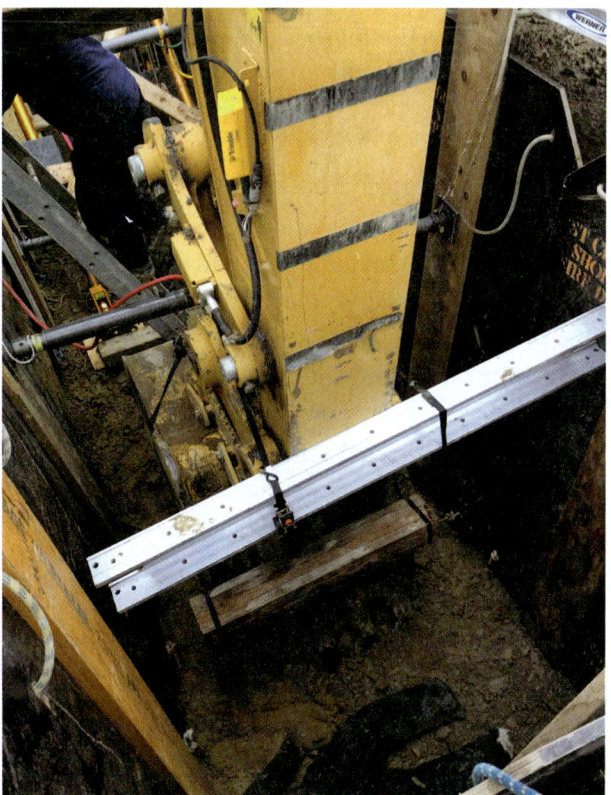

FIGURE A-6 Attach wales to the excavator arm/bucket for excavator shoring.
Courtesy of Ron Zawlocki.

- Directs the positioning of the arm and bucket in the trench. It can be positioned either between the end wall and the victim or beyond the victim. That will depend on both the distance that the victim is from the end wall, the safe position on the lip and the excavator's horizontal reach (boom length).
- Ensures that the bucket is a safe distance from the victim and directs the heavy vehicle operator to dig the teeth of the bucket into the trench floor and then push down, lifting the track or outriggers a few inches off the ground.

Long Wall Panel Team. This team does the following:
- Places lip protection over the victim's area on both long walls
- Installs primary panels to protect the victim's head and chest from additional collapse
- Completes backfill after the backfill strut is in place

Long Wall Shoring Team. This team does the following:
- Measures the trench width and places the ladder
- Installs positioning, backfill, and compliance struts on the primary panel sets on the long walls

End Wall Panel Team. This team does the following:
- Places lip protection on the end wall that is within the collapse zone
- Installs panel(s) on end wall to protect the victim's head and chest from additional collapse
- Places a wale (cut to fit) on the floor parallel to the end wall before installation of long wall struts

Long Wall Shoring Team. This team positions a ladder for entry into the shored primary panel set.

End Wall Panel Team. This team installs and positions (ropes or wale hangers) the wales for the first tier of the end wall struts. The team follows the installation of the first tier of the end wall struts and installs and positions (ropes or wale hangers) the wales for the second tier of end wall struts.

End Wall Shoring Team. This team installs the first tier of end wall struts with two struts from the wale on the end wall panel to the wale on the excavator. The team repeats the previous step for the second tier of end wall struts.

Step 3: Secondary Shoring

The long wall panel team positions panels for secondary shoring on the long walls. The long wall shoring team installs a minimum of two struts per secondary panel set.

Step 4: Complete Shoring

All teams work together to install fillers in gaps between wales and panels. The safe zone is expanded with panels and struts as needed **FIGURE A-7**. Supplemental shoring is installed as needed during extrication operations. All strut bases are nailed. Finally, a shoring performance assessment is completed. All struts are physically inspected twice during the first hour and once per hour thereafter.

End Wall Raker Shore Design

An end wall raker shore design is applicable for end wall protection when an opposing end wall is out of the reach of the longest strut or when the opposing (end) wall is not suitable to resist the potential lateral forces. The raker assembly, which is bolted to a Finn-Form panel, can be easily lowered and controlled with a rope system. It can be safely lowered into place directly over a trapped victim and installed to protect the victim and create a safe working area for rescuers. An end wall raker shore can be assembled on lip

FIGURE A-7 Complete shoring with excavator positioned on the near end wall.
Courtesy of Ron Zawlocki.

protection by a seasoned rescue squad in less than 10 minutes. The shore can be completely installed and pressurized in an additional 10 to 15 minutes. This shore requires that pickets be driven into the trench floor to resist the raker from sliding; it also requires driving pickets through the raker panel into the wall to prevent the strut from walking up the wall.

Typically, the side walls are shored first, leaving enough room between the end wall and the primary shoring struts to allow the end wall raker strut to be lowered into the trench. Hands-on instruction, some specialized equipment, and practice are required for end wall raker proficiency. The application of raker shores for rescue operations at trench end walls up to 8 feet (4.2 m) deep and 8 feet (4.2 m) wide is presented in this section **FIGURE A-8**. An additional 4-foot (1.2-m) high and 8 foot-(4.2-m) wide excavation panel can be added for excavations up to 12 feet (3.7 m) deep.

Application

When the end walls of a trench are farther apart than the struts can span, traditional sheeting and shoring becomes impractical. By using diagonal struts and the geometrical strength of a triangle, rescuers can reduce the risk of cave-in by applying raker shores. The application of rakers at end walls include the following:

- End walls that are not sloped or benched to OSHA standards
- Any end wall with unstable soil that is nearly vertical and is showing signs of movement
- A collapse where the victim's location allows for the installation of an end wall raker

Victim Location

The trapped victim's location and distance from the end wall will determine the techniques and the sequence of long wall and end wall shoring installation. During the shoring size-up, rescuers must determine the most likely secondary collapse area and then shore that area first. If the end wall is determined to be most likely to collapse, the challenge may be removing enough debris (collapsed soil) from the floor to install the end wall raker. In this case a vacuum truck, hydro excavator, and/or air knife that can be operated from a protected trench lip is essential. The long wall shoring can then be installed. However, when the long walls are most likely to collapse, then they must be shored first. If the trapped victim's location is greater than 8 feet (4.2 m) from the end wall but is still within the collapse zone (1.5 times the depth of the trench) the basic raker installation (Option 1) can be used **FIGURE A-9**.

When the trapped victim's location is less than 8 feet (4.2 m) from the end wall, rescuers still must shore the long wall first, but rescuers must also provide a way to lower the assembled raker system into that primary shoring area. Option 2 and Option 3 allow the long walls to be shored first while providing for the lowering and installation of the end wall raker.

Option 1: Basic Raker Installation

This option is utilized when the victim is trapped more than 8 feet (4.2 m) from the end wall. This option allows for primary shoring to be installed in the long walls without interfering with the lowering and installation of the raker system.

Procedure Overview. The teams install the primary panel set on the long walls and install the positioning strut. Backfill is positioned behind the panel(s) and backfill struts and compliance struts are installed as needed. The raker is lowered into place and secondary panels are added to long walls. The trench is entered in order to install ground anchor and wall pickets.

FIGURE A-8 End wall raker shore with long wall shoring in place.
Courtesy of Ron Zawlocki.

Option 2: Strut By-Pass

This option is utilized when the victim is trapped less than 8 feet (4.2 m) from the end wall. Option 2 resolves this problem by installing primary shoring on the long walls to protect the victim area from a secondary collapse while making room for the raker as it is being lowered using the strut by-pass technique. The strut by-pass technique takes more practice and precision, but once it is mastered, it is the fastest way to protect the victim from both end wall and long wall collapse.

Procedure Overview. The teams install the primary panel set on the long walls and install the positioning strut. Backfill is installed behind the panel(s). The backfill strut and compliance strut are installed as needed. The raker is lowered and stopped as it reaches the top strut in the primary panel set **FIGURE A-10**. A by-pass strut is installed just above the top raker struts. The initial top struts are removed from the primary panel set and the team resumes lowering the raker and stops as it reaches the next struts. The by-pass procedure is repeated and the raker is lowered to the trench floor. Secondary panels are added to long walls. The trench is entered to install the ground anchor and wall pickets.

Option 3: Cantilever Wale

This option may be utilized when the victim is trapped less than 8 feet (4.2 m) from the end wall. Installing primary shoring on the long walls would interfere with the lowering and installation of the end wall raker. Option 3 resolves this problem by using cantilever wales to protect the victim area from a secondary collapse while maintaining room for the raker lowering and installation **FIGURE A-11**. Installing three sets of panels, struts, and wales to protect the victim from long wall collapse while proving enough room to lower the end wall raker is more time consuming than the strut by-pass technique, but this option does not require as much precision from the lowering team nor as much coordination with the shoring team.

Procedure Overview. The teams lower wales to the floor on both sides of the trench. The primary panel set is placed over the victim on the long walls and a positioning strut installed. Two additional panel sets (secondary panels) are set beyond the primary panels and positioning struts are installed.

The first set of wales are positioned either suspended on wale hangers or ropes. The most likely area

FIGURE A-9 Primary shoring in place with raker lowered at the end wall.
Courtesy of Ron Zawlocki.

FIGURE A-10 Raker lowered to the initial top strut. The by-pass strut above is being positioned to replace the top strut.
Courtesy of Ron Zawlocki.

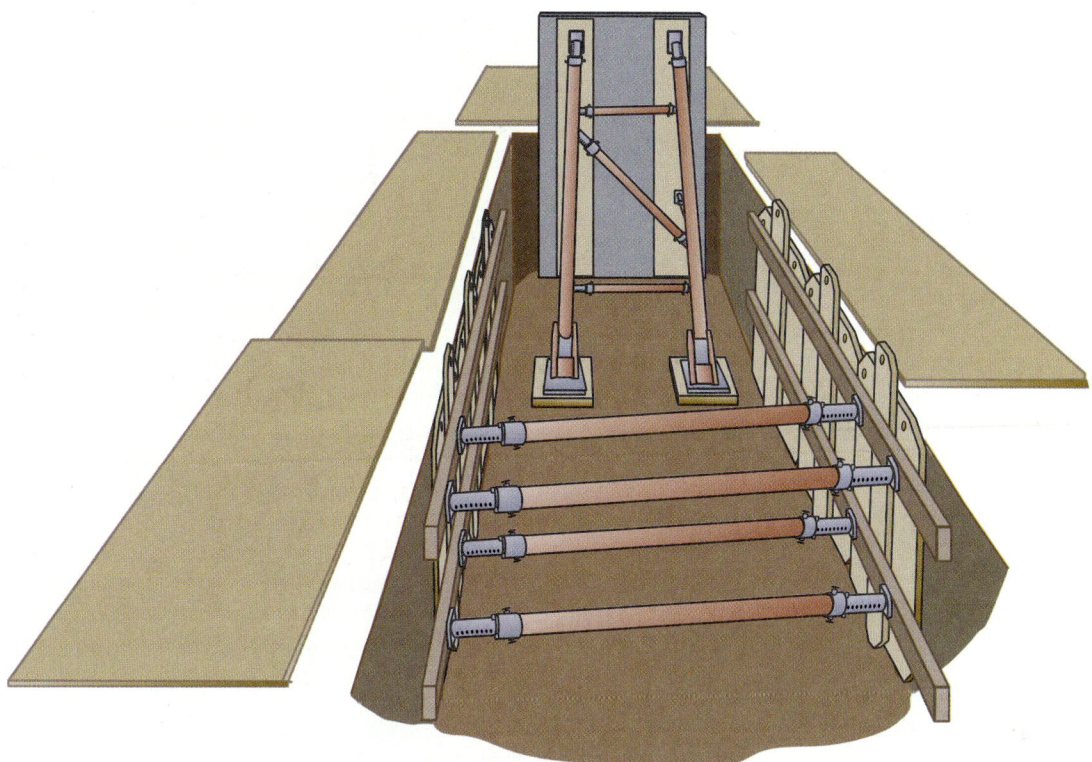

FIGURE A-11 Primary shoring in place with cantilever wales and raker lowered at the end wall.

(top or bottom of the trench wall) to collapse should get the first set of wales. Two struts are installed on the wales at the strongbacks on the secondary panels. The second set of wales are positioned either suspended on wale hangers or ropes. Two struts are installed on the wales at the strongbacks on the secondary panels.

The positioning struts that would interfere with the lowering and installation of the raker are removed. Backfill is installed behind long wall panels as needed and the raker is lowered into place. The trench is entered to install the ground anchor and wall pickets.

End Wall Raker Shoring System Disassembly and Removal

The removal and disassembly of a raker strut from a trench is different than the removal of sheeting and shoring systems. Sheeting and shoring system removal is described in Chapter 7, *Operations Level Trench Rescue Shoring*. The procedure described next is designed to minimize rescuer exposure to potential collapse conditions during the removal and disassembly of the end wall raker shoring system.

Step 1

The long wall shoring team follows the manual removal procedure (see Chapter 7, *Operations Level Trench Rescue Shoring*), to remove any long wall struts that would prevent the machine lift out of the raker strut(s). All other struts are left in place and four nails are placed in each strut base for all struts on long wall panel sets.

Step 2

End Wall Shoring Team. This team connects ropes to prevent 610 struts (angled struts) from expanding while being lifted. A ½-inch (13-mm) diameter prusik (Triple P) is attached to wall pickets. The excavator bucket is positioned in front of raker. Kernmantle ropes (½-inch [13-mm]; prusiks) are used to tie the pickets to the bucket (via the lifting ring).

End Wall Panel Team. This team connects ropes to prevent 610 struts (angled struts) from expanding while being lifted. Lifting ropes are attached to raker systems and tag lines are attached to each side of the raker system. Rescuers are assigned as tenders.

Step 3

End Wall Shoring Team. This team powers up the air system and loosens the collars on the struts. Spin collars are placed 3 to 4 inches (7.6–10.2 cm) away from the strut. All personnel exit the trench and the air pressure is discharged. The hoses are disconnected from the controller. The hoses are rolled and attached to the panels.

Step 4. All teams work together to use the excavator/heavy equipment operator to remove the wall pickets. The bucket is positioned at the lip and the raker is attached to the excavator using lifting lines. Ropes are then connected to the lifting eye on the excavator bucket. The shoring system is removed with the excavator. The panel team and shoring team members are assigned to disassemble the raker system and return the equipment to a state of readiness. The excavator is used to remove (dig out) the sole anchor pickets from the trench floor.

Long Wall Shoring Equipment Removal

Rescuers should follow the manual removal procedure or the machine removal procedure (see Chapter 7, *Operations Level Trench Rescue Shoring*) to remove the shoring equipment on the long walls in the reverse order of installation.

Replacement Shoring

Rescue personnel may arrive on the scene of a trench emergency and be faced with sheeting and shoring that has already been installed but which has safety margins that are unknown or below the recognized rescue shoring standard described in this text. Substandard shoring typically consists of undersized wood struts (cross-braces made from 2 × 4, 2 × 6, or 4 × 4-inch lumber) and may feature construction-grade plywood used as makeshift sheeting, or it may have no sheeting at all (skip shoring and spot shoring). In some cases, this shoring may actually be temporarily preventing additional cave-ins. In those cases, the shoring must be systematically enhanced in phases before the substandard shoring can be removed.

The trench rescue team needs to be able to recognize substandard shoring and must be able to utilize what is in place while reinforcing the safety margin to a suitable rescue standard **FIGURE A-12**. Because substandard shoring is typically in the way of rescue operations, the trench rescuer must be able to adapt and overcome problems that are not usually associated with installing trench rescue shoring in an unshored trench. Replacing substandard shoring requires an additional step called *interim shoring*. Interim shoring improves the level of protection from cave-in while providing room to install standard trench rescue shoring panels, struts, and backfill.

FIGURE A-12 Substandard shoring.
Courtesy of Ron Zawlocki.

In this section, we will evaluate the application of techniques used for replacement shoring. We will discuss the precautions associated with their use and identify a procedure for replacing substandard sheeting and shoring.

Application

Replacing or enhancing substandard sheeting and shoring should be done whenever the risk–benefit analysis concludes that the safety margin provided by the current sheeting/shoring system is inadequate for the entry of rescue personnel. This analysis would include, but not be limited to, the following:

- Wooden struts are not compliant with trench rescue shoring tabulated data.
- Struts are placed in a fashion that does not provide axial loading, or are not perpendicular to the sheeting.
- Struts are not adequately compressed or secured.
- Sheeting is not compliant with trench rescue shoring tabulated data.
- There is no sheeting (skip shoring or spot shoring).

- Backfill of voids created by the trench collapse is inadequate.
- Vertical/horizontal spacing does not provide adequate safety factors.
- Wall angles are less than 70 degrees.

Precautions

Utilizing replacement shoring techniques requires some additional understanding and precautions. Some points to remember include the following:

- The shoring, even if it is substandard, may be providing enough support to prevent a collapse for the time being.
- The removal of substandard struts prior to the installation of replacement shoring may result in trench cave-in.
- Installing and pressurizing new shoring may loosen up the substandard struts and cause them to drop into the trench, injuring the victim. These substandard struts must be supported prior to the installation of replacement struts.
- Lip protection must be in place prior to replacement shoring operations.

Preparation

The first step in preparation to replace substandard shoring, like all other shoring, is to mitigate hazards. The next step is to conduct a shoring size-up. Upon completion of the size-up, the available personnel onsite are assigned into teams and officers. For trenches that require end wall shoring, the authors recommend the following teams and minimum number of team members:

- Panel team (6 members)
- Shoring team (4 members)
- Entry team (4 members)

After team assignment, a clear shoring plan must be developed and briefed to the team.

Replacement Shoring Procedure Walkthrough

Step 1: Interim Shoring

Panel Team. This team does the following:

- Places lip protection over the victim area on both long walls.
- Positions two 2 × 12-inch uprights (per wall) directly across the trench from each other. If plywood was used as a panel in the substandard shoring system, the team installs the uprights over the plywood.
- The inside edges of the uprights must be between 4.5 feet (54 inches; 1.4 m) and 5 feet (60 inches; 1.5 m) apart (to allow panels to be placed between them at a later time).
- Ready a primary panel set at the trench lip to be able to be installed as soon as the substandard struts are removed.

Shoring Team. This team does the following:

- Ties off the substandard struts that are within 4 feet (1.2 m) of the victim on both sides.
  - Ties an overhand bight at the end of each rope and lowers it past the strut.
  - Use a pike pole to reach into the bight from the opposite side of the strut.
  - Brings the bight up to the lip protection and inserts the terminal end of the rope into the bight.
  - Pulls up on the end of the rope as a team member releases and allows the bight to slide down and tighten up around the strut.
  - Secures the terminal ends of all ropes to an anchor point (driven pickets or other suitable anchor).
- Installs two struts on each upright (for each 8 vertical feet [4.2 vertical m]) and slowly increases the strut activation forces to 500 pounds (227 kg) of force. If voids are present behind the uprights, stops adding air pressure if the upright bends before reaching the 500 pounds (227 kg) of force **FIGURE A-13**.
- Removes substandard struts that have loosened.
- Uses tools (pike poles, sledgehammer, 4 × 4-inch timbers, etc.) to dislodge the remaining substandard struts and lift them out of the trench using the ropes attached to them.

Step 2: Primary Shoring

Panel Team. This team does the following:

- Immediately installs a primary panel set between installed uprights to protect the victim's head and chest as soon as the substandard struts are removed **FIGURE A-14**. If plywood was used as a panel in the substandard shoring system, installs the trench rescue panels over the plywood (depending on placement of uprights, two primary panel sets may be necessary to encompass and protect the victim).
- Coordinates backfill with strut installation.

FIGURE A-13 Substandard struts tied off with an interim upright and positioning the struts in place.
Courtesy of Ron Zawlocki.

FIGURE A-14 Primary panel set installed between interim uprights.
Courtesy of Ron Zawlocki.

FIGURE A-15 Primary shoring in place.
Courtesy of Ron Zawlocki.

Shoring Team. This team does the following:
- Installs at least two struts on each primary panel set **FIGURE A-15**
- If voids are present in the primary shoring area, follows the shoring void procedures
- Places a positioning strut
- Places a backfill strut
- Places a compliance strut (as needed)

TIP

If voids are present behind the primary panels, begin with strut activation forces of 500 pounds (227 kg) of force and slowly increase to 1000 pounds (454 kg) of force. Stop adding air pressure if the panel bends and enhance backfill before increasing the strut activation force.

Step 3: Secondary Shoring

Panel Team. This team installs panel sets on both sides of the victim (when possible, it may require placing more uprights and removing substandard struts). The panel team coordinates backfill installation for secondary panels with the installation of secondary struts **FIGURE A-16**.

Shoring Team. This team installs at least two struts on secondary panel sets. If voids are present in the secondary shoring area, the shoring team follows the shoring void procedures.

Appendix A

FIGURE A-16 Secondary shoring panels in place.
Courtesy of Ron Zawlocki.

FIGURE A-17 An angled wall.
Courtesy of Ron Zawlocki.

Step 4: Complete Shoring

The teams work together to expand the safe zone with panels and struts as needed. Supplemental shoring is installed as needed during extrication operations. Finally, the team performs a shoring performance assessment. All struts are physically inspected twice during the first hour and once per hour thereafter.

Angled Walls

Vertical walls (90 degrees) and near vertical walls (70–89 degrees) are conducive to standard rescue shoring practices—rescue panels, struts, and backfill. A vertical or near vertical wall is called a **high-angle wall**. When the angle of the wall or walls is less than 70 degrees, a sliding force greater than the friction resistance between the dirt and panel exists. Additional resistance against this sliding force must be provided if traditional rescue shoring practices are to be used. When the wall angle is less than 70 degrees, it is called a **low-angle wall**. Once the walls reach an approximately 45-degree angle, the chance of collapse is substantially reduced, if not eliminated. A low-angle wall may not need support, but panels may be needed on a low-angle wall to resist forces from an opposing trench wall **FIGURE A-17**.

Installing panels and struts on low-angle walls using traditional rescue shoring practices is dangerous. Any load from the earth transmitted into the shoring system will exert a sliding force and there will be insufficient resistance to counteract it. In this section, we will explore options for the application, precautions, preparation, and procedures needed to strut low-angle walls. We will present two methods of resolving the sliding force that occurs on low-angle walls. The picket method resolves the sliding of panels by installing pickets through predrilled holes in the panels and penetrating the intact trench face a minimum of 36 inches (0.9 m). The counterweight method resolves the sliding of panels by using the weight of heavy equipment that is available onsite (e.g., excavators, backhoes, loaders, etc.). Both methods require the addition of either cleats, nails, or pinned strut bases to prevent sliding at the strut base or panel interface.

Preparation: Determine Strut Insertion Point on Angled Wall

Strut to panel connection points on a low-angle wall (<70 degrees) must be capable of resisting sliding forces. For traditional trench rescue shoring panels, that require the use of **cleats**, and for excavation panels, that requires the use of rail latches or wale stop blocks. All cleats, rail latches, and stop blocks should be installed from the outside of the trench before the panels are installed. The cleats, rail latches, and wale stop blocks must be accurately placed to ensure compliance with the strut spacing found in the tabulated data. They are placed at insertion points on panels that will be installed on the angled walls. The angled wall insertion point chart is designed to provide the insertion points based on the angle of the wall. An example is provided to demonstrate how the angled wall insertion point chart should be utilized.

Utilizing the Angled Wall Insertion Point Chart

First, use an angle finder to determine the angle of the wall. Then determine the strut placement (distance from the trench floor). Finally, find the needed strut distance from the trench floor and the wall angle on the angled wall insertion point chart and determine the insertion point on the angled wall **TABLE A-1**.

For example, if the angled wall is 65 degrees, a strut is needed at 2 feet (0.6 m) above the floor. Find 2 feet (0.6 m) in the *Distance from Trench Floor* Depth column and follow the row to the 65-degree angle (70–60 degrees). The insertion point on the angled wall is at 2' 3" or 27 inches (69 cm). A strut is needed at 6 feet (1.8 m) above the floor. Find 6 feet (1.8 m) in the *Distance from Trench Floor* Depth column and follow the row to the 65-degree angle (70–60 degrees). The insertion point on the angled wall is at 6' 6" or 78 inches (198 cm).

Determine Strut Length

In angled wall conditions, the distance between the walls increases when moving from the bottom of the trench (trench floor) to the top (trench lip). To determine the strut lengths that will be needed to strut the low-angle walls, begin by determining the *baseline strut length*, which is the distance between the original trench walls. If an existing vertical wall exists beyond the collapse, then measure the distance between them and use that distance as the baseline strut length.

If a vertical wall does not exist beyond the collapse, a baseline strut length can be obtained without entering the trench by inserting a straight object (panel, timber, ladder, etc.) into the toe of the trench at the angled wall and standing it vertically. Placing the straight object and measuring the distance must be done while working on a lip bridge. Follow these steps to obtain the baseline strut length:

1. Measure the distance between the object inserted at the toe of the angled wall and the opposite trench wall. Use that distance as the baseline strut length.
2. Use an angle finder to determine the wall angles.
3. Use the angled wall strut length chart for the strut depths and follow the row to the wall angle. For one angled wall, use the baseline strut length and add it to the lengths listed under the wall angle. **TABLE A-2** assumes one vertical wall and one angled wall. If both walls are angled, use the sum of both additional lengths. For two angled walls, use the baseline strut length and add it to the lengths listed under the wall angle.

The Picket Method

This method utilizes pickets driven into holes that have been drilled in the panels and into the intact angled wall(s) to resolve the sliding force. Lower wall angles result in more sliding force, which will require more pickets. For this method to work, the following conditions must be present:

- The soil must be solid and compact (not flowing, running, or excessively fractured)
- The trench must be wide enough for a 42-inch (107-cm) picket to be driven 36 inches (91 cm) into place.

TABLE A-1 Angled Wall Strut Insertion Points

Depth	70-60°	60-50°	50-45°
18' (5.5 m)	20' (6.1 m)	22' (6.7 m)	24'-6" (7.4 m)
14' (4.2 m)	15'-6" (4.7 m)	17' (5.2 m)	19' (5.8 m)
10' (3.0 m)	11' (3.3 m)	12'-3" (3.7 m)	13'-6" (4.1 m)
6' (1.8 m)	6'-6" (2.0 m)	7'6" (2.3 m)	8'-3" (2.5 m)
2' (0.6 m)	2'-3" (0.7 m)	2'-6" (0.8-2.7 m)	3' (0.9 m)

TABLE A-2 Angled Wall Strut Length Chart

Distance from Trench Floor	70° Angle	65° Angle	60° Angle	55° Angle	50° Angle	45° Angle
18' (5.4 m)	add 6'-8" (2.0 m)	add 8'-5" (2.5 m)	add 10'-5" (3.2 m)	add 12'-8" (3.9 m)	N/A	N/A
14' (4.2 m)	add 5'-2" (1.6 m)	add 6'-6" (2.0 m)	add 8'-2" (2.5 m)	add 9'-10" (3.0 m)	add 11'-9" (3.6 m)	add 14' (4.2 m)
10' (3.0 m)	add 3'-8" (1.1 m)	add 4'-9" (1.4 m)	add 5'-10" (1.8 m)	add 7' (2.1 m)	add 8'-5" (2.5 m)	add 10' (3.0 m)
6' (1.8 m)	add 2'-3" (0.7 m)	add 2'-10' (0.9 m)	add 3'-6" (1.0 m)	add 4'-3" (1.3 m)	add 5' (1.5 m)	add 6' (1.8 m)
2' (0.6 m)	add 9" (0.2 m)	add 1' (0.3 m)	add 1'-3" (0.4 m)	add 1'-8" (0.5 m)	add 1'-9" (0.5 m)	add 2' (0.6 m)

Precautions

Angled walls create hazards to rescuers during both the shoring installation process and during the victim extrication process. Rescuers must recognize these hazards and must follow the listed precautions:

- Rescuers working on ladders are exposed to potential collapse while installing pickets, therefore interim struts (temporary) must be installed to provide protection from collapse.
- Interim struts are installed at equal angles to both walls (splitting the angle).

Preparation

Shoring a low-angle wall requires a few additional preparation steps. These steps include the following:

- Measure the distance from the lip of the angled wall(s) to the cave-in pile (debris) or trench floor (along the angled trench face). That distance will be significantly different than the vertical distance from the lip to the cave-in pile or floor.
- Use an angle finder (handheld or app on smart phone) to determine the angle of the wall. This can be done visually from the end of the trench by aligning the trench wall with the angle finder.
- Use the angled wall strut/cleat insertion point chart to determine the insertion points.

SAFETY TIP

If aluminum wales are bolted to engineered plywood, only wale/base connectors that have been tested under angled wall conditions may be used **FIGURE A-18**.

FIGURE A-18 Engineered plywood and bolted-on aluminum wales with the initial pickets in position.
Courtesy of Ron Zawlocki.

- If traditional trench rescue shoring panels are used, nail the cleats at the strut insertion points (in a 14-nails).

Picket Method Walkthrough

Step 1: Trench Size-Up. The purpose of the trench size-up is to gather the information needed to strut the trench. That information includes the six standard assessments listed previously.

Step 2: Primary Shoring. The panel team does the following:

- Installs lip protection.
- Prepares panels for angled walls.
- Attaches ropes to all primary panels and lower into the trench to protect the victims head and chest.
- Installs backfill as needed.
- After the interim struts are installed, the team works from ladders to install the pickets, working from the top down.

The shoring team does the following:

- Measures the trench width and places a ladder.
- Prepares struts and the strut air system.
- If using trench rescue shoring panels, a swivel base (near vertical wall) and a hinge base (angled wall) will be needed for each strut.
- If using excavation panels, a swivel base (near vertical wall) and a rail latch base (angled wall) will be needed for each strut.
- Installs interim struts to protect the rescuers installing pickets.
- Installs positioning, backfill, and compliance struts (as needed) on the primary panel sets.
- Removes interim struts.

Step 3: Secondary Shoring. The panel team does the following:

- Adds secondary panels to both sides of the primary panel set (when possible) to develop a safe zone of at least 12 feet (3.7 m) wide (6 feet [1.8 m] on both sides of the victim)
- Adds pickets and backfill (as needed) to the secondary panel sets following the previous primary shoring procedure

The shoring team does the following:

- Installs interim struts to protect rescuers installing pickets on the secondary panel sets.
- Installs positioning, backfill, and compliance struts (as needed) on secondary panel sets.
- After pickets and backfill are placed, the team brings strut activation forces up to 1000 pounds (454 kg).
- Removes interim struts.
- Sends a shoring team member into the trench along with an entry team member (to perform victim care) to place four nails in each strut base.

Step 4: Complete Shoring. The teams work together to do the following:

- Expand the safe zone with panels and struts as needed.
- Nail all strut bases (four nails per base).
- Install supplemental shoring as needed during extrication operations. Cleats that are 24 inches (61 cm) long must be nailed onto the strongbacks of the supplemental panels used on the low-angle wall.

Finally, a shoring performance assesment is performed, and all struts are physically inspected twice during the first hour and once per hour thereafter.

Counterweight Method

This method uses a cantilever beam design placed on the lip of the angled wall and the weight of an excavator to resolve the sliding force. This concept includes the building of a grillage of beams and girders that will transfer the resistance of the excavator bucket to the tops of the panels on the angled wall. Beams will run perpendicular to the panels, resisting the force of the panels trying to slide up the angled wall. A girder is placed over the beams to transfer the resistance from the bucket to up to three beams (typically primary and secondary shoring length). Lower wall angles result in more force on the pickets. For this method to work, the soil must be compact and there must be sufficient room to place an excavator behind the angled wall.

Precautions

Angled walls create hazards to rescuers during both the shoring installation process and during the victim extrication process. Rescuers must recognize the following hazards and must follow these precautions:

- The tracks of an excavator or the outriggers of a backhoe must be outside of the simple L to minimize the effects of the excavator's weight (surcharge). The bucket (counterweight) should not be closer to the trench face than 3 feet (0.9 m) or one-half the simple L distance.
- Excavator must be in good mechanical condition and be able to support loads without hydraulic fluid leak-down.
- The surcharge load and vibrations associated with positioning an excavator on an unstable trench lip may cause soil failure and collapse.

Preparation

Shoring a low-angle wall using the counterweight method requires a few additional preparation steps. Those steps include the following:

- Measure the distance from the lip of the angled wall(s) to the cave-in pile (debris) or trench floor. That distance will be significantly different than the vertical distance from the lip to the floor.
- Use an angle finder (handheld or app on smart phone) to determine the angle of the wall.
- Use the charts to determine the strut/cleat insertion points, and the angled wall strut length chart **TABLE A-3** and **TABLE A-4**.
- If traditional trench rescue shoring panels are used, nail the continuous cleats at the strut insertion points (five-nail pattern).
- Bolt the aluminum wales to the engineered plywood panels **FIGURE A-19**.
- If aluminum wales bolted to engineered plywood are used, only wale/base connectors that have been tested under angled wall conditions may be used **FIGURE A-20**.

Counterweight Method Walkthrough

Step 1: Trench Size-Up. The purpose of the trench size-up is to gather the information needed to strut the trench. That information includes the standard six assessments detailed previously.

Step 2: Primary Shoring. The entry team first determines the skill level and the emotional stability of the heavy equipment operator. If the heavy equipment operator is suitable for the task, the entry team

FIGURE A-19 Bolting the aluminum wales to engineered plywood.
Courtesy of Ron Zawlocki.

TABLE A-3 Strut/Cleat Insertion Points

Depth	70-60°	60-50°	50-45°
18′ (5.4 m)	20′ (6.1 m)	22′ (6.7 m)	24′-6″ (7.5 m)
14′ (4.3 m)	15′-6″ (4.7 m)	17′ (5.2 m)	19′ (5.8 m)
10′ (3.0 m)	11′ (3.3 m)	12′-3″ (3.7 m)	13′-6″ (4.1 m)
6′ (1.8 m)	6′-6″ (2.0 m)	7′6″ (2.3 m)	8′-3″ (2.5 m)
2′ (0.6 m)	2′-3″ (0.7 m)	2′-6″ (0.8 m)	3′ (0.9 m)

TABLE A-4 Angled Wall Strut Length Chart

Depth	70°	65°	60°	55°	50°	45°
18′ (5.4 m)	add 6′-8″ (2.0 m)	add 8′-5″ (2.6 m)	add 10′-5″ (3.2 m)	add 12′-8″ (3.9 m)	N/A	N/A
14′ (4.3 m)	add 5′-2″ (1.6 m)	add 6′-6″ (2.0 m)	add 8′-2″ (2.5 m)	add 9′-10″ (3.0 m)	add 11′-9″ (3.6 m)	add 14′ (4.3 m)
10′ (3.0 m)	add 3′-8″ (1.1 m)	add 4′-9″ (1.4 m)	add 5′-10″ (1.8 m)	add 7′ (2.1 m)	add 8′-5″ (2.6 m)	add 10′ (3.0 m)
6′ (1.8 m)	add 2′-3″ (0.7 m)	add 2′-10″ (0.6 m)	add 3′-6″ (1.1 m)	add 4′-3″ (1.3 m)	add 5′ (1.5 m)	add 6′ (1.8 m)
2′ (0.6 m)	add 9″ (0.2 m)	add 1′ (0.3 m)	add 1′-3″ (0.4 m)	add 1′-8″ (0.5 m)	add 1′-9″ (0.5 m)	add 2′ (0.6 m)

*After the baseline length is determined use your angle finder to determine the degree of the wall.
*Use the chart above to determine the length of the struts that will be needed at each depth (distance from the trench floor).
*Add the baseline strut length to the lengths in this chart.

FIGURE A-20 Counterweight beams positioned directly over the aluminum wales.
Courtesy of Ron Zawlocki.

develops a communication plan with the operator (see Chapter 7, *Operations Level Trench Rescue Shoring*, for a discussion on working with heavy equipment operators).

Next, the team must determine the suitability of the equipment and mechanical soundness. The heavy equipment operator should push the bucket down in the ground, lifting the front of the machine in the air a few inches. This must occur far away from the trench. Then shut off the equipment with the front in the air to verify that there is no hydraulic fluid leak-down of the arm/bucket over 5 minutes.

After the primary panels are installed on the angled wall, the entry team positions a beam on the trench lip that runs perpendicular to the trench wall and is directly over the panel strongback or vertical aluminum wale (strongback cut flush with top of panel). The team measures 4 feet (1.2 m) (on center) on both sides of the beam and positions additional beams at spots parallel to the first beam (for support of secondary panels). A 7 × 7-inch LVL girder (at least 12 feet (3.7 m) long) is placed perpendicular to and on top of the beams and outside of the exclusion zone. The positioning of the excavator or backhoe is directed outside of the exclusion zone so that the bucket is centered on the girder sitting on the beams.

The panel team installs lip protection and prepares panels for angled walls. The team attaches ropes to all panels and lowers the panels into the trench to protect the victim's head and chest. The panel team also adjusts or cuts the panel(s) and strong-back(s) on the angled wall side so that they are flush with the trench lip.

The shoring team measures the trench width and places ladder. The team prepares struts and the strut air system. The shoring team also installs positioning, backfill, and compliance struts (as needed) on the primary panel sets.

Step 3: Secondary Shoring. The panel team adds panels to both sides of the primary panel set (when possible) to develop a safe zone of at least 12 feet (3.7 m) (6 feet [1.8 m] on both sides of the victim). The secondary panels are slid under the counterweight beam placed with the primary panel. The panel team adds pickets and backfill (as needed) to the secondary panel sets.

The shoring team installs positioning, backfill, and compliance struts (as needed) on the secondary panel sets. After all struts and backfill are in place, the shoring team brings all strut activation forces up to 1000 pounds (454 kg). A shoring team member is sent into the trench along with entry team members to provide victim care and to place two nails in each swivel base.

Step 4: Complete Shoring. The teams work together to expand the safe zone with panels and struts as needed. All strut bases are nailed, and supplemental shoring is installed as needed during extrication operations. Cleats that are 24 inches (61 cm) long must be nailed onto the strongbacks of the supplemental panels used on the low-angle wall. A shoring performance assessment is completed by physically inspecting all struts twice during the first hour and once per hour thereafter.

Extreme Shoring Conditions

Extreme shoring conditions can be defined simply as conditions that exceed the capabilities of prescribed trench rescue shoring. Unfortunately, there is no one size fits all checklist to know when you are encountering extreme conditions. Typically, it is a combination of several factors. These items include the following:

- Moving soil (even after the installation of shoring)
- Flowing soil
- Uncontrollable water
- Shoring that is continually loosening or moving
- Overhanging soil conditions that cannot be shored with typical methods
- Overhanging appurtenances (pavement, equipment, or materials) that cannot be shored or quickly removed
- Unstable spoil piles that are falling into the excavation/trench
- Trench depths greater than 20 feet (6.1 m)
- Excavations greater than 16 feet (4.9 m)
- Excavations in the side of tall, steep, or unstable slopes (hills)
- Failure of an engineered shoring system

Because extreme conditions are uncommon and they often include a combination of items that rescuers are not used to seeing, it is difficult to immediately make a classification upon arrival at a trench emergency scene. However, the sooner that extreme conditions can be identified, the sooner that the needed resources to resolve them can be on the scene. Extreme conditions are often determined during the shoring process. Basically, if shoring is so difficult that nothing is working despite multiple attempts, even with the best trench technicians onsite, then the emergency has extreme conditions. When nothing is working, a higher focus must be put on the safety of the rescuers. This will likely drive the incident commander to alternate solutions.

When extreme conditions exist, the probability of a rescue is reduced, and the likelihood of a body recovery is increased. When that is the case, a shift from rescue to recovery mode will allow for the time needed to bring in alternate methods of protection and hazard mitigation, which includes using construction-based protective systems.

Construction Protective Systems

In extreme conditions, typical trench rescue shoring equipment and techniques cannot provide the needed level of safety. In these cases, professional engineers and construction contractors should be called to assist with the rescue or recovery operation. It is important to remember that construction-based protective systems are meant to be implemented with no one in the trench. That is because there is a relatively high probability of injury or death to someone in the trench resulting from the use of heavy equipment (collapse caused by surcharged loads or being hit by moving equipment). Keep in mind that construction workers do not rescue people for a living, and they do not implement their protective system with people in the trench. Likewise, rescuers are not familiar with the use of construction-based protective systems. Safe operations at trenches with extreme conditions require the use of both trench rescue technicians and construction workers. Combining rescue and commercial shoring techniques requires planning and coordination between rescuers and the construction workers.

It is important to note that construction workers will *not* have knowledge of the T-L method that is presented in this text. Instead, the construction workers and their engineers will consider the maximum potential earth pressures, which are calculated according to field conditions and type of shoring to be used.

Protective systems must be capable of withstanding all intended and reasonably expected loads. Predesigned systems are engineered to withstand certain forces, which are determined by the manufacturer to reach a desired depth of excavation. The decision regarding the type of protective system to use may vary based on the type of rescue operation.

The selection of protective systems for rescue or body recovery operations is based on a set of factors that are evaluated during each operation. These factors may include, but are not limited to, the following:

- Adjacent structures
- Existing hazards
- Soil type
- Water profile and hydraulic table
- Depth and width of the trench
- Purpose of operation (rescue versus utility installations)

Use of Construction Protective Systems at Rescue Scenes

Construction protective systems require the use of heavy equipment and highly skilled heavy equipment operators. Heavy equipment is common at every construction site. When the decision is made to utilize construction protective systems to make the trench safe for rescue or recovery operations, rescuers must work hand in hand with heavy equipment operators. Sometimes the best answer may be utilizing a combination of techniques.

Sloping and Benching Systems

Sloping and **benching** are similar methods used to protect underground construction workers from cave-in. Both techniques utilize the same principle, which removes or minimizes the dangerous vertical sections of the trench walls. The walls are angled to a point where the material will support its own weight and will no longer collapse or flow—the so-called angle of repose or maximum allowable angle. Sloping cuts the excavation back to a relatively smooth slope, while benching is like stair steps, where the steps are generally no more than 5 feet (1.5 m) in height and the line connecting the corners of the steps meets the desired sloping angle. The required slope is dependent on the type of soil present. Determination of the safe angle for a specific soil type is based on soil analysis completed by the contractor's competent person and engineered data (angle chart) or state OSHA guidance. Because benching and sloping remove all the cave-in hazards, it is a safe method, but often, there will not be enough room to implement sloping or benching in developed (urban or suburban) areas in which structures and infrastructure are in close proximity.

Sloping and benching are not typically used during rescue operations, but rather may be more appropriate for body recovery. Because sloping and benching expose a live victim to the risks of falling soil or additional collapse, their proper role in rescue operations must be determined by completing a risk–benefit analysis. If the use of the sloping or benching is deemed necessary, the following issues must be considered:

- A complete soil assessment, including determination of the maximum allowable angle, will require the use of a pocket penetrometer by a competent person. If a soil assessment by a competent is not readily available at a rescue situation, consider the soil to be distressed Type C soil.
- The added surcharge and vibrations from the heavy equipment needed to slope or bench may cause additional collapse during the digging.
- The site needs to have sufficient room for equipment placement to excavate the slope or bench and space for the spoils.
- Sloping and benching are both very time-consuming operations.
- Sloping and benching significantly expand the area of excavation and, when considering spoils, take a lot of room.
- Sloping or benching during rescue operations should be used in conjunction with an isolation device whenever possible.

Sloping and benching can be used in any soil that is not sand or running mud and implemented to just about any depth the equipment can reach. The limitations on their use are based on the soil type and the amount of room around the trench needed to obtain the necessary slope. Using this method for recovery operations will require knowing the limits of the original excavation to avoid digging into the area of the victim.

TIP
Sloping is referred to as "cutting back to the maximum allowable slope"—that is, the point where the material can support its own weight and is not expected to move.

SAFETY TIP
Distressed soil is any soil that shows signs of failure, including collapse, fissures, sloughing, bulging, raveling, and heaving.

Trench Box

Trench boxes are commonly found at trenching sites **FIGURE A-21**. This equipment usually contains steel or aluminum solid side walls and spreaders. Trench boxes come in a variety of lengths, heights, widths, side wall thickness, and weights. Unfortunately, they often spend more time above grade than below grade during underground construction work.

The selection and use of a trench box are based on soil type and trench width and depth. The criteria for use are based on the manufacturer's data and the analysis of a competent person. A trench box is a safe and effective method for protecting construction workers in a trench. Using one at a rescue scene, however, requires a well-developed plan. If the use of a trench box is deemed necessary, the following issues should be considered:

- Installation of a trench box may cause additional collapse. The issues of surcharged (backhoe/excavator) loads, vibrations (heavy equipment), and collision of the trench box with the trench wall must be considered and resolved.

FIGURE A-21 The trench box is a stable piece of commercial equipment used to provide worker protection.
© Jones and Bartlett Publishers. Courtesy of MIEMSS.

- Lowering or dragging a trench box, which can weigh several tons, into the area of a trench containing a victim may cause additional victim trauma or entrapment. The rescue team must have strong indications that the trench box insertion will stay clear of the victim's position and location.
- The gap between the trench wall and trench box sides should be eliminated. Backfill (e.g., air bags, backstruts, timber/soil) may be required to resolve larger gaps.
- The top of the box must either extend above the lip of the trench or any soils above the trench box must be cut back to their angle of repose (or otherwise be protected).
- The bottom of the trench box may be a maximum of 2 feet (0.6 m) from the bottom of the trench, and only if a competent person determines the soil will stand with no possible loss from behind or below the shield.
- The trench box height may not be extended with steel plates unless approved by an engineer.
- The box must be 4 feet (1.2 m) longer than the rescue work area.
- Rescuers may never be outside the protection of the trench box or other protected area of the trench.

Modular Shields

Modular shields are segmental trench boxes that are assembled at the scene and are rated for human safety. They are often present on trailers at the trench site. Modular aluminum or steel shoring is available in a variety of shapes and configurations based on the size of the excavation and the type of soil in which they will be installed. In addition to coming in fixed sizes, some of these units are air or hydraulically adjustable. Modular shields can be purchased with four-wall protection.

Modular systems can be quickly assembled and lowered into the trench for use as rescue isolation devices or safe zones for rescuers. Once the shield is built, it can be lowered into the trench with an excavator. Lightweight aluminum modular shields can be manually installed with ropes or in conjunction with ladder A-frames, gantries, and rope-based mechanical advantage systems. Many modular shields can be stacked, and some can be purchased with legs that can be adjusted to accommodate for uneven trench floor surfaces.

Isolation Devices

The term isolation device is specific to rescue operations. The use of concrete manhole segments, pre-cast concrete catch basins, steel or concrete pipe, and concrete vaults to provide a measure of protection in a trench are not OSHA-compliant techniques. However, their use can provide a protective barrier for a victim involved in a trench accident. Experienced trench rescue personnel working in conjunction with heavy equipment operators may find it necessary to develop isolation devices improvised from material and equipment found at the construction site. The use of this type of isolation device is a temporary or last-resort option.

Using Heavy Equipment to Install a Protective Device

Improper installation of a protective device (trench box, modular shield, or isolation device) can cause additional collapse and/or injury to the victim(s). Placement of a protective device requires a cool, calm, and collected heavy equipment operator. If the device is placed on a victim, it may injure buried (unseen) extremities. Once placed, the device needs to be braced to the trench walls or use backfilling techniques to prevent the device from shifting or sliding if a collapse occurs.

The following walkthrough is an example of a protective device installation procedure. Rescuers *must* have hands-on training and procedures approved by the AHJ for installing an isolation device over a trapped victim. First, the entry team officer meets with the heavy equipment operator to determine their fitness for use during the rescue or recovery. If the equipment operator is suitable, then the entry team officer conducts a briefing with heavy equipment operator, signal person, and rescuers assigned to tag lines.

The heavy equipment operator performs a pre-lift to check rigging, stability, and capacity of the equipment to safely lift and place the isolation device. Rescuers attach tag lines.

The signal person to gives directional orders to the heavy equipment operator.

When the protective device is not moving or swaying, the heavy equipment operator is directed to slowly lower it over the victim while tag line attendees stabilize and guide the isolation device into place.

The protective device must extend above the top of the trench wall and must be large enough for a rescuer to be lowered into the device to begin assisting and freeing the victim. If a rescuer cannot safely enter the protective device, the trench walls may be cut to the maximum allowable angle or sheeted and shored as resources become available. The area between the protective device and the trench walls should be backfilled to prevent collapsing material from moving the device.

Additional considerations include the following:

- The added surcharge and vibrations from the heavy equipment may cause additional collapse.
- If a manhole segment is used, the fixed ladder or rungs inside the ring may hit and injure the victim during installation.
- The device selected must be large enough to fit over the victim with room to spare, yet small enough to fit into the excavation area. If the top of the protective device is below the trench lip, a collapse may turn the device into a coffin.

Soldier and Sheet Piling

Soldier piling and sheet piling are fixed shoring systems that support trench walls. They include sheet piles that can be cantilevered, braced, or tied back to provide ground support **FIGURE A-22**. Uprights made of steel piling or steel sheeting can be pushed into the ground at a depth that is appropriate for the type of soil and depth of the excavation. Piling is installed by pushing it in with a backhoe; drilling a hole to fit the pile; or using an impact, vibrating, or hydraulic hammer. The main advantage of this type of shoring system is that it can easily be installed around existing utilities. The use of soldier piling with lagging at rescue scenes is unusual, because planning and installation is a time-consuming process. It would be used only in the situations where it is the only feasible method to ensure safety. A modified version of sheet piling, using road plate in combination with trench boxes, is occasionally used in the transition from rescue shoring to construction shoring techniques.

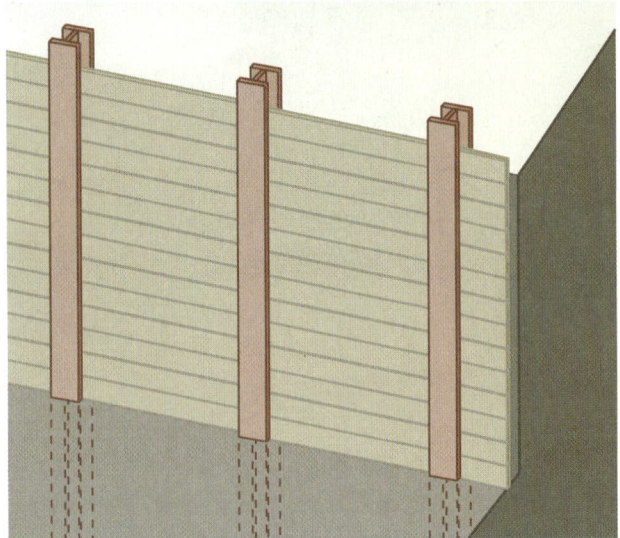

A

B

FIGURE A-22 A. Soldier pile shoring consists of a set of horizontally installed wales held in place by a set of vertically installed piles. **B.** Interlocking steel sheet pile shoring is a commercial technique typically used in deep trenches subject to tremendous compressive forces. It is seldom used at rescue incidents.
B. Courtesy of Chuck Wehrli

Engineered Shoring Systems

The shoring systems that have been presented in in this text describe a trench rescue shoring system that will handle many trench collapse incidents. When extreme conditions are encountered, bringing in an experienced underground contractor, along with their equipment and their professional engineer, is the likely next step. At times, there can be a middle ground where the trench rescue shoring standards can be applied and built, with some customization, but site-specific conditions leave the rescue team in a place where the team is unsure of potential loading or capacity. Either of these two situations will require support from a knowledgeable engineer.

Having an engineer who is familiar with the rescue team, the team's abilities, and the equipment that they possess is important in assisting in these situations. The team will be best served if they can develop a relationship with a local engineer and work with them regularly before the need arises. The T-L method presented in this text is now being covered in the Urban Search and Rescue (USAR) Structures Specialist training classes. Starting with a local USAR Structures Specialist would be a great first step in finding a qualified engineer.

If a rescue team is in an extreme condition situation and becomes involved with an unfamiliar engineer, the experience of the engineer will need to be evaluated. Engineers are a lot like doctors, there are many different specialties, and within those specialties, there can be greatly different levels of experience and ability. Some may be risk tolerant, others may be conservative. An engineer may come to the site to help but may not have much experience dealing with soils and structures.

When it comes to trench rescue shoring, an engineer with a combination of geotechnical (soils) and structural engineering is ideal. A geotechnical engineer would have a strength in determining soil stability and potential soil pressures. A structural engineer would have an expertise in the shoring equipment side. It is also important to understand that structural engineers can have specialties relating to the materials of structures. Many will specialize in steel or concrete, but experience with wood may be less common.

The team will want to size up the engineering resource. Most engineers will *not* have knowledge of the T-L method presented earlier, so that alone cannot be used to gauge the engineer's suitability. The following list contains some questions that can be asked in a nonconfrontational way to help assess the engineer's suitability. The response to these questions should provide an idea of whether the engineer is experienced and practical for trench rescue.

- What do you think might fail next?
- What soil classification are you using for this trench?
- What lateral soil pressure are you going to use?

In addition, the following list contains some questions to ask to help assess the engineer's shoring knowledge:

- Are you familiar with our trench rescue shoring systems?
- Are you familiar with engineered lumber (LVL and arctic birch plywood construction) capacities?
- What kind of cave-in protection do you think will be needed?

KEY TERMS

Benching A method of protecting workers from cave-ins by excavating the side of a trench or excavation to form one or a series of horizontal levels or steps, usually with vertical or near vertical surfaces between levels.

Cleats Strips of wood that are used to support and resist the movement of struts.

Extreme shoring conditions Any soil condition, trench depth, or trench width that exceeds the limits of default trench rescue shoring equipment and techniques.

High-angle wall Trench walls that are vertical or near vertical having angles between 70 and 90 degrees.

Low-angle wall Trench walls with angles less than 70 degrees.

Header image: Courtesy of Ron Zawlocki.

Appendix B
NFPA 1006 and 1670 Correlation Guide

NFPA 1006 *Standard for Technical Rescue Personnel Professional Qualifications, 2021 Edition*		
Awareness	Chapter(s)	Page(s)
12.1	1–4	3–91
12.1.1	1, 2, 3	5, 33–35, 47–49
12.1.1(A)	1, 2	5, 33–35
12.1.1(B)	3	47
12.1.2	2, 3, 4	19–36, 47, 55–57, 61, 73–82
12.1.2(A)	2, 4	19–36, 55–57, 61, 73–82
12.1.2(B)	3	47, 57, 61
12.1.3	1, 2, 3, 4	10–12, 15, 19–36, 47–48, 53–55, 70–73
12.1.3(A)	1, 2, 3, 4	10–12, 15, 19–36, 47–48, 53–55, 70–73
12.1.3(B)	4	87–88
12.1.4	3	45–53
12.1.4(A)	1, 3	5–9, 45–53
12.1.4(B)	3	46–51
12.1.5	2, 3, 4	33–36, 55–57, 66–69, 70–73
12.1.5(A)	2, 3, 4	33–36, 55–57, 66–69, 70–73
12.1.5(B)	4	76
12.1.6	1, 3, 4	7–8, 9–10, 51–53, 73–82

Appendix B

Awareness	Chapter(s)	Page(s)
12.1.6(A)	1, 3, 4	4–5, 51–53, 73–82
12.1.6(B)	3	46–51
12.1.7	3	55–57
12.1.7(A)	3	55–57
12.1.7(B)	4	66–69

Operations	Chapter(s)	Page(s)
12.2	5, 6, 7, 8	94–204
12.2.1	5	95–98, 99–110
12.2.1(A)	5, 7	95–98, 99–110, 133–143
12.2.1(B)	5	103, 104–105
12.2.2	5, 7, 8	98, 99–102, 106, 99–110, 145–147, 155–158, 161–163, 201–202
12.2.2(A)	5, 7, 8	98, 99–102, 106, 99–110, 133–147, 155–158, 161–163, 201–202
12.2.2(B)	5, 7	101, 103, 104–105, 107, 108–110, 157–158
12.2.3	7	145–147, 155–156, 161–163, 178–179
12.2.3(A)	7	133–147, 155–156, 161–163, 178–179
12.2.3(B)	7	145–178
12.2.4	7	144–145, 155–160, 163–174
12.2.4(A)	5, 7	99–102, 106, 144–145, 155–160, 162–174, 177, 178–179
12.2.4(B)	7	157–158, 159–160, 163–165, 166, 167–169, 171–173, 174, 177
12.2.5	8	186–187, 189–194, 197–198, 202
12.2.5(A)	8	186–187, 189–194, 197–198, 202
12.2.5(B)	5, 8	99–102, 189–190, 193–194
12.2.6	8	198, 201–202
12.2.6(A)	8	198, 201–202

Operations	Chapter(s)	Page(s)
12.2.6(B)	8	201–202
12.2.7	6	125
12.2.7(A)	6	125
12.2.7(B)	6	125
12.2.8	6	121, 125
12.2.8(A)	6	121, 125
12.2.8(B)	6	121, 125
Technician	**Chapter(s)**	**Page(s)**
12.3	9, 10	206–271
12.3.1	10	232–253, 256–262
12.3.1(A)	10	232–253, 256–262
12.3.1(B)	10	145–178
12.3.2	10	232–253, 256–262
12.3.2(A)	10	232–253, 256–262
12.3.2(B)	10	143–145
12.3.3	10	253–262
12.3.3(A)	10	253–262
12.3.3(B)	10	257–260
12.3.4	10	253–262
12.3.4(A)	10	253–262
12.3.4(B)	10	257–260
12.3.5	10	233–253, 256–262
12.3.5(A)	10	233–253, 256–262
12.3.5(B)	10	143–145
12.3.6	10	256–262
12.3.6(A)	10	256–262

Technician	Chapter(s)	Page(s)
12.3.6(B)	10	262
12.3.7	10	262–263
12.3.7(A)	10	262–263
12.3.7(B)	10	262–263
12.3.8	9	210–212, 214–219
12.3.8(A)	9	210–212, 214–219
12.3.8(B)	9	211, 214–216
12.3.9	9	207–212, 219–224
12.3.9(A)	9	207–212, 219–224
12.3.9(B)	9	219–225
12.3.10	9	226–228
12.3.10(A)	9	226–228
12.3.10(B)	9	219–222
12.3.11	10	232
12.3.11(A)	10	232
12.3.11(B)	10	232

NFPA 1670 *Standard on Operations and Training for Technical Search and Rescue Incidents*, 2017 Edition

Objective	Chapter(s)	Page(s)
11.1	1–4	3–91
11.2	1–4	3–91
11.2.1	1–4	3–91
11.2.2	1–4	3–91
11.2.3	1–4	3–91
11.2.3(1)	3	42–61
11.2.3(2)	3	52–55
11.2.3(3)	3	50–55

Objective	Chapter(s)	Page(s)
11.2.3(4)	3	55
11.2.3(5)	2, 3	18–36, 47–56
11.2.3(6)	2	18–36
11.2.3(7)	3	57–61
11.2.3(8)	2	18–36
11.2.3(9)	3	55
11.3	5, 6, 7, 8	94–204
11.3.1	5, 6, 7, 8	94–204
11.3.1(1)	5, 6, 7, 8	94–204
11.3.1(2)	5, 6, 7, 8	94–204
11.3.1(3)	5, 6, 7, 8	94–204
11.3.2	7	129–182
11.3.2(1)	7	129–182
11.3.2(2)	7	129–182
11.3.2(3)	7	129–182
11.3.3	5, 6, 7	95–182
11.3.3(1)	6	116–120
11.3.3(2)	7	129–182
11.3.3(3)	5	95–113
11.3.3(4)	8	189
11.3.3(5)	7	129–182
11.3.3(6)	5	96
11.3.3(7)	7	132–134
11.3.3(8)	5	106–110
11.3.3(9)	5	95–113
11.3.3(10)	7	157–158

Objective	Chapter(s)	Page(s)
11.3.3(11)	7	157–158
11.3.3(12)	5, 8	98, 189
11.3.3(13)	8	189
11.3.3(14)	Appendix A	294–296
11.3.3(15)	Appendix A	294
11.3.3(16)	7	129–182
11.3.3(17)	8	186–202
11.3.3(18)	8	186–202
11.4	9, 10	207–264
11.4.1	9, 10	207–264
11.4.1(1)	9, 10	207–264
11.4.1(2)	9, 10	207–264
11.4.2	10	229–264
11.4.3	10	229–264
11.4.3(1)	10	229–264
11.4.3(2)	10	229–264
11.4.3(3)	5	99–100
11.4.3(4)	10	229–264
11.4.3(5)	10	229–264
11.4.3(6)	10	229–264

Appendix C
T-L Method – A Metric Guide

The measurements in this appendix are a guide and it is recommended that if an agency uses the metric system, that a qualified engineer review and verify all tabulated data based on the equipment utilized by the agency.

The T-L method relies on the failure signs observed onsite and uses those visual clues to more closely approximate the actual load that will occur in the short term of a rescue. Specifically, it uses the horizontal distance of trench failure (collapse), or signs of failure such as cracks approximately parallel to the trench lip. The "L" in the T-L method is the distance measured horizontally from the trench lip (or where it used to be) to the farthest collapse or cracking that parallels the trench, or "L distance."

The T-L method calculates the vertical weight of the block of soil bounded by farthest sign of failure (Trench depth × L distance × Length × Assumed soil density of 20.9 kN/m^3; 20.9kN/m^3 is used rather than the 18.9 kN/m^3 average to provide a greater range of soil conditions). The method applies a coefficient of 0.5 to determine lateral earth pressure. The shoring is assumed to be 1.2 × 2.4 meter panels with two struts per panel and the method produces a force where the only variable is the L distance. This provides easy equation that depends entirely on L for systems using 1.2-meter wide panels.

> Lateral force = 16.328 × L (in meters). For simplicity, this is rounded to 16.4 × L [kN]

This equation calculates the force that would act on a 1.2 × 1.2-meter section of panel, which becomes a strut load or a wale load. In cases where more than two struts are being placed on a panel, one would double the result of the equation to come up with a load on the entire panel and divide by the number of supports being placed.

If signs of failure are not visible, or possibly covered by a spoil pile, a failure depth of 0.7 times the trench depth is used to determine the L value. Assigning such a high L value to a trench that has not yet collapsed might seem counterintuitive, but in the case where no failure signs are present, the soil has not given clues and a larger potential soils force must be considered (see **TABLE C-1** for a simplified method to make this determination).

The T-L method utilizes soil pressure formulas and worst-case conditions (weight of 20.9 kN/m^3; active earth pressure coefficient of 0.5). That means a lateral force of about 3.16 kN/m^2

How To Use The T-L Method for Trench Rescue

A tape measure is used to find the simple L from the original trench face (wall) to the farthest point of soil failure. Simple L (SL) is the distance (length) measured in meters from the original trench wall perpendicular to the farthest point of soil failure or cracks/fissures. When there are multiple signs of failure, always use the farthest point as the SL. To use the T-L method for trench rescue, follow these steps:

1. From a safe area on the lip, measure the distance from the original trench wall (face) perpendicular to the farthest failure point.
2. Round that measurement up to the next 1/10th of a meter and utilize that total L distance in the shoring charts, going up to the next higher L value in the chart

Surcharge Loads (ScL)

Surcharge loads (ScL) are spoil piles and construction equipment that are within the area that is between the original trench faces and the farthest point of soil failure. ScL is measured in meters perpendicular to the trench wall. Note that construction equipment can

TABLE C-1 Depth to L Conversion Chart (Metric Version)

Trench Depth (Meter)	SL (Meter)
1.5	1.1
1.8	1.3
2.1	1.5
2.4	1.7
2.7	1.9
3	2.1
3.3	2.3
3.6	2.5
3.9	2.7
4.2	3.0
4.5	3.5
4.8	3.4
5.1	3.6
5.4	3.8
5.7	4.0
6	4.2

TABLE C-2 Surcharge Table (Metric Version)

Surcharge (ScL) Meters within Simple L (SL)

Spoil	Add to SL	Equipment	Add to SL
0.3 within SL	0.3	0.3 within SL	0.3
0.6 within SL	0.3	0.6 within SL	0.6
0.9 within SL	0.3	0.9 within SL	0.9
1.2 within SL	0.6	1.2 within SL	1.5
1.5 within SL	0.9	1.5 within SL	2.4
1.8 within SL	1.2	1.8 within SL	3.4
2.1 within SL	1.5	2.1 within SL	N/A
2.4 within SL	2.1	2.4 within SL	N/A
2.7 within SL	2.7	2.7 within SL	N/A
3.0 within SL	3.0	3.0 within SL	N/A

Note: Total L (L) = Simple (SL) plus Surcharge L (ScL)
Chart is valid for Total L of 6.0 or less.

Determining the Spoil Pile Surcharge

The T-L method includes spoil pile surcharge by approximating the weight of a spoil pile within the SL distance. An equation was developed to determine the equivalent L value to add to the SL. To determine the SL using the table, follow these steps:

1. Measure the amount of spoil pile that is within the SL.
2. Round the measurement up to the next 1/10th of a meter and go to the next higher measurement in the chart.
3. Use this measurement in the surcharge table (spoil column).
4. Table C-2 provides the corresponding value to add to the SL.

Adding the Spoil Pile Surcharge

To determine the total L (L) when a spoil pile is present, use the following steps:

1. Determine the SL. For this example, the SL is 1.2. In addition, 0.6 meters of the spoil pile is within the SL.

include, but is not limited to, excavators, dump trucks, trench boxes, and pipes.

Surcharge loads that are within the area between the original trench face and the farthest point of soil failure can add significant lateral forces to shoring systems. Those additional forces must be added to the SL to obtain the total L (L). The total L (L) is the SL plus the ScL (spoil pile and equipment), if present. The total L is used to select the proper trench rescue shoring from engineered tabulated data charts approved by the authority having jurisdiction.

TABLE C-2 allows the user to easily determine the additional value to add to the SL to account for the ScL and obtain the total L.

2. Use 0.6 meters as the spoil pile measurement and refer to the surcharge table (spoil column) to determine the additional L value that must be added to the SL. In this case, with a measurement of 0.6, the surcharge table determines an additional L value of 0.3, which results in a total L (L) of 1.5 (1.2 + 0.3).

Determining the Equipment Surcharge

Equipment also adds additional weight on the soils adjacent to the trench, increasing lateral earth pressures. Equipment surcharge loads are handled similarly to the spoil load surcharge described previously. To determine the equipment surcharge, follow these steps:

1. Measure the amount of equipment that is within the SL.
2. Round the measurement up to the next 1/10th of a meter.
3. Use the measurement in the surcharge table and go to the next higher measurement (Equipment column).
4. Use the table to locate the corresponding value to add to the SL.

Adding the Equipment Surcharge

The procedure used to calculate equipment surcharge is fairly straightforward. Consider the following example: the SL is 1.8, and 0.9 meters of the equipment is within the SL. Use 0.9 meters as the equipment surcharge L and refer to the surcharge table (equipment column) to determine the corresponding value to add to the SL. In this case, with 0.9 meters of equipment within the SL, the surcharge table tells us to add 0.9 to the SL for a total L (L) of 2.7 (1.8 + 0.9).

Trench Depth to L Conversion

As mentioned previously, when absolutely no signs of failure or distress are apparent, the SL can be taken as 70 percent of the trench depth. **TABLE C-3** provides easy-to-look-up values. It can be used by measuring the trench depth, then reading the corresponding SL equivalent value from the chart.

Tabulated Data for Shoring Equipment

Engineered tabulated data is information that has been developed by a licensed professional engineer and arranged in easy-to-read rows and columns. Valid engineered tabulated data is developed for specific purposes using appropriate theories and assumptions. It is very important to carefully read all the engineer's notes to determine if the tabulated data addresses the type of trench and soil conditions that you are dealing with at the scene. Note that the values provided are general guides only and are based on tabulated data for collapsed trench walls and rescue conditions created by the Michigan Urban Search and Rescue (MUSAR), using the imperial measurement system. Their engineered tabulated data is based on worst case soil (T-L soil) and their shoring charts may be used at all rescue incidents and in most soil conditions.

To interpret and use tabulated data from organizations such as MUSAR, rescuers only need to measure the trench failure point (SL) and calculate the Total L (L) if surcharge loads are present. Then use the charts in **TABLE C-4** through **TABLE C-13** to determine the appropriate shoring component (strut, panel, or wale). If you know what struts, panels, and wales your team carries, it becomes very easy to know a maximum Total L that each component can support.

TABLE C-3 Depth to SL Conversion Table (Metric Version)

Depth to Simple L (SL) Conversion Guide	
Trench Depth	SL Equivalent
1.2 – 2.4 meters	SL-1.8
2.7 meters	SL-1.9
3.0 meters	SL-2.1
3.3 meters	SL-2.4
3.7 meters	SL-2.6
4.0 meters	SL-2.8
4.3 meters	SL-3.0
4.6 meters	SL-3.3
4.9 meters	SL-3.5
5.2 meters	SL-3.7
5.5 meters	SL-3.9
5.8 meters	SL-4.1
6.0 meters	SL-4.2

Note: Total L (L) = Simple L (SL) plus Surcharge L (ScL)

TABLE C-4 Partech Strut Chart (Metric Version)

	Grey Struts		Gold Struts			
Total L	Width Less than or Equal to 2.4 m	Width 2.4 to 3.0 m	Width Less than 3.0 m	Width Less than or Equal to 3.6 m	Width Less than or Equal to 4.2 m	Width Less than or Equal to 4.8 m
L-0.3	1.2 m	1.2 m	1.2 m	1.2 m	1.2 m	1.2 m
L-0.6	1.2 m	1.2 m	1.2 m	1.2 m	1.2 m	1.2 m
L-0.9	1.2 m	1.2 m	1.2 m	1.2 m	1.2 m	1.2 m
L-1.2	1.2 m	1.2 m	1.2 m	1.2 m	1.2 m	1.2 m
L-1.5	1.2 m	1.2 m	1.2 m	1.2 m	1.2 m	1.2 m
L-1.8	1.2 m	1.2 m	1.2 m	1.2 m	1.2 m	0.9 m
L-2.1	1.2 m	1.2 m	1.2 m	1.2 m	1.2 m	0.9 m
L-2.4	1.2 m	1.2 m	1.2 m	1.2 m	1.2 m	0.9 m
L-2.7	1.2 m	1.2 m	1.2 m	1.2 m	1.2 m	0.6 m
L-3.0	1.2 m	0.9 m	1.2 m	1.2 m	1.2 m	0.6 m
L-3.3	1.2 m	0.9 m	1.2 m	1.2 m	0.9 m	N/A
L-3.7	1.2 m	0.9 m	1.2 m	1.2 m	0.9 m	N/A
L-4.0	1.2 m	0.9 m	1.2 m	0.9 m	0.9 m	N/A
L-4.3	1.2 m	0.9 m	1.2 m	0.9 m	0.9 m	N/A
L-4.6	1.2 m	0.6 m	1.2 m	0.9 m	0.9 m	N/A
L-4.9	1.2 m	0.6 m	1.2 m	0.9 m	0.9 m	N/A
L-5.2	1.2 m	0.6 m	1.2 m	0.9 m	0.6 m	N/A
L-5.5	1.2 m	0.6 m	1.2 m	0.9 m	0.6 m	N/A
L-5.8	1.2 m	0.6 m	1.2 m	0.9 m	0.6 m	N/A
L-6.0	1.2 m	N/A	1.2 m	0.6 m	0.6 m	N/A

Reproduced from MUSAR Training Foundation. (2020). *Trench rescue shoring operations guide* (2nd ed.). Author. https://paratech.com/wp-content/uploads/2021/01/MUSAR-SOG-2020.pdf

TABLE C-5 ResQTec Strut Chart (Metric Version)

Strut Model Total L	PxM470 PxM600 PxM880 PxM1400 Width Less than or Equal to 1.8 m	PxM470 PxM600 PxM880 PxM1400 PXM2300 Width Less than or Equal to 2.4	PxM1400 PxM2300 Width Less than 3.0 m	PxM1400 PxM2300 Width Less than or Equal to 3.6 m	PxM2300 Width Less than or Equal to 4.2 m	PxM2300 Width Less than or Equal to 4.8 m	PxM2300 Width Less than or Equal to 5.4 m
L-0.3	1.2 m	1.2 m	1.2 m	1.2 m	1.2 m	1.2 m	1.2 m
L-0.6	1.2 m	1.2 m	1.2 m	1.2 m	1.2 m	1.2 m	1.2 m
L-0.9	1.2 m	1.2 m	1.2 m	1.2 m	1.2 m	1.2 m	1.2 m
L-1.2	1.2 m	1.2 m	1.2 m	1.2 m	1.2 m	1.2 m	1.2 m
L-1.5	1.2 m	1.2 m	1.2 m	1.2 m	1.2 m	1.2 m	1.2 m
L-1.8	1.2 m	1.2 m	1.2 m	1.2 m	1.2 m	1.2 m	1.2 m
L-2.1	1.2 m	1.2 m	1.2 m	1.2 m	1.2 m	1.2 m	1.2 m
L-2.4	1.2 m	1.2 m	1.2 m	1.2 m	1.2 m	1.2 m	0.9 m
L-2.7	1.2 m	1.2 m	1.2 m	1.2 m	1.2 m	1.2 m	0.9 m
L-3.0	1.2 m	1.2 m	1.2 m	1.2 m	1.2 m	0.9 m	0.9 m
L-3.3	1.2 m	1.2 m	1.2 m	1.2 m	0.9 m	0.9 m	0.6 m
L-3.7	1.2 m	1.2 m	1.2 m	1.2 m	0.9 m	0.9 m	0.6 m
L-4.0	1.2 m	1.2 m	1.2 m	1.2 m	0.9 m	0.9 m	0.6 m
L-4.3	1.2 m	1.2 m	1.2 m	1.2 m	0.9 m	0.6 m	0.6 m
L-4.6	1.2 m	1.2 m	1.2 m	1.2 m	0.9 m	0.6 m	N/A
L-4.9	1.2 m	1.2 m	1.2 m	1.2 m	0.9 m	0.6 m	N/A
L-5.2	1.2 m	1.2 m	1.2 m	0.9 m	0.6 m	0.6 m	N/A
L-5.5	1.2 m	1.2 m	1.2 m	0.9 m	0.6 m	0.6 m	N/A
L-5.8	1.2 m	1.2 m	1.2 m	0.9 m	0.6 m	N/A	N/A
L-6.0	1.2 m	1.2 m	1.2 m	0.9 m	0.6 m	N/A	N/A

Reproduced from MUSAR Training Foundation. (2020). *Trench rescue shoring operations guide* (2nd ed.). Author. https://paratech.com/wp-content/uploads/2021/01/MUSAR-SOG-2020.pdf

TABLE C-6 Holomatro Strut Chart (Metric Version)

Total L	Width Less than or Equal to 1.5 m	Width Less than or Equal to 1.8	Width Less than or Equal to 2.4	Width Less than 3.0 m	Width Less than or Equal to 3.6 m	Width Less than or Equal to 4.4 m
L-0.3	1.2 m	1.2 m	1.2 m	1.2 m	1.2 m	1.2 m
L-0.6	1.2 m	1.2 m	1.2 m	1.2 m	1.2 m	0.9 m
L-0.9	1.2 m	1.2 m	1.2 m	1.2 m	0.9 m	0.6 m
L-1.2	1.2 m	1.2 m	1.2 m	1.2 m	0.9 m	N/A
L-1.5	1.2 m	1.2 m	1.2 m	0.9 m	0.6 m	N/A
L-1.8	1.2 m	1.2 m	1.2 m	0.9 m	N/A	N/A
L-2.1	1.2 m	1.2 m	0.9 m	0.6 m	N/A	N/A
L-2.4	1.2 m	1.2 m	0.9 m	0.6 m	N/A	N/A
L-2.7	1.2 m	1.2 m	0.9 m	N/A	N/A	N/A
L-3.0	1.2 m	1.2 m	0.6 m	N/A	N/A	N/A
L-3.3	1.2 m	1.2 m	0.6 m	N/A	N/A	N/A
L-3.7	1.2 m	0.9 m	0.6 m	N/A	N/A	N/A
L-4.0	1.2 m	0.9 m	N/A	N/A	N/A	N/A
L-4.3	1.2 m	0.9 m	N/A	N/A	N/A	N/A
L-4.6	1.2 m	0.9 m	N/A	N/A	N/A	N/A
L-4.9	1.2 m	0.9 m	N/A	N/A	N/A	N/A
L-5.2	1.2 m	0.9 m	N/A	N/A	N/A	N/A
L-5.5	1.2 m	0.6 m	N/A	N/A	N/A	N/A
L-5.8	1.2 m	0.6 m	N/A	N/A	N/A	N/A
L-6.0	0.9 m	0.6 m	N/A	N/A	N/A	N/A

Reproduced from MUSAR Training Foundation. (2020). *Trench rescue shoring operations guide* (2nd ed.). Author. https://paratech.com/wp-content/uploads/2021/01/MUSAR-SOG-2020.pdf

TABLE C-7 Hurst/Airshore Strut Chart (Metric Version)

	Struts with 1 Pin		Struts with 2 Pins		
Total L	Width Less than or Equal to 2.4 m	Width Less than or Equal to 2.4 to 3.6 m	Width Less than or Equal to 1.2 m	Width Less than 2.4 m	Width Less than or Equal to 3.6 m
L-0.3	1.2 m	1.2 m	1.2 m	1.2 m	1.2 m
L-0.6	1.2 m	1.2 m	1.2 m	1.2 m	1.2 m
L-0.9	1.2 m	1.2 m	1.2 m	1.2 m	1.2 m
L-1.2	1.2 m	1.2 m	1.2 m	1.2 m	1.2 m
L-1.5	1.2 m	1.2 m	1.2 m	1.2 m	1.2 m
L-1.8	1.2 m	1.2 m	1.2 m	1.2 m	1.2 m
L-2.1	1.2 m	1.2 m	1.2 m	1.2 m	1.2 m
L-2.4	1.2 m	1.2 m	1.2 m	1.2 m	1.2 m
L-2.7	1.2 m	1.2 m	1.2 m	1.2 m	1.2 m
L-3.0	1.2 m	1.2 m	1.2 m	1.2 m	1.2 m
L-3.3	1.2 m	1.2 m	1.2 m	1.2 m	1.2 m
L-3.7	1.2 m	1.2 m	1.2 m	1.2 m	1.2 m
L-4.0	1.2 m	0.9 m	1.2 m	1.2 m	1.2 m
L-4.3	1.2 m	0.9 m	1.2 m	1.2 m	1.2 m
L-4.6	1.2 m	0.9 m	1.2 m	1.2 m	1.2 m
L-4.9	1.2 m	0.9 m	1.2 m	1.2 m	1.2 m
L-5.2	1.2 m	0.9 m	1.2 m	1.2 m	1.2 m
L-5.5	0.9 m	0.9 m	1.2 m	1.2 m	1.2 m
L-5.8	0.9 m	0.9 m	1.2 m	1.2 m	0.9 m
L-6.0	0.9 m	0.6 m	1.2 m	1.2 m	0.9 m

Reproduced from MUSAR Training Foundation. (2020). *Trench rescue shoring operations guide* (2nd ed.). Author. https://paratech.com/wp-content/uploads/2021/01/MUSAR-SOG-2020.pdf

TABLE C-8 Wood Strut Chart (Metric Version)

0.9 m Maxiumum Vertical Spacing

Strut Wood Type	Trench Width Up To 2.4 Meters
Douglas Fir or Spruce/Pin/Fir	Maximum L-2.4

Note: Maximum L is based on 100 mm x 100 mm wood struts and Ellis screw jacks.
Reproduced from MUSAR Training Foundation. (2020). *Trench rescue shoring operations guide* (2nd ed.). Author. https://paratech.com/wp-content/uploads/2021/01/MUSAR-SOG-2020.pdf

TABLE C-9 Wale/Wood Strut Chart (Metric Version)

2.4 m Maximum Span/1.2 m Maximum Vertical Wale Spacing

Strut Wood Type	Trench Width					
	0.9 m	1.2 m	1.5 m	1.8 m	2.1 m	2.4 m
Douglas Fir	Max L-2.7	Max L-2.7	Max L-2.7	Max L-2.4	Max L-1.8	Max L-1.2
Spruce/Pine/Fir	Max L-2.1	Max L-2.1	Max L-2.1	Max L-2.1	Max L-1.8	Max L-1.2

Notes: Based on based on 100 mm x 100 mm wood struts, Ellis screw jacks, 180 mm x 180 mm LVL wales (21.4 mPA bending strength).
Span is the distance between the struts supporting the wales.
Gaps between the wales and panels at the panel edges and both ends of the wales must be filled with spacers and/or wedges.
Reproduced from MUSAR Training Foundation. (2020). *Trench rescue shoring operations guide* (2nd ed.). Author. https://paratech.com/wp-content/uploads/2021/01/MUSAR-SOG-2020.pdf

TABLE C-10 Panel Chart (Metric Version)

Composite Construction	Maximum Total "L"	Non-Composite Construction	Maximum Total "L"
50 mm x 300 mm LVL/FinnForm	L-6	50 mm x 300 mm LVL/FinnForm	L-1.5
50 mm x 300 mm Wood/FinnForm	L-3.3	50 mm x 300 mm Wood/FinnForm	L-0.9
50 mm x 300 mm Wood/CDX	L-2.4	50 mm x 300 mm Wood/CDX	L-0.3

Notes: Composite construction requires the application of heavy-duty exterior construction grade adhesive that has an ultimate strength of at least 27.6 mPA between the strongback and sheeting. Surface areas must be sanded prior to the application of adhesive. Non-compostive construction includes attaching strongbacks against the sheeting with nails, bolts, and screws. Simply positioning strongbacks against the sheeting is also considered non-composite.
Plywood (FinnForm and CDX) is 19 mm.
Wood (strongback) values are based on #1 Douglas Fir.
LVL strongback values are based on f_b-21.4 mPA laminated veneer lumber.
Reproduced from MUSAR Training Foundation. (2020). *Trench rescue shoring operations guide* (2nd ed.). Author. https://paratech.com/wp-content/uploads/2021/01/MUSAR-SOG-2020.pdf

TABLE C-11 Wale Chart (Metric Version)

1.2 Meter Maximum Vertical Spacing

Wale Type	2.4 Meter Span	3.6 Meter Span
150 mm x 150 mm	L-0.6	L-0.3
200 mm x 200 mm	L-1.5	L-0.6
180 mm x 180 mm LVL	L-3.6	L-1.2
Paratech	L-2.1	L-0.6

Notes: 2 point supported wales.
L values reprsent the maxium allowable Total L.
Span is the distance between the struts supporting the wale.
Notes: For all wales:
Gaps between the wales and panels at the edges and both ends of the wale must be filled with spacers and/or wedges.
Spacing wales at 0.6 meters (vertical) will increase the maximum Total L capcaity by 150% (Max L shown x 1.5).
150mm x 150 mm and 200 mm x 200 mm timber capacticies are based on #1 Douglas Fir.
180 mm x 180 mm LVL capacity based on a bending strength of 21.4 mPA.
Reproduced from MUSAR Training Foundation. (2020). *Trench rescue shoring operations guide* (2nd ed.). Author. https://paratech.com/wp-content/uploads/2021/01/MUSAR-SOG-2020.pdf

TABLE C-12 Buttress Chart (Metric Version)

Composite Panels	1.2 Meter Deep Void	1.8 Meter Deep Void
50 mm x 300 mm LVL/FinnForm 19 mm	Max Total L-1.8	Max Total L-1.2
50 mm x 300 mm Wood/FinnForm 19 mm	Max Total L-0.6	Max Total L-0.3

Note: Capacticies can be increased by placing an upright in front of the strongbacks and installing struts on the upright.

Added Upright	1.2 Meter Deep Void	1.8 Meter Deep Void
50 mm x 300 mm LVL	Added capacity L-0.6	Added capacity L-0.3
150 mm x 150 mm Wood	Added capacity L-1.2	Added capacity L-0.6

Composite Panels	1.2 Meter Deep Void	1.8 Meter Deep Void
200 mm x 200 mm Wood	Added capacity L-3.0	Added capacity L-2.1
180 mm x 180 mm LVL	Added capacity L-6.0	Added capacity L-4.8

Notes: Buttress picket details:
The distance from the void or sign of failure for a L≤1.5 failure shall have pickets installed no less than 1.2 meters from the void or failure point.
L≥1.8 failures shall have pickets installed no less than 75% of the distance of the simple L from the void or failure point.
Pickets must be 25 mm diameter steel driven a minimum of 910 mm into the soil.
Picket spacing shall be a minumum of 150 mm apart unless they are used in a one-piece picket anchor system device.
See the Sole Anchor Picket Chart (Table C-13).
Reproduced from MUSAR Training Foundation. (2020). *Trench rescue shoring operations guide* (2nd ed.). Author. https://paratech.com/wp-content/uploads/2021/01/MUSAR-SOG-2020.pdf

TABLE C-13 Sole Anchor Picket Chart (Metric Version)

Total L	Minimum Pickets
L-0.3	2
L-0.6	3
L-0.9	5
L-1.2	6
L-1.5	8
L-1.8	9
L-2.1	10
L-2.4	12

Notes: This chart included picket requirements for sole anchors used for rakers, buttress, and tension systems.
Pickets must be 25 mm diameter steel driven a minimum of 910 mm into the soil.
Picket spacing shall be a minumum of 150 mm apart unless they are used in a one-piece picket anchor system device.
Reproduced from MUSAR Training Foundation. (2020). *Trench rescue shoring operations guide* (2nd ed.). Author. https://paratech.com/wp-content/uploads/2021/01/MUSAR-SOG-2020.pdf

Glossary

A

Active earth pressure coefficient How much (percentage) of the vertical pressure translates to horizontal pressure pushing outward.

Active soil Soil containing energy as it relates to movement.

Air bags Inflatable bags used for rescue operations available in low pressure (7 psi), medium pressure (14 psi), or high-pressure (100-150 psi) varieties. These systems include pressure regulators, hoses, controllers, and bags.

Air knife Tool that injects air into the soil at approximately 100 pounds per square inch (psi; 7 kg/cm^2) for the purpose of breaking the soil into smaller particles.

Angle of repose The natural angle at which loose particulate products will support their own weight and can be expected not to flow from a standing position.

Atmospheric contaminate The accumulation of substances in the air in quantities that are large enough to produce harmful effects.

Atmospheric monitoring A method of evaluating the ambient atmosphere of a space, including but not limited to its oxygen content, flammability, and toxicity.

Authority having jurisdiction (AHJ) The local authority that draws on the power of law to make rules and regulations for the organization.

B

Backfill strut Strut designed to (1) oppose the force of the backfill and (2) transfer the force of the opposing wall into the backfilled area of the shoring system.

Backhoe A piece of excavating equipment consisting of a digging bucket on the end of a two-part articulated arm (boom and stick).

Beam Long section of wood or metal used by trench teams as shoring or a wale. It can also be used to span the top of the trench from lip to lip to create an overhead anchor point.

Bell pier condition A condition where the bottom few feet of shaft has a larger diameter than the majority of the shaft. This can be caused by soil failure or an oversized cut.

Benching A method of protecting workers from cave-ins by excavating the side of a trench or excavation to form one or a series of horizontal levels or steps, usually with vertical or near vertical surfaces between levels.

Bipod bases High-strength hinged rotating anchor plates for specialized lifting equipment.

Bipod head assembly An engineered, designed head with a rotational capability that provides attachment points for lifting equipment and system stabilizing.

Bipod lift Rescue support struts such as those manufactured by Paratech Rescue Inc. that allow rescuers to build a two-legged structure for overhead lifts. Rigging can be attached to the head of the system to raise a load.

Boom The section of the arm closest to the vehicle.

Bridge lift A lifting tool positioned on or connected to a bridge made of timbers that can be positioned directly above a heavy object in the trench. Rigging is attached to the bridge and the heavy object.

Bridge supports Cribbing used to raise a bridge up over panels or to create a longer throw with the lifting tool when the bridge is not set at grade level. This support can also distribute the load across a larger footprint and/or keep it away from the lip of the trench.

Bucket A specialized container attachment designed with teeth to facilitate digging. They are attached to the second arm (stick) of excavators and backhoes, and they come in a variety of sizes and volumes.

C

Cantilever lip bridges Lip bridges created by placing a ladder on the ground, extending it over the missing lip, placing 2 ×10s over the rungs, and anchoring the far side of the ladder with ladder rung brackets with pickets driven into the ground to allow a rescuer to stand on an overhanging lip bridge.

Cantilever wales (beams) Beams that have two or more supports but extend beyond one end support and end in clear space.

Cave-in The separation of a mass of soil or rock material from the side of an excavation or trench, or the loss of soil from under a trench shield or support system, and its sudden movement into the excavation, either by falling or sliding, in sufficient quantity so that it could entrap, bury, or otherwise injure and immobilize a person.

Center of gravity (COG) The point on a body around which the body's mass is evenly distributed or balanced.

Center pin The middle of the rotating platform on an excavator.

Centrifugal vacuum truck Apparatus that uses a large fan to create suction in the intake line.

Class 3 harness A harness designed with lower body and upper body attachments.

Class III lever Type of lever in which the force is located between the load and the fulcrum.

Class II lever Type of lever in which the load is located between the fulcrum and the force.

Class I lever Type of lever in which the fulcrum is located between the force and the load.

Clay Soil consisting of very small and flat particles; it is the only soil type capable of electrical attraction (cohesion).

Cleats Strips of wood that are used to support and resist the movement of struts.

Cohesion The sticking together of particles of the same substance.

Cohesive soils Soils with very fine particals (cannot be seen with the naked eye) that derive their primary strength via electromagnetism at the particle level.

Compartment syndrome A condition in which increased pressure within a muscle compartment of the arm or leg causes nerve damage due to decreased blood supply.

Competent person An individual, designated by the employer, who is capable of identifying existing and predictable hazards in the surroundings or working conditions that are unsanitary, hazardous, or dangerous to workers, and who is authorized to take prompt corrective measures to eliminate them.

Compliance strut Additional struts placed in order to meet the strut spacing requirements as listed in the tabulated data.

Composite behavior When two members are rigidly connected to act as one larger member, significantly increasing strength.

Confined space ventilator Fan that uses duct tubing to send air to a specific location; can be used with heating elements.

Critical incident stress debriefing (CISD) Process used to debrief rescue personnel after an emotionally charged incident.

Critical lift Any lift in which a human being is below the object, and/or any lift where the object exceeds 75 percent of the lifting or stabilizing tool's rated capacity.

Crush syndrome A condition that is the result of a prolonged entrapment where the victim's body tissue is crushed and circulation to the tissue is restricted.

D

Deep trench A trench that is deeper than 8 feet (4.2 m).

Defensive measures Methods that protect people from specific hazards without directly exposing first responders and rescuers to the hazards.

Dewatering devices Devices that control water from ground seepage, broken water mains, and rainwater runoff (e.g., trash pumps).

Dewatering devices Devices that include well points and pumping systems, that are typically installed by construction crews, and portable pumps (such as trash pumps and mud pumps); used by trench rescue teams.

Ductile materials Materials that bend before they fracture.

Duplex nail Nail that has two shoulders, which allows it to be removed easily.

E

Electric-powered fire department smoke ejector Fan used to provide an adequate flow of fresh air into a trench and hazardous atmosphere out of the trench.

Energy The capacity for doing work and overcoming resistance.

Entrenching tool A small, collapsible version of the larger shovel, designed to be used in situations where space is limited and a regular shovel is too big. Commonly known as a military shovel.

Entry team officer The individual responsible for supervising and managing the entry team assignments.

Entry team The group of rescuers responsible for hazard identification and mitigation, initial patient care, extrication and removal from the trench.

Equipment surcharge load The weight of equipment on the ground that translates into increased horizontal pressure on the trench walls.

Evaluation A systematic determination of scope and magnitude of the emergency using the assessment criteria. Its primary purpose is to determine the incident objectives, strategy, and tactics, and the resources that will be needed to accomplish them.

Excavation Any human-made cut, cavity, trench, or depression in an earth surface formed by the earth's removal. In practical terms, when a hole is more than 15 feet (4.5 m) wide at its base, it is called an excavation. Overall, an excavation is wider than it is deep.

Excavator An engineering vehicle consisting of an articulated arm (boom, stick), bucket, and cab mounted on a pivot (a rotating platform) atop an undercarriage with tracks or wheels.

Extreme shoring conditions Any soil condition, trench depth, or trench width that exceeds the limits of default trench rescue shoring equipment and techniques.

F

Forward staging An area located beyond 100 feet (30 m) from the trench, initially designated by the incident commander for fire, rescue, emergency medical services, and police responders and their vehicles.

Freestanding time The amount of time an excavation is open to the elements.

Friction Measure of the amount of force it takes to move an object across the surface of another object.

G

Granular soils Soils made up of individual grains of soil (can be seen with naked eye) that derive their primary strength from friction between individual grains.

Gravity The force of attraction between two bodies that have mass.

Ground pads Wooden material used to line the trench lip for the purpose of distributing rescuer and equipment load (weight).

H

Hazard control plan A plan developed to address the safety of the trapped victims, bystanders, and all responders by identifying and mitigating all hazards.

Hazards Any agents that can cause harm or damage to humans, property, or the environment.

Heavy equipment lift Type of lift that employs the hydraulic cylinders on an excavator or backhoe to provide the force needed to lift an object. Rigging is attached from the lifting eye of an excavator bucket to the object.

High-angle wall Trench walls that are vertical or near vertical having angles between 70 and 90 degrees.

Hydraulic lifting tools Tools that use cylinders, pumps, and fluid to develop the forces needed to list or push heavy objects.

Hydraulic shores Type of shore (strut and rails) that is lowered into the trench from the top and then expanded by using a connected hydraulic pump.

Hydrostatic pressure Increased pressure caused by the addition of water to the soil profile.

I

Incident commander (IC) The individual responsible for developing the strategic goals for the operation.

Incident command system (ICS) Management system by which emergency personnel are assigned specific responsibilities and areas of supervision.

Inclined plane Angled plane that gains efficiency by reducing required force over time.

Initial actions The immediate actions taken during the first 20 to30 minutes after arrival at the site of an emergency; includes scene management, hazard control, site control, non-entry rescues, summoning resources, and establishing support functions.

Initial stabilization A technique used to prevent or minimize the movement of a load.

Internal angle of friction A measure of the ability of a soil to internally resist shear from an outside loading. When low, it results in a weak soil, and when high it results in a strong soil.

Intersecting trench A trench where multiple trench cuts or legs converge at a single point.

J

Joined corner panels A prefabricated corner panel constructed from two aluminum or steel brackets placed at tiers, secured with ½-inch (13-mm) diameter bolts and nuts to allow two trench rescue panels to be bolted together at a 90-degree angle.

L

Lateral soil pressure The horizontal force produced by soil due to gravity or other loads on the soil.

Lever A simple machine used to move a load. It includes a fulcrum (pivot point), a force, and a load.

Liaison officer Individual responsible for providing information and gathering information from organizations and people outside of the immediate rescue effort.

Lifting eye The factory installed rigging point on a bucket.

Lip bridges A type of lip protection built with girders (timbers), platforms (aluminum, wooden, or fire service ladders with lumber), and bases (sections of timbers placed away from the compromised wall and used to elevate the bridge above the lip).

Lip protection Ground pads or bridges placed on or over the trench lip to distribute the load of rescuers over a greater area, or further away from the trench lip.

Logistics officer Individual responsible for obtaining the appropriate equipment and personnel for deployment by the operations officer.

Low-angle wall Trench walls with angles less than 70 degrees.

L-shaped trench An intersecting trench where one cut or leg converges with another cut or leg resulting in the shape of the letter "L."

M

Manual lifting tools Used in trench rescue, these are typically levers, such as a pry bar, Halligan bar, or pinch bar.

Mechanical advantage The force achieved using a tool, mechanical device, or other equipment.

Mechanical lifting tools Used in trench rescue, these are typically a mechanical ratchet-style jack (Hi-Lift brand is common).

Mechanics A branch of physics that deals with energy and forces in relation to bodies.

Medical officer Individual responsible for establishing a medical control area to treat any on-scene rescuer injuries and to provide for victim care as necessary.

Moment of force The amount of force rotating around the fulcrum times the distance from the fulcrum.

N

Noncomposite behavior When two members are used that act individually.

Nonentry rescue A rescue that can be accomplished without rescuers entering the trench or excavation.

O

Ongoing stabilization A technique that prevents the object from shifting or dropping during the lifting operation. It is the "safety net" as the object is intentionally moved.

Operations officer Individual responsible for overall coordination of the rescue effort and the implementation of the tactical decisions that will make the incident commander's strategy successful.

Outriggers Hydraulically operated stabilizers found on tractor-style backhoes.

Overburden pressure The pressure that the weight of the spoil pile or other object exerts on the trench.

P

Panel set A panel set is any two panels directly across the trench from each other that have or are intended to have struts installed between them.

Panel team officer The individual responsible for supervising and managing the entry team assignments.

Panel team The group of rescuers responsible for installing lip protection, panels, wales, and backfill.

Pile side In a T-trench or L-trench, the side of the trench with the taller pile of dirt that resulted from a trench collapse.

Pneumatic lifting tools Used in trench rescue, these are typically high-pressure air lifting bags.

Pneumatic strut Type of strut that is extended by using compressed air. After extension, the strut either locks itself or is manually locked to prevent a collapse under load.

Positioning strut The first strut installed between the primary panels must be placed in a position that will hold the primary panels in place so that backfilling can begin. The positioning strut must be placed in an area where sections of stable and accessible trench walls exist.

Prescriptive shoring designs Modular rescue shoring designs prepared by a rescue engineer that when assembled per the guidance will safely carry the loads to the indicated level.

Previously disturbed soils Sometimes called fill soils, these are soils that lack cohesiveness because they are broken and/or mixed with other soil types.

Primary panels The first sets of panels that are positioned to protect the victim (centered around the victim's head and chest) from trench collapse.

Primary shoring The initial panels and struts installed to protect the victim from trench collapse.

Primary staging An area designated by the incident commander and located at least 300 feet (91 m) from the trench, for the staging of trench rescue resources.

Probes A non-metallic rod that is pushed into the cave in spoil pile to determine the buried victim's location.

Public information officer (PIO) Individual who provides information to the media regarding incident activities.

R

Rapid intervention team (RIT) Team established and equipped to handle everything from intervention for a secondary collapse or a medical emergency involving a member of the rescue team.

Rehabilitation An area that provides for rescue personnel rotation to address medical monitoring and fluid replacement needs.

Rescue area The area located immediately surrounding the rescue site.

Rescue plan The tactical portion of the incident action plan.

Rescue team logistics officer (RTLO) An individual responsible for tracking and determining the availability of trench rescue team equipment during the emergency.

Rescue team manager (RTM) A member of the trench rescue team who manages and supervises the activities of the trench rescue team.

Risk An exposure to a hazard or hazards that will likely result in injury, health problems, or death.

Risk versus gain analysis A decision made based on a hazard identification and situation assessment that weighs the risks likely to be taken against the benefits to be gained for taking those risks.

Rotational failure A scoop-shaped collapse that starts back from the trench lip and transmits itself to the trench wall in a half-moon shape; also called slough failure.

S

Safety officer Individual responsible for all aspects of operations that deal with safety and health of the rescue personnel.

Safe zone Area within the trench that is fully shielded by adequate trench rescue shoring.

Sand Particles that can be seen with the naked eye; particles are larger than silt and clay.

Scabs Small pieces of lumber, usually 2 × 4 inches, that are used to secure timber struts to panels by boxing-in the strut end and other connective devices in place.

Screw jack strut Type of shore or strut end that is tightened by a thread and yoke assembly. It is also sometimes referred to as a pipe jack when used in conjunction with varying lengths of pipe.

Secondary shoring Panels, struts, and wales (as needed) installed to enhance the area being protected by the primary shores.

Self-contained breathing apparatus (SCBA) A respirator that supplies breathing air to the user from an air source that is independent of the environment. It is designed to be carried by the user.

Severe environmental conditions Operations involving frozen soil, running soil (e.g., gravel, sand, liquid), severe weather (e.g., heavy rain, wind, lightning, or flooding), or nighttime (dark) operations.

Sheeting The portion of the protective system designed to hold back running debris.

Shoring performance assessment An inspection of in-place shoring that verifies continuous safety by regularly verifying the shoring integrity.

Shoring system An assembly of shoring components consisting of panels, wales, struts, backfill, etc., to provide a complete protective system.

Shoring team The group of rescuers responsible for installing struts in a shoring system.

Shoring team officer The individual responsible for supervising and managing the shoring team assignments.

Silt Particles sized between sand particles and clay particles that form a weak soil. It will feel slightly gritty when rubbing between fingers.

Simple L (SL) The distance (length) measured in feet (meters) from the original trench wall perpendicular to the farthest point of soil failure or cracks/fissures.

Single-point shores A nonstandard shoring type that should be used only as a secondary, not primary, means of trench shoring and should only be used when conventional trench rescue shoring cannot be properly installed.

Size-up Determination and analysis of information used to develop a rescue plan of action for an incident.

Slough failure The loss of part of the trench wall starting at an area back from the trench lip and extending down into the trench wall.

Soil density The weight of a standard unit of soil, typically pounds per 1 cubic foot of soil.

Span of control The number of workers whom a supervisor can manage based on the type of work being performed.

Spoil pile A pile of excavated soil next to the excavation or trench.

Spoil pile slide The result of excavated earth that is placed too close to the lip and subsequently falls into the trench.

Spot shoring Shoring without the use of panels or strongbacks.

Staging officer In a rescue situation, the person responsible for positioning and accounting for resources that are not immediately assigned.

Standard precautions Precautions (e.g., handwashing, gloves, gowns, masks) that cover protection of both the provider and the victim from exposure to blood and other body fluids, and/or airborne products that could pass from the rescuer to the victim or from the victim to the rescuer.

Stick The section of a backhoe or excavator arm that carries the bucket.

Strongbacks The 2 × 12-inch lumber components that are 10 or 12 feet (3.0 or 3.7 m) in length on the trench rescue panel that transmit forces along the vertical plane of the trench wall.

Strong side The side of the trench without the obstruction.

Strut A compression element used in the support of structures, excavation openings, or other loads.

Strut activation force A measurement of the total force that the strut exerts. The total force is calculated by multiplying the strut surface area (square inches) by the pressure (psi).

Strut pressure The amount of pressure that the air or hydraulic system sends to the strut. It is measured in psi and can be seen on the system's gauge.

Struts The component in the protective system that transfers the force from one side of the trench to the other.

Supplemental shoring Shoring consisting of cut-down sections of panels and struts used in conjunction with full trench rescue shoring panel and strut systems.

Supplied air breathing apparatus (SABA) A respirator that supplies breathing air to the user from an air source that is independent of the environment, supplied from a remote source. It is not designed to be carried by the user.

Surcharge loads (ScL) Spoil piles and equipment that are within the area between the original trench walls and the farthest point of soil failure. Measured in feet (meters) perpendicular to the trench wall.

Swing area The dangerous area or radius around which a rotating cab can reach.

T

Technical rescue safety officer (TRSO) A trained safety officer that is a certified trench rescue technician.

Tieback Term used to describe tension-based stabilization systems.

Timber strut A wood column with a compression strength of not less than 1500 psi (105.5 kg/cm) that transfers load from one side of a trench to the other side.

T-L method A method designed for use by trench rescuers to obtain rapid and accurate determinations of actual earth pressures and lateral forces based on the visible soil failure conditions.

T-L soil conditions (T-L soil) A classification of soil that assumes the worst-case conditions are present.

Toe failure A slough failure that occurs at the bottom of the trench, where the floor meets the wall.

Total L (L) The simple L (SL) plus the surcharge loads (ScL [spoil pile and equipment]), if present.

Traditional sheeting and shoring The use of 4 × 8-foot (1.2 × 2.4-m) sheet panels, with a strongback attachment, supplemented by a variety of conventional shoring (strut) options such as hydraulic, screw, and/or pneumatic shores.

Trench A narrow excavation (in relationship to its length) made below the surface of the ground. In general, the depth is greater than the width, but its width measured at the bottom does not exceed 15 feet (4.5 m).

Trench boxes Boxes that consist of steel or aluminum side walls and spreaders that are assembled at the scene used to provide worker protection; also called modular shields.

Trench floor The bottom of the trench.

Trench lip The area 2 feet horizontal and 2 feet vertical (0.6 m and 0.6 m) from the top edge of the trench face.

Trench rescue panel A combination of plywood sheeting and lumber strongback, used in shoring.

Trench rescue shoring The use and application of prescriptive shoring techniques designed to resolve the collapse conditions at the majority of trench rescue situations.

Trench walls The vertical or inclined earth surfaces formed as a result of excavation work.

T-shaped trench An intersecting trench where one cut or leg converges with another cut or leg resulting in the shape of the letter "T."

Type A soil A strong soil with an unconfined compressive strength of over 1.5 tons per square foot

Type B soil A medium strength soil with an unconfined compressive strength between 0.5 and 1.5 tons per square foot.

Type C soil A weak soil with an unconfined compressive strength less than 0.5 tons per square foot.

U

Unconfined compressive strength (UCS) The force, or load per unit area, as calculated with a penetrometer or other device and stated numerically in tons per square foot, that determines the point at which a soil will fail in compression.

Uniform shape A single, rigid symmetrical object with constant density. The balance point (center of mass) is located at the object's geometrical center (centroid).

V

Vacuum system Equipment that attaches to most municipal vacuum trucks that uses a standard hose with couplings and specially designed nozzles to create vacuum pressure to expedite the removal of soil.

Varying soil profiles Multiple layers of different soils found at various levels in the trench or excavation wall.

Vector A force having a specific magnitude and direction.

Victim packaging The application of devices (e.g., wristlets, cinch rings, harnesses, and litters) on the victim for removal from the trench.

Victim survivability profile Determination based on a thorough risk–benefit analysis and other incident factors that address the potential for a victim to survive or die with or without rescue intervention.

W

Wale A beam used horizontally in a trench shoring system to provide resistance to struts when voids or other obstructions are present.

Wale hangers Pipe clamps that are preassembled that can hang on the top of a panel and support a wale at a specific height for placing struts.

Wall shear collapse A collapse that occurs when a section of soil loses its ability to stand and falls into the trench along a mostly vertical plane.

Water table The top of the water surface in the saturated part of an aquifer.

Weak side The side of the trench with the obstruction.

Wedge failure A failure that usually occurs in intersecting trenches, in which an angled section of earth falls from the corner of two intersecting trenches.

Work Distance times force.

Working length The length of the crib minus the required overlap (minimum 4 inches [10 cm] on both sides).

Working load limit (WLL) The maximum allowed weight (load). It is a percentage of the "break strength" that is often expressed as a ratio. Example: A chain that breaks at 10,000 pounds has a value of 2,500 (25%) at a 4 to 1 factor of safety.

Wristlets Webbing or other material designed to wrap around the wrist for the purpose of providing a lifting or pulling attachment point.

Index

Note: Page numbers followed by *f* and *t* indicate figures and tables, respectively.

A
activation force, strut, 148–149
active earth pressure coefficient, 25
active soil, 40
advanced life support (ALS), 239
advanced trench rescue shoring, 274–297. *See also* angled walls;
    end wall rescue shoring; extreme shoring conditions;
    replacement shoring
air bags, 78–79, 79f, 220, 232
    backfill, 174–175
    bridge lift, 225
air knife, 192, 203
alloy steel chain slings inspection, 219
American National Standards Institute (ANSI), 68
angle of repose, 27, 40
angled box crib, 216, 216f
angled walls, 287–292
    baseline strut length, 288
    counterweight method, 290–292
    picket method, 288–290
    strut length, 288
atmospheric contaminate, 113
atmospheric hazard, 97–98. *See also* severe environmental conditions
atmospheric monitoring, 16, 102, 103f
atmospheric testing equipment, 99–100
authority having jurisdiction (AHJ), 7, 16, 43, 197
awareness level, 7

B
backfill, 78–80
    air bags, 78–79, 79f
    backshores, 79
    buttress, 79
    options, 78–80
    soil, 79
    strut, 162, 183
    wood, 79–80
backhoe, 227, 228f, 232
backshores, 79
baseline strut length, 288
basic life support (BLS), 239
beam, 225, 232
bell pier condition, 30, 40
benching systems, 294, 297
biological hazards, 97

bipod bases, 226, 232
bipod head assembly, 226, 232
bipod lifts, 224, 232
    bases, 226
    capacities, 226
    components, 226
    hand signals for, 226
    head assembly, 226
    performing, 225–226
blown out trench walls, 36–39
boom, 227, 232
box crib system, 214–215
bridge lifts, 224, 232
    air bag bridge lift, 225
    capacities, 225
    hand signals for, 226
    performing, 225
bridge supports, 225, 232
briefing, 98
bucket, 227, 232
buttress, 79, 175–176
    chart, 137, 143t, 306, 312t

C
cantilever lip bridges, 243, 272
cantilever wales, 272, 282–283
carabiner, 219
cave-in, 16
    incidents, 186–187
center of gravity (COG), 210–212, 232
center pin, 228, 232
centrifugal vacuum truck, 191, 191f, 203
chainsaw, 85
choker hitch, 222, 223f
class 3 harness, 187, 188f, 203
class I lever, 209, 232
class II lever, 209, 232
class III lever, 209, 232
clay, 9, 22, 23f, 40
cleats, 297
closed void, 154, 154f
Code of Federal Regulations (CFR), 18
cohesion, 26, 40
cohesive soils, 26, 40
collapse mechanics, 9, 17–39

collapse patterns, 32. *See also* trench collapse
collapse situations, 5
collect/distribute loads, 131–132
command, 117–118
command post, 44
compartment syndrome, 200, 203
competent person, 5, 16
compliance strut, 163, 183
composite behavior, 73, 91
compression-based stabilization, 214–216
    angled box crib, 216, 216f
    box crib system, 214–215
    cribbing angled objects, 216
    cribbing details, 215–216
    cribbing rules, 215
confined space ventilator, 102, 102f, 106, 113
construction protective systems, 293–297
construction shoring, 129–131. *See also under* trench rescue shoring
correlation grid, 298–303
counterweight method, 290–292
cracks (fissures), 36
cribbing, 214
cribbing angled objects, 216
    shims, 216
    wedges, 216
critical angles, 222–224
critical incident stress debriefing (CISD), 125
critical lift, 207, 233
cross-shore design, 275
crush syndrome, 199–200, 203
cutting station, 86–90, 87–88f
cutting tool protection, 69

D

deep trenches, 8f, 237–238, 260–263, 264–267f, 272
    application for, 260
    complete shoring, 263
    personnel for, 261
    precautions for, 260
    preparation tasks for, 260
    primary shoring, 261
    secondary shoring, 262–263
    shoring plan for, 261
        size-up, 261
    shoring team, 262
    walkthrough of shoring, 261–263
default shoring, 131
defensive measures, 55–56, 62
defensive mitigation actions for underground utilities, 56–57
department of public works (DPW), 193
dewatering devices, 85–86, 86f, 91, 100, 104–105f, 113
    sewer line breaks, 102–106
    water line breaks, 102–106
digging trenches, 4f
dispatch information, 46
disturbed soils, 34
ductile materials, 210, 233
duplex nails, 84, 85f, 91
dynamic soil, 129, 130f

E

electric control equipment, 101
electric-powered fire department smoke ejector, 101, 113
emergency medical services (EMS), 198

end wall raker shore design, 280–281
end wall raker shoring system, 283–284
    disassembly and removal, 283–284
        panel team, 283
        shoring team, 283
end wall rescue shoring, 275–284
    application, 275–276, 278
    end wall raker shore design, 280–281
    excavator shore design, 278, 278f
    excavator shore walkthrough, 279–280
    long wall shoring equipment removal, 284
    precautions, 275–276, 279
    preparation, 276, 279
    victim location, 281–283
        basic raker installation, 281
        cantilever wale, 282–283
        strut by-pass, 282
    walkthrough, 276–278
        complete shoring, 278, 278f
        primary shoring, 277
        secondary shoring, 277
        trench size-up, 276–277
energy, 207, 233
engineered shoring systems, 297
enhanced hazard mitigation, 113
entrenching tool, 84, 91
entry operations, 189
entry shoring, 161–162, 162f, 167–174
    pneumatic shores with entry operations, 169–171f
    pneumatic struts without entering the trench, 167–169f
    timber shores, installing, 171–172, 172–173f
        inside wales, installing, 172–174
entry team, 119, 125
    duties, 189–190
    officer, 119, 125, 219, 233
equipment, 64–90. *See also* personal protective equipment (PPE)
    basics, 64–90
    importance of, 65–66
    surcharge load, 34, 40, 136, 306
excavation, 3, 3f, 16
excavator, 207, 227, 228f, 233
    hand signals, 179f, 228, 229f
    shore design, 278
    shore walkthrough, 279–280
exhaust ventilation, 109
extreme shoring conditions, 292–297
    construction protective systems, 293–297
        isolation devices, 295
        modular shields, 295
        sloping and benching systems, 294
        trench box, 294–295
        use at rescue scenes, 293
    engineered shoring systems, 297
    heavy equipment to install a protective device, 295–296
extrication, 189–190
eye protection (safety glasses), 68

F

Federal Emergency Management Agency's (FEMA), 116
firefighters, 132–137
firefighting turnout gear, 69
fissures, 134, 134f
foot protection (boots), 68
forward staging, 44, 62

freestanding time, 40
friction, soil, 25–26, 212, 233

G

granular soils, 40
gravity, 40
ground pads, 70, 91

H

hammers and nails, 84–85
hand protection (gloves), 67–68
hazard assessment, 47–48
hazard control equipment, 99–102
    atmospheric testing equipment, 99–100
    dewatering devices, 100
    electric control equipment, 101
    ventilation equipment, 101–102
hazard control plan, 98, 113, 119, 125
    briefing, 98
    hazard mitigation PPE, 98
hazard identification, 95–98
hazard management, 55–57
    defensive measures, 55–56
    defensive mitigation actions for underground utilities, 56–57
        natural gas line break, 56
        sewer line break, 56
        underground electrical wires, 56
        water line break, 56
    unbroken utility lines, 56–57
hazard mitigation, 99–113. See also severe environmental conditions; potential offensive mitigation techniques; ventilation
    enhanced, 113
    gas, 106
    hazardous materials, 107
    ongoing, 113
    PPE, 98
    underground electrical wires, 106
hazardous atmospheres, 12
hazards associated with heavy objects, 212–213
hazards, types of, 95–98
    atmospheric hazard, 97–98
    biological hazards, 97
    hazardous materials, 97
    physical hazards, 97
    severe environmental conditions, 97
    traffic, 96–97
    trench collapse, 95–96
    utilities, 96
    water, 97
head protection (helmet), 66–67
hearing protection, 69
heavy equipment, 226–231
    communications, 227
        face-to-face, 227
        hand signals, 227
        radio communications, 227
    coordinating the use of, 226–231
    equipment operator, 227
    hand signals needed for lifting, 228–231
    lifts, 224, 233
    terminology, 227–228
high-angle wall, 287, 297
Holomatro strut, 137, 140t, 306, 309t
    219

horizontal force, 25
Hurst/Airshore strut, 137, 141t, 306, 310t
hydraulic lifting tools, 220, 233
hydraulic shores, 82, 91
hydraulic struts, 82–83, 148, 148f, 150
hydro vacuum trucks, 192, 192f
hydrostatic pressure, 23, 41

I

immediately dangerous to life and health (IDLH), 69, 99
incident action plans, 45–46
    developing, 45–46
    large-scale IAPs (written), 45
    medium-scale IAPs (verbal), 45
    size-up, 46
    smaller-scale IAPs (mental), 45–46
    tactical sheets, 45
incident command system (ICS), 43–44, 63, 116–120, 118f. See also logistics
    command, 117–118
    liaison officer, 118
    operations officer, 117, 118f
    public information officer (PIO), 118
    safety officer, 118
    size-up and expanding, 117–120
    staging officer, 117, 118f
    technical rescue safety officer, 118
incident commander (IC), 44, 62
incident management system (IMS), 193
incident management tools, 120–121
incident termination, 121–125
inclined plane, 208, 233
initial actions, 42–61, 63. See also hazard management; nonentry rescue; scene management; site control; summon resources; victim self-rescue
    evaluation, 51–52
        awareness level incidents, 51
        operations level incidents, 51
        specialist level, 51
        technician level incident, 51
    site for incoming resources, preparing, 55
initial assessments, 47
initial stabilization, 213, 233
initial ventilation, 106
injury mechanism, 186–187
    cave-in incidents, 186–187
    incidents without a cave-in, 187
inside wales, installing, 172–174
interim shoring, 284
internal angle of soil friction, 26, 26f, 41
intersecting trenches, 8f, 237, 272
interview techniques, 47
isolation devices, 295

J

joined corner panels, 243, 272

L

L-shaped trench, 237, 273
L-trench shoring plan, 239–240, 243–244, 243f, 247–250f
    primary shoring, 244
    secondary shoring, 245–246
    shoring size-up, 244
    walkthrough of shoring L-trench, 244–246

ladders, 85
large-scale IAPs (written), 45
lateral soil forces, 132–137
lateral soil pressure, 41
levers, 208–209, 233
liaison officer, 118, 125
lifting and stabilizing heavy objects, 86
lifting capacities, 228
lifting eye, 228, 233
lifting mechanics, 207–210
    basic techniques, 210
    inclined plane, 208
    lever, 208
    mechanical advantage, 207
    pulleys, 209–210
    rope-based systems, 210
lifting plan, 219
lifting techniques, 220–222. *See also* heavy equipment; load assessment; load stabilization; overhead lifting techniques
    struts, 216–217
    surface-based lifting techniques, 220
        tool inspection, 221–222
    tool capacities, 220–221
        hydraulic, 221
        manual, 220
        mechanical, 220, 221*f*
        pneumatic, 220
    wheel chocks, 217
lip bridge, 71, 71*f*, 91
lip failures, 134, 134*f*
lip protection, 70–72, 157–160, 183, 225, 233
    2-foot wide ground pads, 72
    4-foot wide ground pads, 72
    best practices, 72
    lip bridge, 71–72, 71*f*
lip shear, 32, 32*f*
    slide pattern, 32*f*
    topple pattern, 32*f*
lip shear failures, 28–29
litter basket, 202
load assessment, 210–213
    center of gravity (COG), 210–212
    ductile materials, 210
    forces acting on the load, 212
    heavy objects, hazards associated, 212–213
    material type, 210
    weight, 210
load stabilization, 213–214. *See also* compression-based stabilization
    mechanics of, 213–214
    stabilization plan, 214
    tension-based stabilization, 217–218
logistics, 118–120
    entry team, 119
    hazard control plan, 119
    logistics officer, 118, 125
    medical team officer, 120
    operations, 119–120
    panel team, 120
    rapid intervention team (RIT), 120
    rescue team logistics officer (RTLO), 119
        entry, 119
        panel, 119
    shoring, 119
    rescue team manager, 119
    shoring team, 120
low-angle wall, 297

M

machine removal, 177, 178
machine tear out, 178
manual lifting tools, 220, 233
manual removal, 177–178
mechanical advantage, 207, 233
mechanical lifting tools, 220, 233
mechanics, 207, 233
medical team officer, 120, 125
medium-scale IAPs (verbal), 45
Michigan Urban Search and Rescue (MUSAR), 147
mid-wall failures, 134, 134*f*
modular shields, 295
moisture content, soil, 27
moment of force, 210, 233

N

National Fire Protection Association (NFPA), 6–9, 51
    NFPA 1670, trench rescue levels, 7–9
        awareness, 7
        operations, 7–8
        technician, 8–9
National Incident Management System (NIMS), 116
natural gas control equipment, 102
natural gas line break, 56
non-entry shoring, 162–163
noncollapse situations, 5
noncomposite behavior, 73, 91
noncontact voltage testers, 101
nonentry rescue, 57–61, 57*f*, 63, 187–190

O

Occupational Safety and Health Administration (OSHA), 3, 18, 68, 187. *See also* sample OSHA safety measures
    OSHA-29 CFR 1926.651 and 1926.652, 19
    OSHA CFR 1926 Subpart P, Excavations, 5–6
ongoing hazard mitigation, 113
ongoing stabilization, 213, 233
open void, 154, 154*f*
operational zones, establishing, 53–54
operations, 119–120
operations level, 7–8
    trench rescue shoring, 127–182. *See also* trench rescue shoring
operations officer, 117, 118*f*, 126
OSHA-29 CFR 1926.651 and 1926.652, 19
OSHA CFR 1926 Subpart P, Excavations, 5–6
outriggers, 228, 233
overburden pressure, 25, 27, 41
overhead lifting techniques, 222–225
    choker hitch, 222, 223*f*
    critical angles, 222–224
    tools, 226
    for trench rescues, 224–225
        bipod lifts, 224
        bridge lifts, 224
        heavy equipment lifts, 224
    vertical hitch, 222, 223*f*

P

panel chart, 137, 142t, 306, 311t
panel installation, 163–167
    one-side set, 163, 163–165f
    two-side panel installation, 166
panel ropes, 75–76
panel set, 162, 183
panel team, 120, 126
panel team officer, 120, 126
Paratech strut, 137, 138t, 162, 306, 307t
personal protective equipment (PPE), 55, 64–90, 67f. *See also* tools and appliances
    arm, 66
    for entry team operations, 189
    eye protection (safety glasses), 68
    firefighting turnout gear, 69
    foot protection (boots), 68
    hand protection (gloves), 67–68
    head protection (helmet), 66–67
    leg, 66
    personal-issue PPE, 66
    team-issue PPE, 66
    torso, 66
    for trench rescue, 66–69
physical hazards, 12–14, 97
picket method, 84, 288–290
pike poles, 84, 84f
pile side, 240, 273
pipe jacks, 82
planned excavations, 130
planning, 130–131
    planned excavations, 130
    unplanned events, 130
pneumatic lifting tools, 220, 233
pneumatic struts, 81, 91, 149, 149f, 218
    pressure, 149–150
    spacing, 150
polyester round sling inspection, 219
portable on demand storage (PODS), 193
positioning strut, 162, 183
potential offensive mitigation techniques, 102–103
    atmospheric monitoring, 102
    dewatering devices, 102–106. *See also individual entry*
pre-entry briefing, 189
prescriptive shoring designs, 144–145, 183
previously disturbed soils, 34, 41
primary panels, 163, 184
primary shoring, 147, 183, 244
primary staging, 44, 63
probes, 199, 203
public information officer (PIO), 118, 126
pulleys, 209–210

R

rapid intervention team (RIT), 86, 120, 126
rehabilitation, 126
removal methods, 177–178
    machine removal, 177
    machine tear out, 178
    manual removal, 177–178
replacement shoring, 284–287
    application, 284–285
    precautions, 285
    preparation, 285
    procedure walkthrough, 285
        interim shoring, 285
        panel team, 285
        primary shoring, 285–286, 286f
        shoring team, 285
    secondary shoring, 286–287
rescue area, 8, 16
rescue plan, 238–239, 273
    path for removal, 239
    pre-entry briefing, 239
    risk versus benefit analysis, 238
    shoring performance assessment, 239
    shoring plan, 238–239
    victim release, 239
rescue team logistics officer (RTLO), 119, 126
rescue team manager (RTM), 119, 126
rescue ventilation, 106, 107, 110f
resist loads, 132
resource assessment, 50–51, 50f
respiratory protection, 69
Resqtech struts, 137, 139t, 162
    chart, 306, 308t
rigging equipment, inspecting, 219
    alloy steel chain slings inspection, 219
    carabiner, 219
    hook, 219
    polyester round sling inspection, 219
    rigging hardware, 219
    shackle, 219
    synthetic web slings, 219
    turnbuckle, 219
    wire rope inspection, 219
rigging hardware, 219
rope-based systems, 210
roping panels, 76
rotational failure, 29–30, 30f, 41
    clays, 30
    cohesive silts, 30
    deep layers of weak clay soils, 30
    shallow groundwater, 30
    soil types likely to fail in rotation, 30

S

safe zone, 144, 183
safety culture, development, 66. *See also* personal protective equipment (PPE)
safety officer, 118, 126
sample OSHA safety measures, 11–15
    collapse (cave-in), 11
    fall hazards, 11
    falling objects, 11
    hazardous atmospheres, 12
    physical hazards, 12–14
    slippery conditions, 11, 11f
    utility hazards, 11–12
sand, 9, 22, 23f, 41
scabs, 151, 183
scene lighting, 85
scene management, 43–51
    classifying the emergency, 47
    command post, 44
    hazard assessment, 47–48
    ICS chart, 44f
    incident action plans (IAPs), 45–46. *See also individual entry*

incident command system (ICS), 43–44
initial assessments, 47
interview techniques, 47
resource assessment, 50–51
situation assessment, 46. *See also individual entry*
staging area, 44–45
standard operating guidelines (SOGs), 45–46
trench assessment, 49–50
victim assessment, 48–49, 49*f*
screw jack struts, 82, 91, 150–151
secondary shoring, 147, 183
self-contained breathing apparatus (SCBA), 55, 69, 91, 98
severe environmental conditions, 107–113, 270, 273
sewer and water control equipment, 100–101
sewer line break, 56
shackle, 219
sheet piling, 296
sheeting, 72, 92
shielding systems, 155–156, 155*f*
shims, 216
shoring intersecting trenches, 239–259. *See also* L-trench shoring plan; T-trench shoring plan
    L-trench shoring plan, 243–244, 243*f*
    personnel for, 243
    plan application for, 242
    precautions for, 242
    preparation to stabilize, 242
    shoring size-up, 244
    standard equipment for, 243
    trench–specific shoring equipment, 243
        cantilever lip bridges, 243
        joined corner panels, 243
        wale hangers, 243
        wales, 243
    walkthrough of shoring L-trench, 244–246
        primary shoring, 244
        secondary shoring, 245–246
shoring performance assessment, 239, 270–271, 273
shoring plan, 238–239
shoring size-up for intersecting trench, 244
shoring system, 72, 92
    disassembly and removal, 176–182
        excavator hand signals, 179*f*
        machine removal, 177, 178
        machine tear out, 178
        manual removal, 177–178
        personnel, 179–182
        plan, 178
        removal methods, 177–178
shoring team, 120, 126
shoring team officer, 120, 126
shoring voids, 154–156, 174–176
    air bag backfill, 174–175
    assessments, 154–155
    buttress, 175–176
    closed void, 154, 154*f*
    open void, 154, 154*f*
    soil backfill, 176, 176*f*
    wood backfill, 175
        size, 154
        void type, 154
        wall angle, 154
shovels, 84, 84*f*
shut down traffic, 53

silt, 9, 22, 23*f*, 41
single point shore, 147, 183, 270, 273
site control, 53–55
    action zone, 54, 54*f*
    hot zone, 54, 54*f*
    operational zones, establishing, 53–54
    shut down traffic, 53
    warm zone, 54, 54*f*
site for incoming resources, preparing, 55
situation assessment, 46
    dispatch information, 46
size-up, 46, 63, 116–117
sledge hammers, 84
slides, 28
slippery conditions, 11, 11*f*
sloping systems, 294
slough failure, 29, 41
smaller-scale IAPs (mental), 45–46
soil, 79. *See also individual entries*
    forces associated with, 22–25
        horizontal force, 25
        vertical force, 22–25
    types, 9, 22, 29
        clay, 9, 22, 23*f*
        sand, 9, 22, 23*f*
        silt, 9, 22, 23*f*
soil backfill, 176, 176*f*
soil classification, 6, 133
    Type A-25, 6
    Type B-45, 6
    Type C-60/C-80, 6
    OSHA, 19–22
        C-60, 20*t*, 22
        C-80, 20*t*, 22
        soil types, 20*t*
        Type A, 20*t*, 21
        Type B, 20*t*, 21
        Type C, 20*t*, 22
soil density, 41
soil entrapment, 190–194. *See also* vacuum trucks
soil forces, 133–134
soil mass, 23, 23*f*
soil mechanics, 22–27
soil strength, 25–27
    angular-shaped particles, 25
    cohesion, 26
    friction, 25–26
    internal angle of friction, 26
    moisture content, 27
    rounded-shaped particles, 25
    unconfined compressive strength (UCS), 27
soldier piling, 296, 296*f*
sole anchor picket chart, 137, 143*t*, 306, 312*t*
span of control, 126
spoil pile, 12, 16
    slide, 27–28, 41
    surcharge, 135–136, 305–306
spot shoring, 238, 269–270, 273
stabilization equipment, inspecting, 218–219
    lever hoist inspection, 218–219
    pneumatic struts, 218
    rigging equipment, inspecting, 219
    synthetic rope, 218
    tiebacks, inspecting, 218–219

stabilization equipment, inspecting (*Continued*)
    webbing inspection, 218
stable rock, 21, 21f
staging area, 44–45
    primary staging, 44
    staging officer, 45
staging officer, 45, 63, 117, 118f, 126
standard operating guidelines (SOGs), 45–46
standard precautions, 197, 203
static soil, 129, 129f
stick (arm), 227, 233
strong side, 163, 183
strongbacks, 72, 92
strut activation force, 148, 183
strut by-pass, 282
strut length, 288
strut pressure, 148, 153, 183
struts, 16, 80–83, 92, 148–150, 216–217. *See also* timber struts
    activation force, 149
    best practices for, 81
    hydraulic struts, 82–83, 150
    pneumatic struts, 81, 149
        placement, 150
        pressure, 149–150
        spacing, 150
    screw jack struts, 82, 150–151
    strut force, 148–149
    timber struts, 81–82, 151
summon resources, 52–53
    Tier, 1 response, 52
    Tier 2 resources, 52–53
superimposed (surcharge) loads, 34–35
supplemental sheeting, 238
supplemental shoring, 147, 183, 238, 263–269, 273
    application for, 267–268
    precautions for, 268
    preparation for, 268
    utilizing, 263–269
    walkthrough of installing, 268
supplied air breathing apparatus (SABA), 69, 92
supply ventilation, 108f
surcharge loads (ScL), 305
surface-based lifting techniques, 220
swing area, 228, 233
synthetic rope, 218
synthetic web slings inspection, 219

T

T-L method, 133, 304–312
    in action, 137
    equipment surcharge, 306
    spoil pile surcharge, 305–306
    surcharge loads (ScL), 305, 305t
    trench depth to L conversion, 306
    for trench rescue, 134–135, 305
T-L soil, 41
T-shaped trench, 237, 239–240, 273
T-trench shoring plan, 246–250, 252–256f
    primary shoring, 250–251
    secondary shoring, 251–252
    walkthrough of, 250–252
    tabulated data for shoring equipment, 137–143, 138–143t, 306–312
        buttress chart, 137, 143t, 306, 312t
        Holomatro strut, 137, 140t, 306, 309t
        Hurst/Airshore strut, 137, 141t, 306, 310t
        panel chart, 137, 142t, 306, 311t
        Paratech strut, 137, 138t, 306, 307t
        Resqtec struct, 137, 139t, 306, 308t
        sole anchor picket chart, 137, 143t, 306, 312t
        wale chart, 137, 143t, 306, 312t
        wale/wood strut chart, 137, 142t, 306, 311t
        wood strut chart, 137, 142t, 306, 311t
tactical sheets, 45
tape measure and angle finder, 83–84
team-issue PPE, 66, 69
    cutting tool protection, 69
    hearing protection, 69
    respiratory protection, 69
technical rescue safety officer (TRSO), 118, 126
technician level trench rescue shoring, 8–9, 235–271. *See also* deep trench; shoring intersecting trenches; spot shoring; supplemental shoring, utilizing
    deep trenches, 237–238
    intersecting trenches, 237
    rescue plan, 238–239. *See also individual entry*
    severe environmental conditions, 270–271
    sheeting and shoring for, 238
    spot shoring, 238
    supplemental sheeting and shoring, 238
    victim removal review, 271
tension-based stabilization, 217–218
tieback, 217, 233
    inspecting, 218–219
timber shores, installing, 171–172, 172–173f
timber struts, 81–82, 92, 151–152
    activation forces, 151–152
    placement, 151
toe failure, 30, 30f, 41
tools and appliances, 83–90
    chainsaw, 85
    dewatering devices, 85–86, 86f
    duplex nails, 84, 85f
    entrenching tool, 84
    hammers and nails, 84–85
    ladders, 85
    lifting and stabilizing heavy objects, 86
    pickets, 84
    pike poles, 84, 84f
    rapid intervention team (RIT) equipment, 86
    scene lighting, 85
    shovels, 84, 84f
    sledge hammers, 84
    strut collar locking tools, 83
    tape measure and angle finder, 83–84
    utility control, 86
    ventilation equipment, 85
    victim removal equipment, 86
toothpaste clays, 36
topples, 28
traditional sheeting and shoring, 238, 273
traffic, 96–97
transfer loads (struts), 132
trench, 3, 16
    assessment, 49–50
        collapse, 49
        depth, 49
        shape, 49
        width, 49

boxes, 8, 16, 155, 184, 294–295
and excavation hazards, 10–15, 11f
trench collapse, 4–5, 27–32, 95–96
    collapse patterns, 32
    collapse situations, 5
    conditions and factors that lead to, 33–35
    lip shear failures, 28–29
    noncollapse situations, 5
    overburden pressure, 27
    rotational failure, 29–30. *See also individual entry*
    signs of impending, 35–39
        undercut trench walls, 36–39
        visible bulging on walls or floors, 36
        visible cracks, 36
        water, 36
    slides, 28
    slough failure, 29
    spoil pile slide, 27–28
    topples, 28
    types of, 27
    wall shear collapse, 30–31, 31f
    wedge failures, 31–32
        disturbed soils, 34
        previously disturbed soils, 34
        severe environmental conditions, 33–34
        superimposed (surcharge) loads, 34–35
        varying soil profiles, 34
        vibration, 35
        water, 33
        water table, 33
trench depth to L conversion, 136–137, 306
trench floor, 9, 16
trench incident, managing, 115–125. *See also* incident command system (ICS)
    size-up, 116–117
trench lip, 7, 16
trench rescue, 5f
    OSHA and, 5–6
trench rescue equipment, 147–152. *See also* struts
trench rescue panels, 72–76, 92, 152–153, 184
    best practices for panels, 74–75
    panel materials, 74
    panel ropes, 75–76
    roping panels, 76
    working of panels, 73–74
trench rescue response systems, 9–10, 10f
trench rescue shoring, 127–182, 184. *See also* entry shoring; shoring system disassembly and removal; shoring voids
    construction shoring and, comparing, 129–131
    entry shoring, 161–162
    equipment, 70–83. *See also* backfill; lip protection; struts; trench rescue panels; wales
    equipment surcharge, 136
    essentials, 131–132
    history of, 129–132
        planning, 130–131
        soil conditions, 129
        time, 130
    lateral soil forces and firefighters, 132–137
    non-entry shoring, 162–163
        backfill strut, 162
        compliance strut, 163
        panels, 162
        positioning strut, 162
    panel installation, 163–167
    practices of, 156–160
        lip protection, 157–160
    prescriptive shoring designs, 144–145
    principles of, 131–132
        collect/distribute loads, 131–132
        resist loads, 132
        transfer loads (struts), 132
    procedure for shoring plan, 145–147
    shoring plan, 145
    soil classification, 133
    soil forces, 133–134
    spoil pile surcharge, 135–136
    surcharge loads (ScL), 135
    T-L method, 133
    tabulated data for shoring equipment, 137–143
    trench depth to L conversion, 136–137
        collapse potential, assessing, 146
        complete shoring, 147
        lip, assessing, 146
        primary shoring, 147
        secondary shoring, 147
        shoring performance assessment, 147
        trench, assessing, 146
        trench size-up, 145–147
        victim(s), assessing, 146
        voids, assessing, 146
trench ventilation, 106–107
trench walls, 9, 16
turnbuckle, 219
Type A-25 soil, 6
Type A soil, 20t, 21, 41
Type B-45 soil, 6
Type B soil, 20t, 21, 41
Type C-60/C-80 soil, 6
Type C soil, 20t, 22, 41

U

unbroken utility lines, 56–57
unconfined compressive strength (UCS), 27, 41
undercut trench walls, 36–39
underground electrical wires, 56, 106
uniform shape, 211, 233
unplanned events, 130
unreinforced masonry (URM), 72
utilities, 96
utility control, 86

V

vacuum systems, 191–193, 203
vacuum trucks, 191–193
    centrifugal vacuum truck, 191, 191f
    hydro vacuum trucks, 192, 192f
    response, 193–194
varying soil profiles, 34, 41
vector, 41
ventilation, 106–107
    concerns, 106
    confined space, 106
    equipment, 85, 101–102
        confined, 102, 102f
        natural gas control equipment, 102
    exhaust, 109

ventilation (*Continued*)
    initial, 106
    rescue, 106–107, 110f
    supply, 108f
    trench, 106–107
vertical force, 22–25
vertical hitch, 222, 223f
vibration, 35
victim assessment, 48–49, 49f
victim care and extrication, 185–202. *See also* non-entry rescue; soil entrapment; vacuum trucks; victim self-rescue, 187–190
    entry operations, 189
    entry team duties, 189–190
    extrication, 189–190
    PPE for entry team operations, 189
    pre-entry briefing, 189
victim care considerations, 194–202
    litter basket, 202
    providing victim care, 197
    removing victim from trench, 201–202
    special considerations, 199
    victim assessment and initial care, 197
    victim care involving a collapse, 198–199
    victim harness, 202
    victim packaging equipment, 202
    victim stabilization, 198
victim packaging, 201, 203
victim removal equipment, 86
victim self-rescue, 57–61
victim stabilization, 198
victim survivability profile, 194, 203
visible cracks, 36

W

wales, 76–78, 92, 131, 153–154, 184, 243, 273
    best practices for, 76–78
    chart, 137, 143t, 306, 312t
    hangers, 78, 243, 273
    ropes, 78
wale/wood strut chart, 137, 142t, 306, 311t
wall shear, 32
    collapse, 30–31, 31f, 41
    topple pattern, 32f
water, 33, 97
    line break, 56
    table, 33, 41
weak side, 163, 184
webbing inspection, 218
wedge failures, 31–32, 41
    lip shear, 32, 32f
    rotational failure pattern, 32, 33f
    slough-in (undercut) pattern, 32, 33f
    toe failure pattern, 32, 33f
    wall shear, 32
    wedge failure pattern, 32, 33f
wedges, 216
weight, 210
wheel chocks, 217
wire rope inspection, 219
wood, 79–80
    backfill, 175
    strut chart, 137, 142t, 306, 311t
work, 207, 233
working length, 215, 233
working load limit (WLL), 219, 233
wristlets, 188, 203